Contributions to Environmental Sciences & Innovative Business Technology

Contributions to Environmental Sciences & Innovative Business Technology (CESIBT) is an interdisciplinary series of peer-reviewed books dedicated to addressing emerging research trends relevant to the interplay between Environmental Sciences, Innovation, and Business Technology in their broadest sense. This series constitutes a comprehensive up-to-date interdisciplinary reference that develops integrated concepts for sustainability and discusses the emerging trends and practices that will define the future of these disciplines.

This series publishes the latest developments and research in the various areas of Environmental Sciences, Innovation, and Business Technology, combined with scientific quality and timeliness. It encompasses the theoretical, practical, and methodological aspects of all branches of these scientific disciplines embedded in the fields of Environmental Sciences, Innovation, and Business Technology.

The series also draws on the best research papers from EuroMid Academy of Business and Technology (EMABT) and other international conferences to foster the creation and development of sustainable solutions for local and international organizations worldwide. By including interdisciplinary contributions, this series introduces innovative tools that can best support and shape both the economical and sustainability agenda for the welfare of all countries, through better use of data, a more effective organization, and global, local, and individual work. The series can also present new case studies in real-world settings offering solid examples of recent innovations and business technology with special consideration for resolving environmental issues in different regions of the world.

The series can be beneficial to researchers, instructors, practitioners, consultants, and industrial experts, in addition to governments from around the world. Published in collaboration with EMABT, the Springer CESIBT series will bring together the latest research that addresses key challenges and issues in the domain of Environmental Sciences & Innovative Business Technology for sustainable development.

Adarsh Kumar · Neelu Jyothi Ahuja ·
Keshav Kaushik · Deepak Singh Tomar ·
Surbhi Bhatia Khan
Editors

Sustainable Security Practices Using Blockchain, Quantum and Post-Quantum Technologies for Real Time Applications

Springer

Editors
Adarsh Kumar
School of Computer Science
University of Petroleum and Energy Studies
Dehradun, Uttarakhand, India

School of Computer Science
College of Computing, Mohammed VI
Polytechnic University
Benguerir, Morocco

Keshav Kaushik
School of Computer Science
University of Petroleum and Energy Studies
Dehradun, Uttarakhand, India

Surbhi Bhatia Khan
School of Science, Engineering
and Environment
University of Salford
Manchester, UK

Department of Electrical and Computer
Engineering
Lebanese American University
Byblos, Lebanon

Neelu Jyothi Ahuja
School of Computer Science
University of Petroleum and Energy Studies
Dehradun, Uttarakhand, India

Deepak Singh Tomar
Department of Computer Science
and Engineering
Maulana Azad National Institute
of Technology
Bhopal, Madhya Pradesh, India

ISSN 2731-8303 ISSN 2731-8311 (electronic)
Contributions to Environmental Sciences & Innovative Business Technology
ISBN 978-981-97-0087-5 ISBN 978-981-97-0088-2 (eBook)
https://doi.org/10.1007/978-981-97-0088-2

This Springer imprint is published by the registered company Springer Nature Singapore Pte Ltd.
The registered company address is: 152 Beach Road, #21-01/04 Gateway East, Singapore 189721, Singapore

Preface

Blockchains and the emergence of quantum technologies are two of the most significant emerging technical developments now underway that are demonstrating an ever-increasing spread both in academic and industrial domains. The problem related to the quantum threat and the execution of post-quantum signatures in a blockchain platform is a hot topic in today's scientific community because blockchains have remarkably progressed in recent years and have found a variety of applications with the assumption that they will drastically increase their safety. The most recent quantum computers are capable of decoding secret encrypted keys and opening the network to attack by solving very difficult classical cryptography equations. In contrast to the billions of years of calculation required by conventional computer equipment, they can nearly instantly answer complicated mathematical problems. Researchers support the creation of innovative methods that incorporate encryption technology in the post-quantum future. Quantum computing, post-quantum cryptography, quantum attack-resistant blockchain, quantum blockchains, interdisciplinary solutions, and real-world use cases are just a few of the themes that are taken into account in the book Advances in Quantum Blockchain with Real-Time Applications. A distributed database that is based on quantum information theory and computing is known as a quantum blockchain. It is decentralized, encrypted, and distributed. The necessity for quantum versions of conventional blockchains and current developments in quantum computing are the causes of this. Quantum blockchain is one of several technologies still in the process of standardization, including algorithms, platforms, models, applications, systems, and databases. There is now more study being done in this area as a result of the development of quantum information theories and computation.

Chapters in this book are organized as follows.

Chapter 1 titled *Blockchain-Based Privacy-Protected Reputation Model for Internet of Vehicles* presented the potential to revolutionize transportation infrastructure using IoT when traditional reputation management approaches face privacy concerns and reliance on centralized servers. This work proposes a reputation model using blockchain technology for anonymity, decentralization, and encryption to improve analysis accuracy.

Chapter 2 titled *Sustainable Blockchain and Supply Chain Management: Integration Issues, Challenges, Potentials and Applications* explores the role of blockchain in supply chain operations, focusing on its principles of decentralization, immutability, and cryptographic security. It highlights the benefits of blockchain integration, such as increased transparency, simplified transactions, traceability, and improved collaboration. The work covers topics like supply chain digitization, smart contracts, data privacy, and decentralized applications. It also addresses integration issues like technological complexities, prices, laws, and governance. The work also looks ahead to future themes like IoT and AI integration, regulatory challenges, and scalability. It emphasizes the need for organizations to embrace blockchain for competitive advantage and supply chain innovation.

Chapter 3 titled *Zero-Knowledge Proofs in Blockchain-Enabled Supply Chain Management* discussed that the global supply chain is undergoing significant changes due to technological advancements and manufacturing growth. Blockchain technology is being used to enhance transparency, traceability, and trust in supply chain management. Zero-knowledge proofs (ZKPs) are powerful cryptographic tools that can validate goods, certifications, and transaction records without exposing sensitive information. ZKPs maintain data integrity, prove provenance, and prevent fraudulent activities by identifying anomalies in supply chain data. This article provides a comprehensive analysis of blockchain-based supply chain management, identifying technical considerations, scalability problems, and interoperability challenges.

Chapter 4 titled *A Sustainable Blockchain and Asymmetric Broadcast Encryption-Based Secure E-Healthcare System* discussed the increasing value of healthcare records has led to the need for electronic healthcare records (EHRs). However, current management methods create a single point of failure and expose users to security risks. Handling electronic health records from the internet of medical things poses significant challenges due to sensitive information. The COVID-19 pandemic has exacerbated these issues, necessitating the use of blockchain technology to enhance health record management. Blockchain technology has sparked research into transaction schemes focusing on privacy preservation. However, the public nature of blockchain exposes transaction information, introducing privacy risks. A proposed solution is a framework of broadcast encryption and a specific instance for conditional data access control. This approach aims to overcome limitations and ensure secure and private handling of healthcare transactions within the blockchain ecosystem.

Chapter 5 titled *Computational Analysis of Interference for Sustainable Operation of Wireless Networks* investigates quantum co-channel and adjacent channel interference in IEEE 802.11 networks to improve network performance. The Riverbed modeler tool was used to monitor and compute interference levels. The study found improper wireless channel distribution as the main cause of excessive interference. Further, the role and importance of quantum and post-quantum computing are discussed for wireless networks.

Chapter 6 titled *Advanced Techniques for Digital Evidence Preservation: The Power of Blockchain and Machine Learning* discussed that cybercrime is a rapidly growing global issue, with attackers entering cyber systems every seven seconds.

Log systems are crucial for handling events, but detecting anomalies is challenging due to constant changes and mutability. A novel model using Blockchain addresses this problem by integrating machine learning models and a Distributed Immutable Storage system. The model achieves 99.7% accuracy in detecting network anomalies using HTML, CSS, and Flask Framework on the Heroku cloud environment.

Chapter 7 titled *Quantum Resilience and Distributed Trust: The Promise of Blockchain and Quantum Computing in Defense* discussed that blockchain technology and quantum computing have revolutionized various industries, particularly the defence and armed forces. These technologies enhance safety and efficiency, providing new opportunities in various sectors. Blockchain technology ensures secure communication and data storage, while quantum computing principles are applied to military and defense-related tasks. Real-world scenarios benefit from the combination of blockchain for secure data storage and transmission in military networks and quantum computing for research and development activities. The future of blockchain and quantum computing is uncertain, but their integration aims to contribute to sustainable security practices for real-time applications. The chapter addresses ethical concerns, such as privacy breaches and potential military use, and provides a forward-looking perspective on the potential impact of these technologies on privacy and security. This comprehensive analysis provides valuable insights for researchers and professionals interested in improving safety and productivity in various industries.

Chapter 8 titled *Integration of Blockchain and the Internet of Things in Healthcare Sector* discussed that the blockchain-based IoT healthcare Industry 4.0 systems are being compared due to their security and privacy benefits. IoT technology simplifies data collection, accelerates information exchange, and increases knowledge processing. Blockchain architecture offers a secure platform for data collection, delivery, and retention. Its attributes include decentralized governance, security, privacy, transparency, and inviolability. Integrating IoMT and blockchain simplifies healthcare personnel's duties in diagnosing and securely storing patient data. Wearable sensors and IoMT networks enable collaboration between patients, medical professionals, and caregivers. Blockchain technology can address cybersecurity challenges, such as malware attacks and DDoS attacks, and secure digital health records (DHR) while maintaining privacy. It could transform next-generation Healthcare IoT applications as a technical improvement.

Chapter 9 titled *Quantum-Resistant Cryptography to Prevent from Phishing Attack Exploiting Blockchain Wallet* discussed that blockchain technology provides a decentralized, distributed digital ledger for various applications, including cryptocurrency, supply chain management, financial services, healthcare, and voting systems. However, phishing attacks are exploiting this technology to trick users into impersonating the wallet. To protect users from phishing attacks, users should follow best practices like verifying authenticity, being cautious of hyperlinks, using secure website connections, and raising awareness. Quantum computing can help prevent phishing attacks by providing secure key generation, robust authentication, and resilient encryption. Both blockchain and quantum computing have the potential to bring significant advancements in securing against phishing attacks.

Chapter 10 titled *Revolutionizing Military Technology: How the Fusion of BlockChain and Quantum Computing is Driving in Defense Application* discussed that blockchain is a revolutionary technology that revolutionizes society by storing data in a decentralized network. It can help combat cyber threats by reducing weak links and enhancing compromised weaknesses. Quantum technology, which uses quantum principles, could have significant implications for military detection, encryption, and correspondence, as well as legislative oversight and approvals. Both blockchain and quantum technologies have been tested in military applications, but their development is still in its early stages.

Chapter 11 titled *Building Resilient Digital Forensic Frameworks for NoSQL Database: Harnessing the Blockchain and Quantum Technology* discussed that digital forensics involves gathering, examining, and presenting digital evidence from devices like computers, smartphones, and cameras. It is used to investigate cybercrimes, retrieve deleted data, verify document veracity, and locate digital information sources. Database forensics investigates database system logs for criminal activity or security breaches. Blockchain and quantum computing are used in digital forensics to find patterns, encrypt data, and ensure data accuracy. However, quantum-safe preservation techniques and other challenges are necessary. A six-phase forensic investigation framework is provided for wide-column NoSQL databases, including MongoDB, CouchDB, and Cassandra.

Chapter 12 titled *Revolutionising Gait Analysis with Blockchain Technology: Enhancing the Privacy and Security* discussted that the Gait analysis is crucial for diagnosing and treating conditions like multiple sclerosis and Parkinson's disease. Traditional methods can be time-consuming, costly, and damaging to patient data. Blockchain technology offers an efficient, transparent system for storing and exchanging patient information. It collects data from wearable devices using a decentralized framework, ensuring immutability and immutability. Integration with blockchain offers benefits like enhanced efficiency, traceability, and reduced risks. However, challenges include infrastructure investment, complex integration, and ethical concerns.

Chapter 13 titled *A Blockchain-Enabled Framework to Implement Supply Chains in Digital Land Registry in the Kingdom of Saudi Arabia* explored a sustainable supply chain framework model for implementing a digital land registry in Saudi Arabia, focusing on the Land Administration Domain Model (LADM). It evaluates existing land registries and discusses the use of blockchain in the supply chain digital land registry model. The study uses an exploratory approach, collecting data through interviews with real estate organizations.

Chapter 14 titled *Blockchain for Secure Payments: A Bibliometric Review* discussed that the blockchain is gaining interest as a potential technology for peer-to-peer transactions, as traditional online financial transactions are vulnerable to cyber-attacks. A bibliometric analysis of 311 documents on blockchain for secure transactions revealed major themes for research in secure payments using blockchain, highlighting the importance of secure transactions for banks and non-bank entities.

Chapter 15 titled *Formalized Overview of ZX-Calculus, the Notion of Completeness Clifford Computation-Based and One Representative Application Case* explored the ZX-calculus, a branch of logic and category theory, as a representation of quantum algorithms. It covers ZX-diagrams, spiders, symmetries, transposes, spider fusion, π-commutation, color changing, algebraic structures, connectivity, advantages, and how a ZX-diagram is translated into a quantum circuit.

This book provides the best learning resource on sustainable security practices in the domain of blockchain, Quantum, and Post-Quantum technologies dealing with the real-time applications. This book highlights the Blockchain and Quantum-based techniques for Security Practices in Banking Applications. Moreover, the applications of Blockchain, Quantum and Post-quantum techniques for the protection of digital assets in healthcare will also be discussed in this book. The book will stress the integration of Blockchain and Quantum techniques in defence, military applications, and supply chain management. In addition, the book also focuses on the blockchain's application in secure messaging and keyless secure infrastructures. This book will be beneficial for academicians, researchers, students, and technical persons in the domain of Blockchain and Quantum Computing.

Dehradun, India — Adarsh Kumar
Dehradun, India — Neelu Jyothi Ahuja
Dehradun, India — Keshav Kaushik
Bhopal, India — Deepak Singh Tomar
Manchester, UK — Surbhi Bhatia Khan

Contents

Blockchain-Based Privacy-Protected Reputation Model for Internet of Vehicles

1

Nikhil Kumar Singh, Deepak Singh Tomar, and Rishi Kumar Singh

Abstract

The Internet of Vehicles (IoV) is a promising emerging technology that could significantly contribute to the development of high-level transportation infrastructure. However, there are constraints to the traditional reputation management approach in the IoV sector. These include a small detection radius, the possibility of privacy issues, and the use of unstable centralised reputation servers. This chapter provides a reputation model for the IoV that makes use of blockchain technology to guarantee anonymity, so addressing these concerns. Traditional IoV reputation systems have a major flaw due to their reliance on a single reputation server. The integrity and dependability of reputation assessments are at risk due to the presence of potential vulnerabilities in these systems, which could allow for unauthorised data alteration or theft. Additionally, one could argue that the current IoV reputation method does not offer sufficient protection for user anonymity. The evaluation of car reputations frequently requires the inclusion of personal information, such as driving habits and location data. Lastly, it is important to note that there exists a distance threshold beyond which the aforementioned reputation technique may encounter difficulties in accurately detecting and identifying vehicles that exhibit malevolent behaviour. The aforementioned proposal employs a distributed reputation update method based on blockchain technology in order to address these issues. This strategy

N. K. Singh (✉)
CSE Department, Indian Institute of Information Technology, Bhopal, M.P., India
e-mail: nikhilsinghmanit@gmail.com

D. S. Tomar
CSE Department, Maulana Azad National Institute of Technology, Bhopal, M.P., India

R. K. Singh
Electrical Engineering Department, Maulana Azad National Institute of Technology, Bhopal, M.P., India

A. Kumar et al. (eds.), *Sustainable Security Practices Using Blockchain, Quantum and Post-quantum Technologies for Real Time Applications*, Contributions to Environmental Sciences & Innovative Business Technology,
https://doi.org/10.1007/978-981-97-0088-2_1

enhances the precision of reputation analysis by obviating the necessity for a centralised reputation server. Furthermore, the utilisation of a multi-key, fully homomorphic encryption methodology is employed for both the encryption and computation of evaluation data, thereby substantially mitigating the potential for privacy breaches. In order to mitigate the possibility of rogue vehicles eluding detection through reputation upgrades, a novel adaptive retroactive time interval adjustment approach has been devised. The simulation results indicate that the proposed approach effectively preserves user privacy while maintaining high detection rates for malicious vehicles, and also minimises the occurrence of false positives. Furthermore, the model exhibits a detection rate that surpasses gold-standard techniques by 32%.

Keywords

Internet of vehicles • Reputation model • Blockchain • Privacy protection • Distributed update • Homomorphic encryption • Adaptive retrospective time interval • Malicious behaviour

1.1 Introduction

In recent years, there has been significant interest in vehicular ad hoc networks (VANETs) among both academic and industrial communities. This interest has been driven by the widespread adoption of fifth-generation communication technology (5G) and the rapid advancement of in-vehicle wireless communication technology (Xu et al. 2020, 2019; Atallah et al. 2014). This can be attributed to their significant capacity to enhance traffic efficiency and enhance road safety. Reputation systems play a crucial role inside vehicular networks, serving as important components for upholding communication security among vehicles.

Reputation systems can be divided into two categories: centralised and distributed, depending on the underlying structure of computational data. Reputation updates and storage are handled by a single independent third party, known as the reputation centre, in centralised reputation schemes (Sun et al. 2021; Xi et al. 2022; Tian et al. 2020). However, centralised solutions are vulnerable to problems like unreliable central servers and user privacy breaches because VANETs are open, decentralised, and extremely dynamic. The Authors (Tan et al. 2011; Song et al. 2020; Atallah et al. 2015) suggested building a distributed reputation system using blockchain technology. Although these systems present a number of workable options, they ignore one crucial fact: malignant intelligent VANETs may adjust the amount of harmful behaviour over particular time periods to avoid detection by the reputation system (Wei et al. 2021; Talpur and Gurusamy 2022).

During the process of reputation updates, some vehicles may hesitate to provide interactive evaluations (Xiao et al. 2020; Cheng et al. 2020). This hesitation arises from the potential risks of malicious eavesdropping on the evaluation data during its transmission and the possibility of semi-trusted roadside infrastructure leaking evaluation data, leading to vehicles neglecting their evaluation obligations. Evaluation data may contain sensitive information such as vehicle user IDs, interests, interaction patterns, and evaluation habits (Ahmed et al. 2021; Wang et al. 2022; Ding et al. 2020; Adil et al. 2022). The leakage of evaluation data can greatly harm user interests and even put users in danger. To address these issues, this paper proposes a privacy-preserving vehicular network reputation model based on a blockchain architecture. This model achieves distributed and trusted vehicle reputation updates using blockchain technology (Kumar et al. 2019) and employs a multi-key fully homomorphic encryption algorithm (Ghane et al. 2021) to protect the privacy of evaluation data during collaborative computation with partial secret keys by road side units (RSUs) and mobile edge servers (MES). By calculating the credibility of vehicle's historical reputation and assessing the security of the communication environment, the dynamic adjustment of the retrospective time interval is achieved.

This paper proposes a privacy-preserving vehicular network reputation model based on a blockchain architecture to accomplish distributed and trustworthy vehicle reputation updates while protecting the privacy of evaluation data. Motivating this research is the potential of vehicular ad hoc networks (VANETs) to improve road safety and traffic efficiency, where reputation mechanisms are essential for ensuring vehicle communication security. Existing centralised solutions are susceptible to issues such as unreliable central servers and violations of privacy. The proposed model employs blockchain technology for distributed reputation updates and a multi-key, completely homomorphic encryption algorithm to secure evaluation data. In addition, the model incorporates dynamic adjustment of the retrospective time interval by evaluating the prior reputation's veracity and the communication environment's security. The ultimate goal of this paper is to enhance the effectiveness and security of reputation mechanisms in VANETs. The primary contributions of this paper are as follows:

1. Proposal for a decentralised and trustworthy reputation model based on a blockchain architecture, minimising the impact of centralised server issues on reputation availability.
2. Design of a method for encrypting evaluation data to protect privacy while allowing for the collaborative computation of encrypted feedback.
3. Introduction of an adaptive adjustment strategy for the retrospective time interval, preventing intelligent malicious vehicles from eluding detection based on reputation update attributes.

1.2 Related Work

1.2.1 Application of Blockchain in Vehicular Network Reputation Mechanism

Since Satoshi Nakamoto's 2008 proposal of Bitcoin, blockchain, the technology underlying the Bitcoin protocol, has been extensively adopted in VANETs for establishing reputation mechanisms (Huang et al. 2019; Hussain et al. 2021). In addition to crowd sensing (Chen and Ho 2019), reputation management (Li et al. 2022), and vehicle insurance claims (Xiao et al. 2022), blockchain technology has been applied to other areas of vehicular networks. Literature (Huang et al. 2019) introduced a blockchain-based Anonymous Reputation System (BARS) that accomplishes transparency in vehicle reputation authentication and revocation by establishing twin blockchains (CerBC and RevBC) to enhance the security of vehicular network reputation processes. As a means of protecting vehicle confidentiality, public keys are employed as communication pseudonyms. Literature (Cheng et al. 2020) presented a blockchain-based trust management strategy for automotive networks, guaranteeing vehicle privacy security with an RSU-led anonymous concealing zone algorithm and preventing identity information leakage with digital certificates as pseudonyms. Author (Safavat and Rawat 2021) presented a blockchain-based identity verification and trust management approach that evaluates the trustworthiness of automobiles and data in order to detect and manage deceptive information in order to combat potential malicious vehicles. Literature (Ullah et al. 2023) proposed a blockchain-based distributed trust management scheme to facilitate secure message sharing and privacy protection among vehicles in VANETs, addressing the problems of internal vehicle broadcast of forged messages and leakage of vehicle trajectory data. This scheme assigns a unique encryption fingerprint to each vehicle using a Certificate Authority (CA) to eradicate the link between public keys and actual identities, thereby protecting the identity of vehicle nodes against attacks. In conclusion, the adoption of blockchain technology for vehicular network reputation assessment has become a trend, but it confronts a number of technical obstacles, such as enhancing detection accuracy, system efficiency, and privacy protection.

1.2.2 Research on Privacy Protection Techniques in Reputation Mechanisms

As the significance of privacy protection techniques for the secure operation of VANETs increases, The Authors (Nan et al. 2022; Su et al. 2017; Sarker et al. 2023; Cao et al. 2021; Li et al. 2022) have proposed a variety of privacy protection schemes suited to various vehicular network use cases.

The literature (Cao et al. 2021) presents a proposed technique that utilises multi-secret sharing reputation and binary asymmetric polynomial to achieve asynchronous secret reconstruction. In this scheme, each participant is assigned a shared partition of a univariate polynomial. During the process of covert reconstruction, the explicit disclosure of the shared partition of information is avoided in order to prevent the disclosure of shared information. In the field of literature, a study (Li et al. 2022) introduced a way for managing trust and privacy in automobile networks that is adaptive in nature. The purpose of this strategy is to address the trade-off between reputation and privacy. This method facilitates the recognition of network nodes by vehicles, taking into account their identities or levels of trust, and allows for the concealment of their identities or levels of trust in situations where privacy is of paramount significance. The utilisation of Private Set Intersection (PSI) employing Bloom filters (BF) has been documented in the literature (Wei et al. 2021). This approach aims to strike a harmonious equilibrium between trust management and privacy preservation during the distribution of emergency messages. As a result, the BTMPP approach was developed. The authors of the study proposed the lightweight privacy-preserving trust evaluation (LPPTE) method with the aim of facilitating distributed data fusion in road vehicle safety applications. This methodology seeks to achieve a more optimal equilibrium between trust evaluation and privacy protection, as documented in the existing literature (Sun et al. 2021). The academic literature (Wang et al. 2022) introduces a vehicle reputation system known as Trust Vote, which utilises crowdsourcing and allows vehicles to hide their review scores when engaging in interactions. The authors of the literature (Adil et al. 2022) have proposed a cloud-assisted protocol, known as Privacy-Preserving Vehicular Feedback (PPVF), for safeguarding the privacy of feedback data in vehicular networks. The proposed approach utilises homomorphic encryption and data aggregation methodologies to assess the reputation of a specific vehicle while preserving the confidentiality of its feedback data. Conventional homomorphic encryption techniques, as employed in the aforementioned schemes, possess the limitation of exclusively encrypting data using a single key. Additionally, the compute centre retains the sole possession of the private key, which must exhibit a high level of resilience. This implies that feedback data has the potential to be transmitted to other vehicles that possess the same key.

1.3 Vehicular Network Reputation Model Under Blockchain Architecture

1.3.1 System Architecture

As shown in Fig. 1.1, the designed vehicular network reputation model consists of four entities: Trusted Authority (TA), MES, RSUs, and communicating vehicles.

- Trusted Authority (TA): TA is primarily responsible for generating entity keys and public parameters within the system, as well as maintaining the required source data for various vehicles. In this system, TA is dependable.

- MES: Each MES has its own coverage area, which is referred to as a management domain. Several neighbouring RSUs are part of the same MES management domain based on their geographical characteristics. In the suggested scheme, MES nodes function as blockchain deployment nodes and are in charge of the generation and verification of reputation blocks. As peer nodes, MESs can use consensus techniques to write updated vehicle reputation and related parameters into the public categorised ledger.
- RSUs: RSUs, as vehicle network infrastructure, are often located along the roadside and have specific data processing capabilities. RSUs communicate with cars using Dedicated Short-Range Communication (DSRC) technology, which allows for wireless communication. RSUs have a communication range of 500–1,000 m (Liu et al. 2015). RSUs are presumed to be semi-trusted in the strategy.
- Communicating Vehicles: Vehicles, as direct users of the system, are outfitted with On-Board Units (OBUs) with effective communication ranges ranging from 50 to 300 m (Liu et al. 2022). Transient Protection Devices (TPDs) are also used in vehicles to save keys. This approach divides communicative vehicles into two types: target vehicles and evaluating vehicles.

1.3.2 Reputation System

1.3.2.1 System Initialisation

The Trusted Authority (TA) selects large prime numbers p and q, satisfying the conditions $p' = (p - 1) \times 2^{-1}$, and $q' = (q - 1) \times 2^{-1}$ where the bit lengths of p and q meet the security parameter k, $\updownarrow(p) = \updownarrow(q) = k$. The TA calculates the system parameters N = pq and = $\mathrm{lcm}(p-1, q-1) \times 2^{-1}$, where $\mathrm{lcm}(p-1, q-1)$ is the least common multiple of p-1 and q-1. A random number $a \in Z^*_{N^2}$ is chosen, where $Z^*_{N^2}$ represents the multiplicative group modulo N^2. The TA then computes $g = -a^{2N}$ to obtain the generator g of order $(p - 1) \times (q - 1) \times 2^{-1}$, and defines the function $P(x) = (x - 1)N^{-1}$.

The TA executes the key rekey generation algorithm $\mathrm{Key}_{\mathrm{rekey}}$ (λ) to randomly split the strong private key SK = λ into i parts. Let $\mathrm{SK}^{(i)} = \lambda_i (i = 1, 2)$ represent the set of weak private keys for communication entities (MES and RSUs), where $\lambda_1 + \lambda_2 \equiv 0 mod \lambda and \lambda_1 + \lambda_2 \equiv 1 mod N^2$. Based on registration information and system parameters, the TA generates public–private key pairs $\left(\mathrm{pk}_i, \mathrm{sk}_i\right) = \{(N, g, h_i), \theta_i\}$ for vehicles, where $\theta_i \in [1, N/4]$ is a randomly chosen number, and $h_i = g^{\theta_i} mod N^2$.

1.3.2.2 Evaluation Data Processing

This module mainly achieves data encryption and decryption operations by combining the distributed two-trapdoor public-key cryptosystem (DT-PKC) algorithm (Kumar et al. 2019). The main steps are as follows:

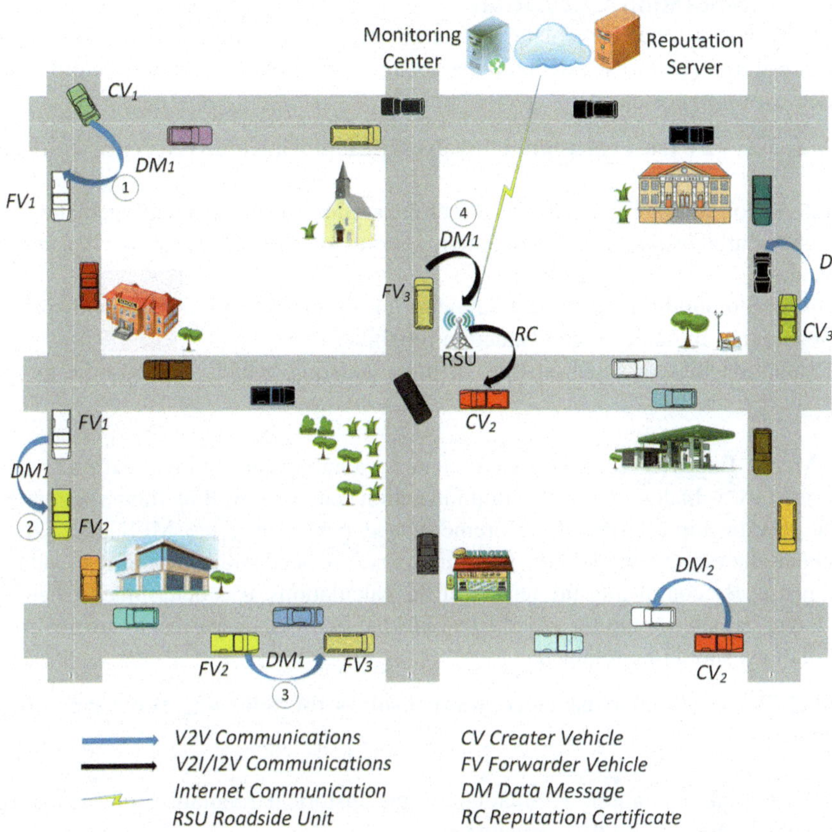

Fig. 1.1 Reputation update model of internet of vehicles

(1) **Data Encryption** For the data (where Z_N is the additive group of integers modulo N), select a random number to calculate, obtaining the data ciphertext, where

$$C_{i,1} = g^{\varepsilon\theta_i}(1 + fN) mod N^2, C_{i,2} = g^{\varepsilon} mod N^2. \tag{1.1}$$

(2) **Data Decryption** For the ciphertext data$[f]_{\mathrm{pk}_i}$, MES uses the private key for pre-decryption:

$$\mathrm{HFD}_{\mathrm{SK}^{(1)}}\left([f]_{\mathrm{pk}_i}\right) : \mathrm{CT}_i^{(1)} = \left(C_{i,11}\right)^{\lambda_1} = g^{\varepsilon\theta_i\lambda_1}(1 + fN\lambda_1) mod N^2 \tag{1.2}$$

$$\text{Obtain plaintext data } f = P\left(\mathrm{CT}_i^{(1)} \cdot \mathrm{CT}_i^{(2)}\right).$$

1.3.2.3 Reputation Calculation

In the proposed scheme, vehicle reputation is primarily composed of V2V trust and global trust, and trust calculation is obtained by synthesising multiple different sources of reference data.

V2V Trust: V2V trust calculation involves the following four reference sources:

- Interaction proof: Evaluation of direct interaction records between vehicles.
- Behavioural evaluation: Quantitative scoring of the behaviour of the target vehicle.
- Spatio-temporal relationship: Evaluation of the impact of temporal and spatial positioning of vehicles on evaluation reliability.
- Evaluation bias: Evaluation bias of the evaluating vehicle towards the target vehicle.

The Mobile Edge Server (MES) accumulates evaluation data and interaction proofs uploaded by vehicles within its communication range, as well as their spatiotemporal relationships. Using the aforementioned parameters, the MES and RSUs calculate the target vehicle's trustworthiness and dependability during the evaluation phase. By contrasting the results of the calculations, it is determined whether the target vehicle should be added to a blacklist or whether the results of the trust calculations should be retained.

Global Trust: global trust calculation involves the following three reference sources:

(i) Historical reputation: Evaluation of the historical reputation of the target vehicle updated by the evaluator.
(ii) Reputation fluctuation curve: Trajectory of changes in vehicle reputation within a sliding time window.
(iii) Reputation bias: Deviation of the target vehicle's historical reputation cached by the MES.

The MES calculates the global trust value and reliability of the target vehicle by retrieving local cached data and relevant parameters. If the reputation fluctuation curve is within the predefined range and below the detection threshold, the trust calculation results are retained. Otherwise, the target vehicle will be added to the local blacklist.

1.3.3 Attack Model

Malicious behaviours of vehicles include but are not limited to the following four types:

1. Selfish behaviour: Vehicles only enjoy VANETs network resources without providing services to other vehicles.
2. Switching attack: Malicious vehicles manipulate their reputation cyclically to surpass the detection threshold and bypass the reputation system's detection.
3. Collusion attack: Vehicles form alliances to maliciously boost their own or others' reputations. They may give unreasonable ratings to target vehicles to lower their reputations.
4. False information attack: Malicious vehicles deceive driving vehicles or edge infrastructure by sending false messages.

1.4 Privacy-Preserving Reputation Model for Vehicular Networks

1.4.1 Reputation Model Modelling

Let X denote the set of vehicles, where each vehicle $x \in X$ is characterised by its behaviour data D and interaction data I. We define a reputation model as a function $f : X \to [0, 1]$ that maps a vehicle to its reputation value.

In the designed reputation system, the updating of vehicle reputation is based on the calculated trust values, which consist of two main components: global trust and V2V trust. The reliability of the evaluation parameters is represented by the weight function w, where w can be either $\mathrm{w}_{\mathrm{global}}$ for global trust or $\mathrm{w}_{\mathrm{V2V}}$ for V2V trust, with $\forall_{X\in(D,I)} w_X \in [0, 1.0]$.

The calculation of vehicle reputation can be expressed as follows:

$$R_{\mathrm{u}}(m, n, t) = \frac{\omega_D \cdot \mathrm{DT}_{\mathrm{gl}}(m, n, t) + \omega_I \cdot \mathrm{IT}_{\mathrm{pv}}(m, n, t)}{\sum_{X\in(D,I)} \omega_X}. \tag{1.3}$$

In the equation 1.3 ω_D represents the spatial risk factor of the evaluation; ω_I represents the distance between the evaluating vehicle and the target vehicle; ω_X is the defined initial attenuation distance, which means that evaluations from vehicles within the range of the target vehicle are not affected (Wang et al. 2019), as shown in Fig. 1.2.

$$\alpha_I(\eta_{\mathrm{r}}) = \partial \cdot \rho. \tag{1.4}$$

1.4.2 Calculation Process of Reputation

As shown in Fig. 1.3, the evaluator uses four data sources, using three calculation models to calculate the reputation value of the target vehicle. The details of the process are shown in Algorithm 1.

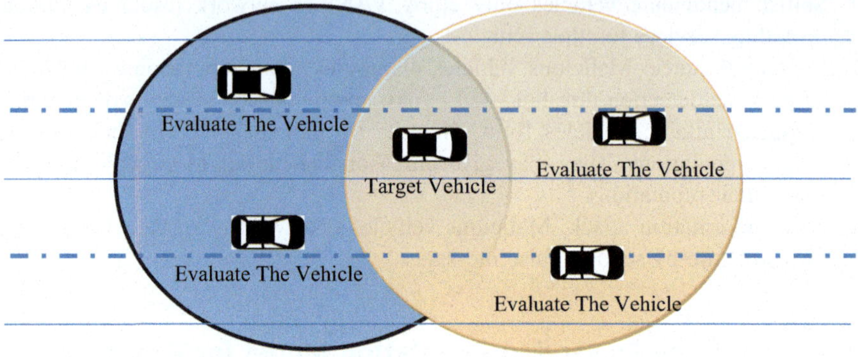

Fig. 1.2 Location distribution among vehicles

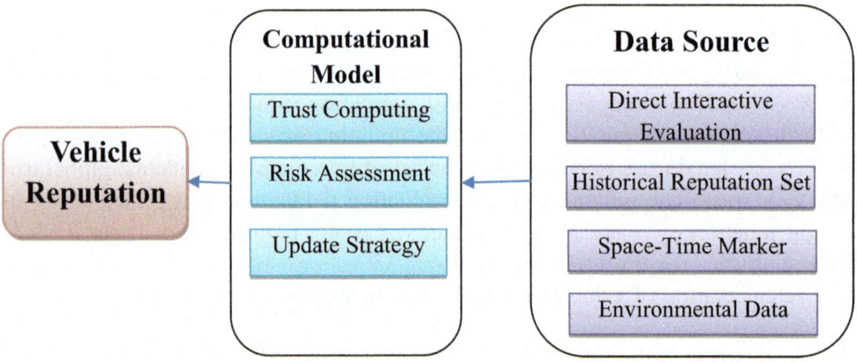

Fig. 1.3 Calculation process of vehicle reputation evaluation

Algorithm 1 Target Vehicle Reputation Value Calculation Algorithm (CAT)

(1) Input r_i, f_j
(2) Output $R_{\mathrm{u}}(m, n, t)$.
(3) Neighbouring vehicles encrypt and upload evaluation data f_j:

$$[f_j]_{\mathrm{pk}_i} = \{C_{i,1} = g^{\varepsilon\theta_i}(1 + f_j N) mod N^2, C_{i,2} = g^{\varepsilon} mod N^2\}. \tag{1.5}$$

(4) After collecting the evaluation data, MES calculates the risk coefficient function, and calculates the global trust of the target vehicle

$$\mathrm{DT}_{\mathrm{gl}}(m, n, t) = \frac{\sum_{\mu_{\mathrm{r}} \in \psi_i(m,n,t)} {}^{\alpha}D^{(\mu_{\mathrm{r}}) \cdot r_i}}{\sum_{\mu_{\mathrm{r}} \in \psi_i(m,n,t)} {}^{\alpha}D^{(\mu_{\mathrm{r}})}}. \tag{1.6}$$

(5) MES calculates and encrypts and cooperates with RSUs to calculate vehicle V2V trust (refer to Algorithm 2).

$$\mathrm{IT}_{\mathrm{pV}}(m,n,t) = \frac{\sum_{\eta_{\mathrm{r}} \in \phi_j(m,n,t)} \alpha_I(\eta_{\mathrm{r}}) \cdot f_j}{\sum_{\eta_{\mathrm{r}} \in \phi_j(m,n,t)} \alpha_I(\eta_{\mathrm{r}})} \tag{1.7}$$

(6) MES calculates the weight coefficients of the evaluation parameters:

$$\omega_X = h_{\mathrm{mc}}(m,n,t) \cdot h_{\mathrm{dc}}(m,n,t) \tag{1.8}$$

(7) According to all obtained parameters, MES calculates the reputation value of the target vehicle

$$R_{\mathrm{u}}(m,n,t) = \frac{\omega_D \mathrm{DT}_{\mathrm{gl}}(m,n,t) + \omega_I \cdot \mathrm{IT}_{\mathrm{pv}}(m,n,t)}{\sum_{X \in (D,I)} \omega_X} \tag{1.9}$$

(8) In the process of calculating vehicle reputation, to protect the privacy and confidentiality of evaluated vehicles and data, evaluated vehicles preprocess the data and submit it after merging with pseudonyms. Select random number confidential data to obtain the evaluation ciphertext.

$$[f_j]_{\mathrm{pk}_O} = \{C_{Q,1}, C_{Q,2}\}. \tag{1.11}$$

Reliability Calculation: Due to potential insufficient interactions among vehicles on the road, evaluators may not be able to collect sufficient evaluation data effectively, or there may be malicious vehicles providing false evaluations. As a result, the data collected by evaluators may have significant biases. This leads to unreliable updates of vehicle reputation. This paper introduces a reliability assessment method to mitigate this detrimental impact. Through reliability assessment, it is possible to effectively filter out evaluation data that is higher or lower than normal, thereby defending against collusion attacks on vehicles.

- **Rating Reliability** During the reputation update period, when there are fewer participating evaluators with lower reputation, the evaluations they provide may be unreliable, resulting in unusable reputation updates. The calculation principle of rating reliability is similar to expectations in statistics, which can measure the quantity and credibility of the computed data (Wang et al. 2019):

$$h_{\mathrm{mc}}(m,n,t) = 1 - \exp\left[-\tau \sum_{X \in (I,D)} \alpha_X(\cdot)\right]. \tag{1.12}$$

- **Indirect Reliability** In VANETs, the data obtained by evaluators may contain errors and biases, leading to inaccurate reputation updates. Indirect reliability is similar to bias in statistical data and can measure the differences between reference data.

$$h_{dc}(m,n,c) = 1 - \frac{\sum_{X\in(I,D)} \alpha_X(\cdot)|\Delta H|}{\sum_{X\in(I,D)} \alpha_X(\cdot)} \tag{1.13}$$

1.4.3 Adaptive Adjustment of Backtracking Time

The backtracking time interval Δt is calculated and adjusted by the Mobile Edge Server (MES) based on the communication security of the monitored road environment. In the equation: Δt represents the unit backtracking time interval, N_fake denotes the number of false messages sent by road entities, N_road represents the number of road traffic entities, and S_road represents the security level of the road communication environment.

$$\Delta t_{\text{Time}} = \frac{\Delta t_0 G_{\text{se}}}{1 + \sum_{x\in E_n} P(x)} \tag{1.14}$$

To quantify the security of the communication environment, considering that the number of interacting neighbours, reputation, and evaluation bias may affect the evaluator's assessment, the reliability calculation results are used (Wang et al. 2019).

$$G_{\text{se}} = \sum_{X\in(D,I)} L_X \text{Sr}. \tag{1.15}$$

In the equation, Sr represents the sum of vehicle reliabilities, α_i represents the fixed weight coefficient, and $\alpha_i = 1$ indicates that different vehicles have the same overall impact on the security of the communication environment.

$$\Delta t = \begin{cases} \Delta t_{\text{Min}}, \Delta t_{\text{Time}} < \Delta t_{\text{Min}}; \\ \Delta t_{\text{Max}}, \Delta t_{\text{Time}} > \Delta t_{\text{Max}} \\ \Delta t_{\text{Time}}, \text{Other}. \end{cases} \tag{1.15}$$

To ensure that the reliability of the vehicle varies with different time periods during the update process, increasing the backtracking time interval allows for dynamic adjustment of the update to prevent malicious vehicles from controlling the reputation cycle through low-cost switch attacks.

1.4.4 Reputation Dynamic Update Based on Blockchain

A consortium blockchain-based distributed management structure is used to achieve the on-chain reputation value and secure sharing of vehicle reputations across various infrastructure regions. This structure uses the practical Byzantine fault tolerance (PBFT) consensus mechanism. The MES calculates the most recent vehicle reputation during the reputation update process by retrieving evaluations uploaded by all Roadside Units (RSUs) within its communication range as well as historical reputations of vehicles that have been saved. The transaction pool is where the MES maintains all changed reputation values. Because there isn't a stable central node in VANETs to manage the blockchain, the system is designed to have numerous accounting nodes and one regulatory node.

When it is time to update reputation values on the blockchain, the MES selects one of a few pre-selected accounting nodes, mediated by the regulatory node, as shown in Fig. 1.4. Currently, the network as a whole elects one node to act as the bookkeeper (Rao et al. 2023). An MES node will be permitted to gather transaction records from the transaction pool into a newly generated block once it has been designated as the accountant. When this block is broadcast, the network nodes take in both the block header and body. Commonly seen in block headers are elements like BlockID, MESID, Time, etc. The relevant data is stored in the block's body using the Merkle tree data structure, which is hashed using the H(•) function. The root of the Merkle tree (hash tree) is contained in the block header (Punithan and Seo 2013). Even if only the hash value of the Merkle tree root is retained, the block's data cannot be changed because of the hash function's irreversibility. The block's PrevHash field stores the hash value of the preceding block, while the timestamp field contains information on the block's genesis time.

The major steps of the Byzantine fault-tolerant block consensus process that MES nodes use are as follows. The accounting nodes of the network will inform all other nodes whenever a new block is created. Each node verifies a new block's authenticity by examining its header for essential details like PrevHash and Time. The node will communicate its signature and the verification outcome if the verification is successful. Every node checks its own verification results against those of other nodes and alerts the network if there are any inconsistencies. Before the block is linked after the current block, updating the status of the targeted vehicle, two-thirds of the Mobile Edge Servers must concur on it.

1.5 Performance Analysis

1.5.1 Theoretical Analysis

Table 1.1 contrasts four aspects of the proposed model to existing vehicular reputation models. Unlike traditional vehicular reputation update models, the proposed model not only takes into account the historical reputation and related parameters accumulated from long-term interactions with target vehicles, but also incorporates

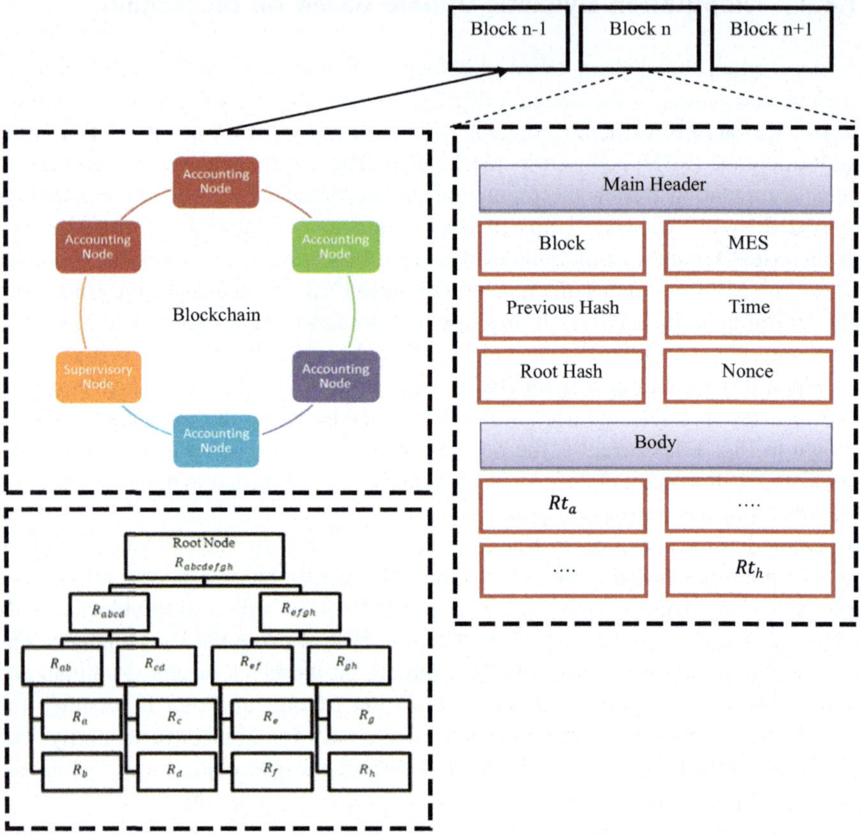

Fig. 1.4 Storage architecture of blockchain

short-term V2V direct evaluation data provided by neighbouring vehicles, which can more accurately reflect the true reputation of the target vehicle. Comprehensively measuring the overall reliability of spatiotemporal risk and trust calculations for target vehicle reputation evaluation makes the evaluator's reputation evaluation of the target vehicle more objective. Concerning vehicle reputation update methods, the proposed model integrates more secure and dependable blockchain technology, thereby avoiding potential single-point failures, unreliability, and privacy data leakage issues associated with centralised reputation servers being attacked.

The suggested model introduces adaptive modification of retrospective time during reputation update cycles, successfully preventing malevolent vehicles from starting switch attacks. Link attacks are somewhat prevented by the scheme's use of frequently changed pseudonyms for vehicle communication. By putting the vehicles on a blacklist, selfish activity that exclusively uses the networking services for automobiles for its own gain is addressed. All interactions are logged

Table 1.1 Comparison with other reputation update models in VANETs

Scheme	Reputation update source	Core algorithm	Reputation update method
Talpur and Gurusamy (2022)	Historical data, direct interaction	0–1 Encoding, FNV Hash	Centralised database
Cheng et al. (2020)	Direct interaction	Distributed K-Anonymous Blockchain	False message attacks, Switch attacks
Wang et al. (2022)	Direct interaction, trustworthiness	Homomorphic Addition, Roadside Units	False message attacks, Collusion attacks
Huang et al. (2019)	Direct interaction	Blockchain	False message attacks
Adil et al. (2022)	Direct interaction	Paillier Encryption (Li et al. 2022), Central Cloud	Selfish behaviour, False Message attacks
Proposed approach	Historical data, Direct interaction, Parameter risk, Trustworthiness	DT-PKC, Blockchain	Link attacks, Selfish behaviour, Switch attacks, Collusion attacks, False message attacks

and analysed as a vehicle's reputation is being updated, protecting it from false message attacks by penalising cars that send misleading information. The development of a data reliability evaluation module filters faulty parameters to stop collusion attacks. The proposed strategy can withstand several attacks compared to existing plans, hence boosting system security and resilience.

1.5.2 Experimental Simulation and Analysis

1.5.2.1 Experimental Setup

The studies were carried out by combining simulation software with Python, similar to conventional vehicular network reputation techniques. In order to replicate vehicle paths realistically, SUMO, an open-source traffic simulation programme, was used. It is frequently used in studies on vehicle networks (Punithan and Seo 2013; Liu et al. 2017; Zheng et al. 2022).

RSUs were placed at 6 intersections and road segments throughout the simulation scenario. Each intersection had a single MES. R2V's communication distance was set at 800 m, whereas V2V's communication distance was 400 m (Liu et al. 2015). There were 300 car nodes in the experiment, and the chance of feedback was 30%. The deployment was performed on a computer with a 3.20 GHz AMD processor and 16 GB of memory. The average results were from 50 simulation runs in order to reduce potential variations. Table 1.2 displays the primary simulation variables.

Table 1.2 Simulation parameters of the vehicle communication scene

Parameter	Value
Simulation area	3.0×3.0 km^2
Simulation time	300 s
R2V communication distance	800 m
V2V communication distance	400 m
Number of vehicles on the road	[30, 300]
Average speed	[20, 120] km/h
Malicious vehicle ratio	[0, 35] %
Initial reputation	0.6

1.5.2.2 Performance Metrics

To evaluate the detection performance of the reputation model against malicious vehicle behaviour in VANETs, performance metrics such as detection rate, false alarm rate, and time overhead are defined, following the reputation security standards (Adil et al. 2022).

(1) **Detection Rate (DPr)** This metric measures the performance of the model in detecting malicious vehicles in different scenarios:

$$\mathrm{DP_r} = \frac{N_{\mathrm{mv}}}{N_{\mathrm{tm}}} \quad (1.16)$$

where Nmv represents the number of malicious vehicles detected and excluded by the model, and Ntm is the total number of malicious vehicles present in the simulation.

(2) **False Alarm Rate (FAr)** This metric evaluates the probability of false alarms in the designed reputation model:

$$\mathrm{FA_r} = \frac{N_{\mathrm{f}}}{N_{\mathrm{f}} + N_{\mathrm{mv}}} \quad (1.17)$$

Here, Nf represents the number of normal or malicious vehicles incorrectly classified as malicious or normal.

(3) **Time Overhead (Tc)** This metric measures the overall time overhead for vehicle reputation calculation. It evaluates the operational efficiency of the proposed model:

$$T_{\mathrm{c}} = T_{\mathrm{e}} + T_{\mathrm{p}} + T_{\mathrm{d}} \quad (1.18)$$

where Te is the time required for data encryption, Tp is the time required for data processing, and Td is the time required for data decryption.

1.5.2.3 Experimental Simulation Analysis

Rate of malicious vehicle detection in various environments: The suggested model's capacity to detect malevolent cars as the number of vehicles on the road (Nv) changes is shown in Fig. 1.5. The testing findings show that, in situations of heavy traffic, the detection rate for malevolent cars can surpass 99%. The detection rate can get as high as 90% in situations with little traffic (50 to 100 vehicles). This is because there are more evaluating cars and data sources available, which raises the cost of combating malicious vehicles and makes it more challenging to execute effective attacks. By using the previous reputation trajectory and pertinent parameters stored by MES when the number of directly interacting vehicles diminishes, the proposed approach somewhat ensures the regular updating of vehicle reputation.

Vehicle speed's effect on the effectiveness of malicious vehicle detection (Fig. 1.6): It is evident that the detecting performance decreases to some extent when the vehicle speed rises. When the car is moving faster than 100 km/h, this effect becomes more obvious. This is due to the fact that when speed increases, a vehicle's unit detection and connection times shorten. However, the proposed model still performs detection at a sufficient level even at faster speeds.

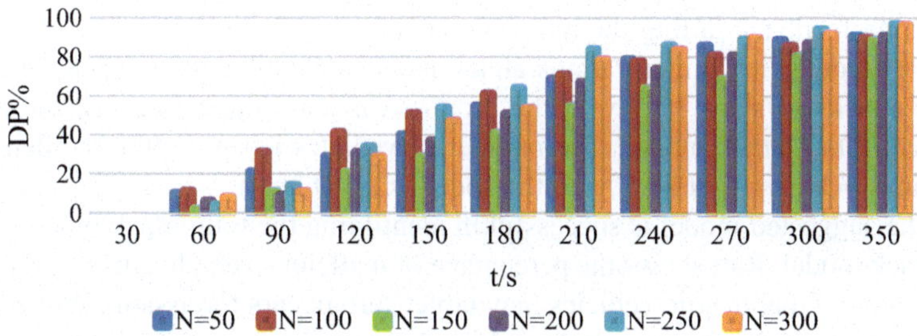

Fig. 1.5 Detection performance under different vehicle densities

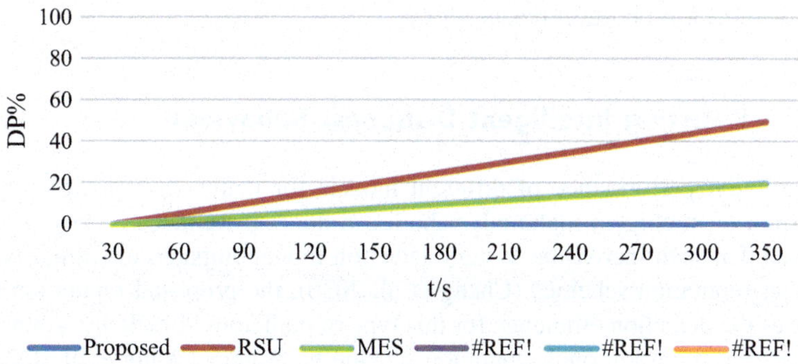

Fig. 1.6 Detection performance at different vehicle speeds

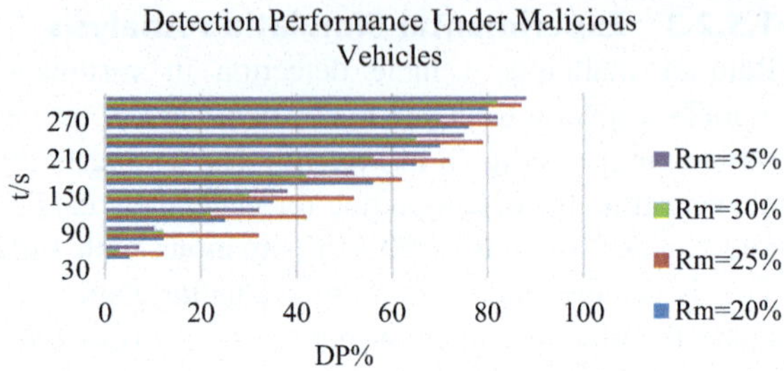

Fig. 1.7 Detection performance under different proportions of malicious vehicles

A shorter connection time implies fewer interactions. In a limited number of interactions, malicious vehicles may not launch attacks, and the reduced detection time further increases the convergence time of detection.

The suggested model's detection performance varies over time as the proportion of malicious cars (Rm) in the road environment increases, as shown in Fig. 1.7. According to the experimental findings, all harmful vehicles are correctly detected when the percentage of malicious vehicles hits 20%. The suggested model keeps a detection effectiveness of 90.35 per cent after running for 300 s, even at a proportion of 35 per cent. Due to an increase in the quantity of malicious vehicles, there has been a modest decline in detection performance, which has led to certain evaluation feedback failing the risk assessment.

Rate of detection under various attack models: Simulations are run for a variety of attacks to show how well the proposed model defends against the attacks detailed in the attack model. The number of vehicles is set at 300, the density of malicious vehicles is at 35%, and the speed is at 60 km/h.

If the suggested model is successful in identifying the four attack types listed in the attack model. Ratt shows the percentage of malicious vehicle attacks, indicating the likelihood that hostile vehicles may target certain cars for assault. As the attack proportion rises, the experimental results indicate a certain degree of decline. When the attack proportion exceeds 60%, this effect is more noticeable. The suggested approach maintains a detection rate of at least 50% for selfish behaviour, switch assaults, false message attacks, and cooperation attacks by malevolent vehicles when the attack proportion reaches 90%.

1.5.3 Resisting Intelligent Dishonest Behaviour

The detection performance of different models for malicious vehicles when the proportion of intelligent malicious vehicles in the road vehicles is 25%. It can be observed that with increasing system detection time, compared to similar types of vehicular reputation schemes (Cheng et al. 2020), the proposed reputation model improves the detection efficiency for this type of malicious vehicle by around 10%. Compared to the traditional reputation scheme by Author (Wang et al. 2022), the proposed model achieves an improvement of over 30% in detection efficiency. This is because the proposed model can adaptively adjust the retrospective time

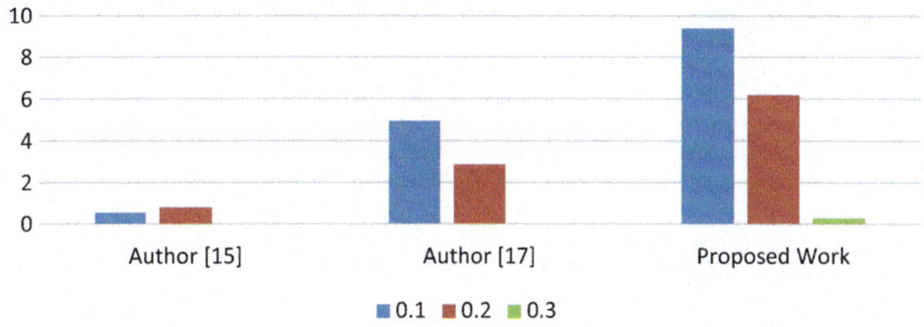

Fig. 1.8 False alarm rate

window for reputation updates based on environmental factors and historical reputation trajectory, significantly avoiding intelligent malicious vehicles bypassing the detection system through periodic or non-periodic low-cost reputation changes.

1.5.4 False Alarm Rate of the Reputation Model

Figure 1.8 compares the false alarm rates of the schemes proposed in Wang et al. (2022); Adil et al. 2022), and this study when the vehicle speed and quantity in the road environment remain constant, but the proportion of malicious vehicles varies. In the figure, PFA represents the false positive rate, and PFN represents the false negative rate. it can be observed that compared to the schemes proposed in Wang et al. (2022) and Adil et al. (2022), the proposed model exhibits lower false alarm rates when malicious vehicles are present.

1.5.5 Time Overhead

When adopting a higher security level of 1,024-bit encryption keys, Fig. 1.9 shows the comparison of calculation time overhead (tc) suffered by cars, RSUs, and MES during system operation. It can be shown that as the number of interacting feedback vehicles grows, so does the amount of time needed to compute each entity. The computation time overhead for the vehicle end is 0.0165 s when the simulated vehicle count reaches 300 on the road, which is reasonable. This is due to the fact that this model achieves single-key processing for vehicle users, avoiding additional complexity in key creation, in contrast to conventional privacy-preserving reputation methods. Through protocol optimisation, the suggested technique minimises pointless data homomorphic encryption steps. The additional computational burden caused by the demands for privacy protection can be effectively reduced by using the proposed technique.

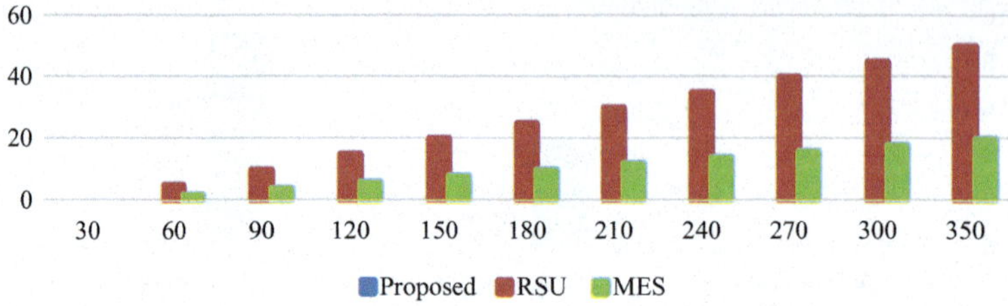

Fig. 1.9 Time Overhead

1.6 Conclusion

This chapter presents a privacy-protected reputation model based on a blockchain architecture to address the issues of an untrusted centralised reputation server, risks to user privacy, and restricted detection scope in standard vehicle network reputation systems. The DT-PKC technique is used for ciphertext data processing, which reduces the danger of user privacy leaks. The use of blockchain technology eliminates a central reputation server's sole control of car appraisal data. The results of the experiments show that the proposed system has advantages in terms of malicious vehicle identification performance, false positive rate, and computing overhead. Further study on consensus techniques, smart contract design, and on-chain data updates should be performed in the future. The goal is to improve the solution's security and efficiency while also maintaining the accuracy of the reputation model detection.

References

Adil M et al (2022) Three byte-based mutual authentication scheme for autonomous internet of vehicles. IEEE Trans Intell Transp Syst 23(7):9358–9369. https://doi.org/10.1109/TITS.2021.3114507

Ahmed GA, Sheltami TR, Mahmoud AS, Imran M, Shoaib M (2021) A novel collaborative IoD-assisted VANET approach for coverage area maximization. IEEE Access 9:61211–61223. https://doi.org/10.1109/ACCESS.2021.3072431

Atallah R, Khabbaz M, Assi C (2014) A novel drive-thru internet channel access scheme. IEEE Commun Lett 18(6):1047–1050. https://doi.org/10.1109/LCOMM.2014.2320903

Atallah RF, Khabbaz MJ, Assi CM (2015) Modeling and performance analysis of medium access control schemes for drive-thru internet access provisioning systems. IEEE Trans Intell Transp Syst 16(6):3238–3248. https://doi.org/10.1109/TITS.2015.2440447

Cao M, Zheng L, Jia W, Liu X (2021) Joint 3D reconstruction and object tracking for traffic video analysis under IoV environment. IEEE Trans Intell Transp Syst 22(6):3577–3591. https://doi.org/10.1109/TITS.2020.2995768

Chen L-W, Ho Y-F (2019) Centimeter-grade metropolitan positioning for lane-level intelligent transportation systems based on the internet of vehicles. IEEE Trans Industr Inf 15(3):1474–1485. https://doi.org/10.1109/TII.2018.2854901

Cheng J et al (2020) Location prediction model based on the internet of vehicles for assistance to medical vehicles. IEEE Access 8:10754–10767. https://doi.org/10.1109/ACCESS.2019.2957785

Ding N, Meng X, Xia W, Wu D, Xu L, Chen B (2020) Multivehicle coordinated lane change strategy in the roundabout under internet of vehicles based on game theory and cognitive computing. IEEE Trans Industr Inf 16(8):5435–5443. https://doi.org/10.1109/TII.2019.2959795

Ghane S, Jolfaei A, Kulik L, Ramamohanarao K, Puthal D (2021) Preserving privacy in the internet of connected vehicles. IEEE Trans Intell Transp Syst 22(8):5018–5027. https://doi.org/10.1109/TITS.2020.2964410

Huang J et al (2019) A crowdsource-based sensing system for monitoring fine-grained air quality in urban environments. IEEE Internet Things J 6(2):3240–3247. https://doi.org/10.1109/JIOT.2018.2881240

Hussain S, Mahmud U, Yang S (2021) Car e-Talk: an IoT-enabled cloud-assisted smart fleet maintenance system. IEEE Internet Things J 8(12):9484–9494. https://doi.org/10.1109/JIOT.2020.2986342

Kumar S, Singh K, Kumar S, Kaiwartya O, Cao Y, Zhou H (2019) Delimitated anti jammer scheme for internet of vehicle: machine learning based security approach. IEEE Access 7:113311–113323. https://doi.org/10.1109/ACCESS.2019.2934632

Li L, Gan J, Ji X, Qu X, Ran B (2022) Dynamic driving risk potential field model under the connected and automated vehicles environment and its application in car-following modeling. IEEE Trans Intell Transp Syst 23(1):122–141. https://doi.org/10.1109/TITS.2020.3008284

Li Z, Yang X, Wang C, Ma K, Jiang C (2022) Crowd-learning: a behavior-based verification method in software-defined vehicular networks with MEC framework. IEEE Internet Things J 9(2), 1622–1639. https://doi.org/10.1109/JIOT.2021.3107581

Liu Y, Wang L, Chen H-H (2015) Message authentication using proxy vehicles in vehicular ad hoc networks. IEEE Trans Veh Technol 64(8):3697–3710. https://doi.org/10.1109/TVT.2014.2358633

Liu Y, Wang Y, Chang G (2017) Efficient privacy-preserving dual authentication and key agreement scheme for secure V2V communications in an IoV paradigm. IEEE Trans Intell Transp Syst 18(10):2740–2749. https://doi.org/10.1109/TITS.2017.2657649

Liu X, Yang Z, Hou J, Huang W (2022) Dynamic scene's laser localization by NeuroIV-based moving objects detection and LiDAR points evaluation. IEEE Trans Geosci Remote Sens 60:1–14, Art no 5230414. https://doi.org/10.1109/TGRS.2022.3184962

Nan Z, Jia Y, Ren Z, Chen Z, Liang L (2022) Delay-aware content delivery with deep reinforcement learning in internet of vehicles. IEEE Trans Intell Transp Syst 23(7):8918–8929. https://doi.org/10.1109/TITS.2021.3087833

Punithan MX, Seo S-W (2013) King's graph-based neighbor-vehicle mapping framework. IEEE Trans Intell Transp Syst 14(3):1313–1330. https://doi.org/10.1109/TITS.2013.2260746

Rao PM, Jangirala S, Pedada S, Das AK, Park Y (2023) Blockchain integration for IoT-enabled V2X communications: a comprehensive survey, security issues and challenges. IEEE Access 11:54476–54494. https://doi.org/10.1109/ACCESS.2023.3281844

Safavat S, Rawat DB (2021) On the elliptic curve cryptography for privacy-aware secure ACO-AODV routing in intent-based internet of vehicles for smart cities. IEEE Trans Intell Transp Syst 22(8):5050–5059. https://doi.org/10.1109/TITS.2020.3008361

Sarker HS, Qiu C, Uehara H, Zhang K (2023) Brake-signal-based driver's location tracking in usage-based auto insurance programs. IEEE Internet Things J 10(12), 10172–10189. https://doi.org/10.1109/JIOT.2023.3237759

Song L, Sun G, Yu H, Du X, Guizani M (2020) FBIA: a fog-based identity authentication scheme for privacy preservation in internet of vehicles. IEEE Trans Veh Technol 69(5):5403–5415. https://doi.org/10.1109/TVT.2020.2977829

Su Z, Hui Y, Luan TH, Guo S (2017) Engineering a game theoretic access for urban vehicular networks. IEEE Trans Veh Technol 66(6):4602–4615. https://doi.org/10.1109/TVT.2017.2700049

Sun G, Song L, Yu H, Du X, Guizani M (2021) A two-tier collection and processing scheme for fog-based mobile crowdsensing in the internet of vehicles. IEEE Internet Things J 8(3):1971–1984. https://doi.org/10.1109/JIOT.2020.3015967

Talpur, Gurusamy M (2022) DRLD-SP: a deep-reinforcement-learning-based dynamic service placement in edge-enabled internet of vehicles. IEEE Internet Things J 9(8):6239–6251. https://doi.org/10.1109/JIOT.2021.3110913

Tan WL, Lau WC, Yue O, Hui TH (2011) Analytical models and performance evaluation of drive-thru internet systems. IEEE J Sel Areas Commun 29(1):207–222. https://doi.org/10.1109/JSAC.2011.110120

Tian Z, Gao X, Su S, Qiu J (2020) Vcash: a novel reputation framework for identifying denial of traffic service in internet of connected vehicles. IEEE Internet Things J 7(5):3901–3909. https://doi.org/10.1109/JIOT.2019.2951620

Ullah MAK, Kumar N, Abdullah AM, AlSanad AA, Noor F (2023) A conditional privacy preserving heterogeneous signcryption scheme for internet of vehicles. IEEE Trans Veh Technol 72(3):3989–3998. https://doi.org/10.1109/TVT.2022.3220041

Wang et al (2019) NDNVIC: named data networking for vehicle infrastructure cooperation. IEEE Access 7:62231–62239. https://doi.org/10.1109/ACCESS.2019.2913656

Wang Z, Liu L, Wang L, Wen X, Jing W (2022) Privacy-protecting and reputation-based participant recruitment scheme for IoV-based MCS. IEEE Internet Things J 9(22):22490–22500. https://doi.org/10.1109/JIOT.2021.3138131

Wei F, Zeadally S, Vijayakumar P, Kumar N, He D (2021) An intelligent terminal based privacy-preserving multi-modal implicit authentication protocol for internet of connected vehicles. IEEE Trans Intell Transp Syst 22(7):3939–3951. https://doi.org/10.1109/TITS.2020.2998775

Xi N, Li W, Jing L, Ma J (2022) ZAMA: a ZKP-based anonymous mutual authentication scheme for the IoV. IEEE Internet Things J 9(22):22903–22913. https://doi.org/10.1109/JIOT.2022.3186921

Xiao Z et al (2020) Blockchain and IoT for insurance: a case study and cyberinfrastructure solution on fine-grained transportation insurance. IEEE Trans Comput Soc Syst 7(6):1409–1422. https://doi.org/10.1109/TCSS.2020.3034106

Xiao T, Chen C, Pei Q, Song HH (2022) Consortium blockchain-based computation offloading using mobile edge platoon cloud in internet of vehicles. IEEE Trans Intell Transp Syst 23(10):17769–17783. https://doi.org/10.1109/TITS.2022.3168358

Xu W, Shi W, Lyu F, Zhou H, Cheng N, Shen X (2019) Throughput analysis of vehicular internet access via roadside WiFi hotspot. IEEE Trans Veh Technol 68(4):3980–3991. https://doi.org/10.1109/TVT.2019.2900032

Xu W, Guo S, Ma S, Zhou H, Wu M, Zhuang W (2020) Augmenting drive-thru internet via reinforcement learning-based rate adaptation. IEEE Internet Things J 7(4):3114–3123. https://doi.org/10.1109/JIOT.2020.2965148

Zheng et al (2022) Digital twin empowered heterogeneous network selection in vehicular networks with knowledge Transfer. IEEE Trans Veh Technol 71(11):12154–12168. https://doi.org/10.1109/TVT.2022.3190941

Sustainable Blockchain and Supply Chain Management: Integration Issues, Challenges, Potentials and Applications

2

Nishant Gupta and Ankit Kumar Jain

Abstract

Blockchain has improved openness, efficiency, and confidence in the global supply chain ecosystem. This chapter investigates blockchain's critical role in supply chain operations, beginning with an analysis of its key principles: decentralization, immutability, and cryptographic security. It then digs into the advantages of blockchain integration, which include more transparency, simplified transactions, greater traceability, and improved collaboration among supply chain players. This section's chapters address a wide range of blockchain integration topics, including supply chain digitization, smart contracts, data privacy, and decentralized applications. Real-world case studies demonstrate effective blockchain use in a variety of industries. Furthermore, it covers integration issues such as technological complexities, prices, laws, and governance, providing useful insights and best practices. The chapter also looks ahead to future themes including Internet of Things (IoT) and Artificial Intelligence (AI) integration, regulatory challenges, and scalability. It emphasizes the need of organizations embracing blockchain to achieve a competitive advantage and promote supply chain innovation. This chapter is a comprehensive resource for professionals, researchers, and students interested in learning about and using blockchain integration in supply chain management. It takes a comprehensive approach to the subject, delivering practical insights, case studies, and future perspectives to inspire and enable organizations to use blockchain technology for supply chain optimization and transformation.

N. Gupta · A. K. Jain (✉) · A. K. Jain (✉)
National Institute of Technology, Kurukshetra 136119, India
e-mail: ankitjain@nitkkr.ac.in

A. Kumar et al. (eds.), *Sustainable Security Practices Using Blockchain, Quantum and Post-quantum Technologies for Real Time Applications*, Contributions to Environmental Sciences & Innovative Business Technology,
https://doi.org/10.1007/978-981-97-0088-2_2

Keywords

Artificial Intelligence • Blockchain • Data security • Decentralization • Governance models • Internet of Things • Regulatory considerations • Scalability • Smart contracts • Supply chain digitization • Supply chain management • Traceability • Transparency

2.1 Introduction

2.1.1 Overview of Supply Chain Management

Figure 2.1 shows how supply chain management (SCM) coordinates and manages all operations involved in the production, acquisition, transformation, and distribution of goods and services from the point of origin to the end user. It comprises several parts: suppliers, manufacturers, distributors, retailers, and customers. SCM entails the planning, execution, and control of the flow of commodities, information, and cash along the supply chain network (Copacino 1997).

2.1.1.1 Importance of Efficient Supply Chain Management

Effective supply chain management is critical for organizations to achieve a competitive advantage in today's global marketplace. It lets businesses optimize processes, cut expenses, increase customer happiness, and adapt swiftly to market demands. Increased inventory management, faster manufacturing processes, increased collaboration with suppliers and partners, and improved customer service are all benefits of effective SCM (Roy et al. 2022).

2.1.1.2 Challenges in Traditional Supply Chain Management

Several difficulties impede traditional supply chain management's efficiency and efficacy. A lack of transparency, restricted traceability, information asymmetry, manual and paper-based processes, lengthy reconciliation and settlement procedures, and the existence of intermediaries are among these problems (Roy et al. 2022). As illustrated in Fig. 2.2, traditional SCM systems frequently need help to give real-time visibility into the movement of items, resulting in delays, inaccuracies, and increased costs.

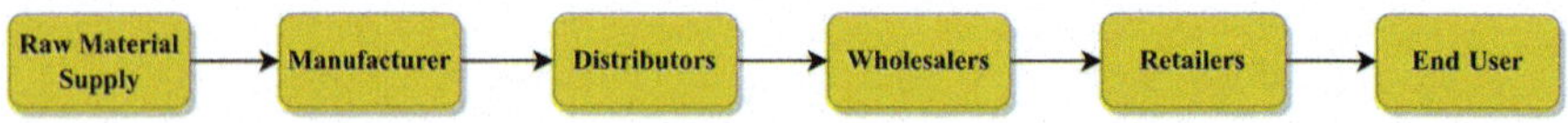

Fig. 2.1 Typical supply chain

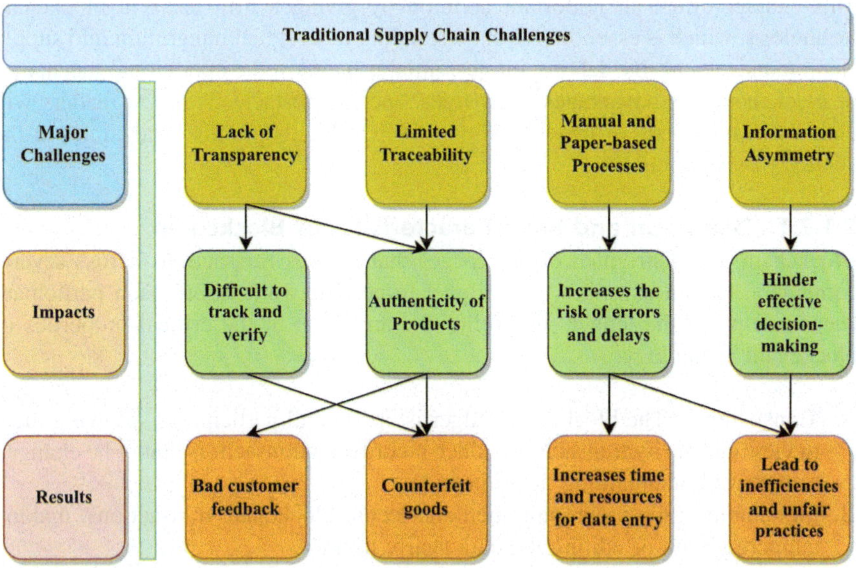

Fig. 2.2 Challenges in the traditional supply chain

2.1.2 Introduction to Blockchain Technology

Blockchain technology has emerged as a disruptive force, affecting businesses worldwide. In this chapter, we will look at the core principles and workings of blockchain technology, establishing the framework for a thorough knowledge of its application in supply chain management. This chapter tries to achieve many critical objectives:

1. **Understanding Blockchain Fundamentals:**
   - Investigating blockchain's key characteristics, such as decentralization, transparency, immutability, and cryptographic security.
   - The structure of blocks and the consensus processes that support blockchain networks are described in detail.
2. **Working of Blockchain:**
   - Investigating blockchain's operating mechanisms, from transaction verification to block generation.
   - Investigating how consensus techniques such as proof-of-work and proof-of-stake help to secure networks.
3. **Key benefits of blockchain technology:**
   - Uncovering the transformational benefits of blockchain integration in supply chain management.
   - Highlighting advancements in supply chain transparency, traceability, efficiency, and security.

This chapter offers an important addition by giving a firm basis in blockchain technology, which is essential for understanding its eventual integration into supply chain management. By delving into essential principles, the operational mechanics of blockchain, the advantages it provides, and the obstacles it poses, readers will obtain a comprehensive understanding of blockchain's promise and limits in the context of supply chains.

2.1.2.1 Definition and Key Characteristics of Blockchain

Blockchain is a distributed digital ledger that records transactions across several computers, known as nodes. It is a peer-to-peer network where each participant keeps a copy of the ledger. The following are some of the critical properties of blockchain technology:

A. Transparency: The blockchain ledger is open to all participants, allowing them to view and verify transactions. Once recorded, a transaction cannot be changed or removed, maintaining data integrity.
B. Immutability: Cryptographic methods secure blockchain transactions, making tampering with or editing the data nearly tricky.
C. Decentralization: Blockchain is a decentralized network and it eliminates the need for central control.

2.1.2.2 How Blockchain Works: Distributed Ledger, Consensus Mechanisms, and Immutability

Blockchain uses a distributed ledger, a constantly expanding list of information known as blocks (Zheng et al. 2017). Each block comprises a set of validated transactions linked together using cryptographic hashes to form a chain, as shown in Fig. 2.3. As illustrated in Fig. 2.4, the distributed ledger is kept across numerous network nodes, ensuring redundancy and transparency.

Consensus methods are critical in the blockchain. They are the algorithms or rules that govern how transactions are approved and recorded on the blockchain (Zheng et al. 2017). These measures assure network participant agreement and prevent fraudulent or conflicting transactions from being recorded in the ledger (Nakamoto 2008).

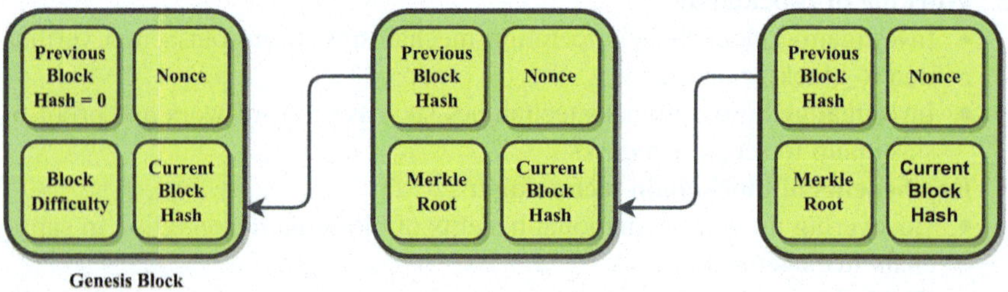

Fig. 2.3 Chain of blocks in a blockchain

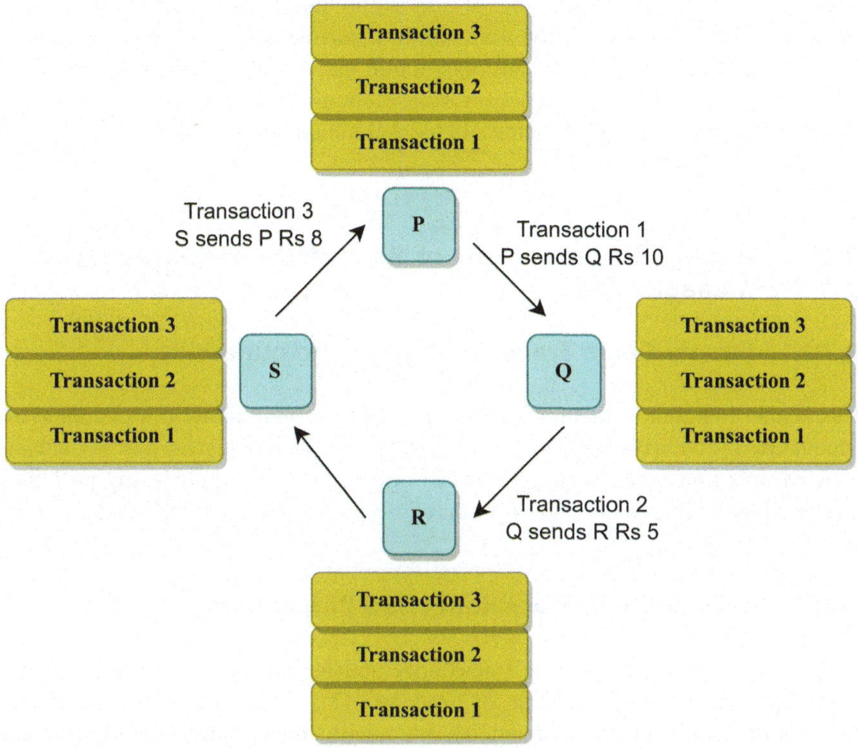

Fig. 2.4 Distributed ledger stored across multiple nodes in a blockchain technology

Blockchain's immutability is a crucial property. It is nearly hard to change or cancel a transaction once it has been recorded on the blockchain. Table 2.1 summarizes the specific benefits that organizations can gain from blockchain integration.

To facilitate effective learning, this chapter is structured as follows. Section 2.2 explains potential applications of blockchain in supply chain management.

Table 2.1 Benefits of blockchain integration in supply chain management

Benefits	Description
Increased transparency	Blockchain allows supply chain data visibility and traceability in real-time
Enhanced efficiency	Processes have been streamlined, transactions have been automated, and paperwork has been decreased
Improved trust and security	Immutable records, cryptographic security, and reduced fraud risks are all advantages of immutable records
Enhanced collaboration	Data exchange that is seamless, enhanced coordination, and stakeholder trust

Section 2.3 discusses blockchain's role in supply chain transparency. Section 2.4 presents blockchain integration for supply chain security and risk mitigation. Section 2.5 discusses blockchain integration for supply chain efficiency and cost savings. Section 2.6 describes various adoption considerations. Section 2.7 presents future trends and challenges in blockchain integration and finally Sect. 2.8 concludes the chapter.

2.2 Potential Applications of Blockchain in Supply Chain Management

2.2.1 Supply Chain Transparency and Accountability

Blockchain integration in supply chain management may allow transparency and accountability by recording every transaction and occurrence on the blockchain. This ensures transparency and eliminates information discrepancies by providing stakeholders with credible and up-to-date information (Mann et al. 2018).

2.2.2 Supply Chain Traceability and Provenance

Blockchain can create accurate traceability and legitimacy by documenting the movement and historical records of products. This procedure allows stakeholders to verify items' origin, authenticity, and quality, decreasing counterfeiting and assuring compliance with legal standards (Mann et al. 2018).

2.2.3 Supply Chain Automation and Smart Contracts

Smart contracts, which are code-based contracts with preset terms, can automate and streamline supply chain procedures. Smart contracts lower transaction costs, improve efficiency, and reduce errors and disputes by eliminating intermediaries and automating payment settlements, procurement, and order fulfillment.

2.2.4 Supply Chain Finance and Payment Solutions

Blockchain technology enables novel supply chain finance options, such as real-time payments and smart contracts for financing arrangements. Organizations can streamline payment processes, reduce dependency on traditional banking institutions, and enable speedier cross-border transactions by utilizing cryptocurrencies or tokenization. Table 2.2 depicts the various applications of blockchain integration in supply chains.

Table 2.2 Use cases of blockchain integration in supply chain management

Use cases	Description
Product traceability	Throughout the supply chain, track and verify the authenticity and provenance of products
Supplier management	Verify the credentials of suppliers, monitor compliance, and manage supplier data
Inventory management	Inventory levels, demand predictions, and tracking are all visible in real-time
Smart contract payment	Payment settlements should be automated based on predetermined supply chain milestones

Finally, Sect. 2.1 of this chapter presents an overview of supply chain management and introduces blockchain technology as a possible answer to the issues that traditional supply chain systems confront. It emphasizes blockchain's core properties, benefits, and possible applications in improving transparency, traceability, security, and efficiency in supply chain management.

2.3 Blockchain Technology and Its Role in Supply Chain Transparency

2.3.1 Enhancing Transparency in Supply Chains

- Transparency Challenges in Traditional Supply Chains

Because of the complexity and fragmentation of global trade, traditional supply chains frequently lack transparency. The supply chain involves multiple stakeholders, including suppliers, manufacturers, distributors, and retailers, making tracking and tracing the movement of items and information difficult. Furthermore, information silos, manual processes, and a lack of visibility into supplier networks all add to the obscurity of traditional supply chains (Hellani et al. 2021).

- Blockchain's Role in Improving Transparency

Blockchain technology provides a decentralized and transparent environment for transaction recording and verification. Each transaction recorded on the blockchain is tamper-proof, maintaining data integrity. This transparency allows all stakeholders to obtain real-time and trustworthy information regarding commodity movement, transactions, and events throughout the supply chain (Hellani et al. 2021). Blockchain helps stakeholders to make educated decisions, eliminate risks, and increase accountability by offering visibility into each stage of the supply chain as shown in Fig. 2.5.

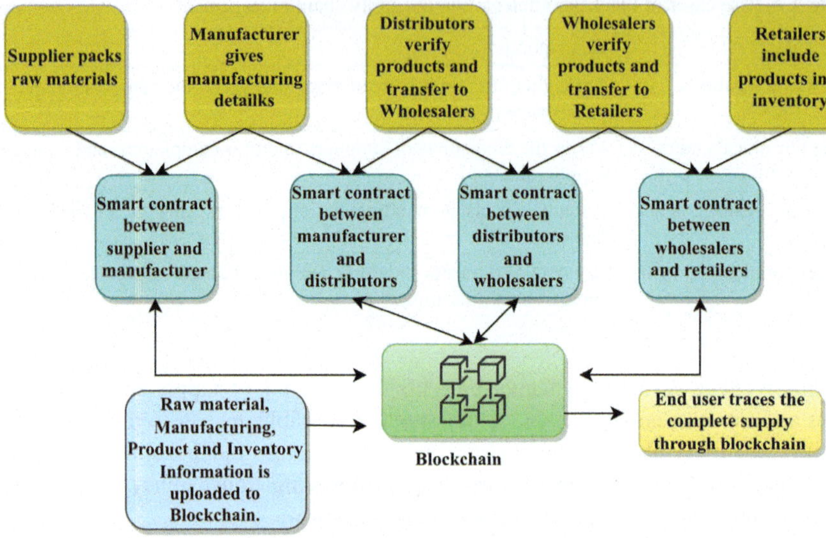

Fig. 2.5 Blockchain-based supply chain

- Real-time Tracking and Traceability

Each transaction, such as a change in ownership or location, is time-stamped and documented, producing a record that can be audited. This enables stakeholders to track the origin, provenance, and condition of the items, assuring authenticity and decreasing the risk of counterfeiting or fraudulent activities (Mann et al. 2018; Hellani et al. 2021). Real-time tracking also enables faster detection and resolution of supply chain bottlenecks or disturbances.

2.3.2 Blockchain-Based Traceability and Provenance

This subsection presents the blockchain-based traceability and provenance. Details are presented as follows.

- Importance of Traceability and Provenance

Traceability refers to the ability to track and trace a product or component from its creation to the ultimate consumer.

- Supply Chain Use Cases for Traceability and Provenance

Several sectors have used blockchain to improve traceability and provenance. Blockchain, for example, can enable customers to trace the origin of products, verify organic or fair-trade certificates, and assure compliance with safety laws in the food business. Blockchain can track the transit of drugs from manufacturers to patients in the pharmaceutical business, decreasing the risk of counterfeit or inferior medications. Similarly, blockchain can verify the authenticity and ownership history of high-value objects in the luxury goods industry, ensuring consumers.

2.3.3 Smart Contracts and Supply Chain Automation

This subsection presents the smart contracts and supply chain automation integration possibilities. Details are presented as follows.

- Smart Contract automation process

Smart contracts run independently and have predetermined terms and conditions expressed in code. When specific criteria are met, these contracts automatically execute and enforce the agreed-upon terms (Wang et al. 2019). Figure 2.6 shows how intelligent contracts operate on the blockchain, using its transparency, immutability, and decentralized nature. They do away with intermediaries, laborious processes, and formal legal agreements, expediting supply chain operations and lowering expenses (Mohanta et al. 2018).

- Benefits of Smart Contracts in Supply Chain Management

Smart contracts offer several benefits in supply chain management:

A. Automation: Smart contracts automate critical supply chain activities like procurement, order fulfillment, and payment settlement (Hamledari and Fischer 2021). This decreases the need for manual involvement, shortens transaction times, and lessens the likelihood of errors or disagreements.

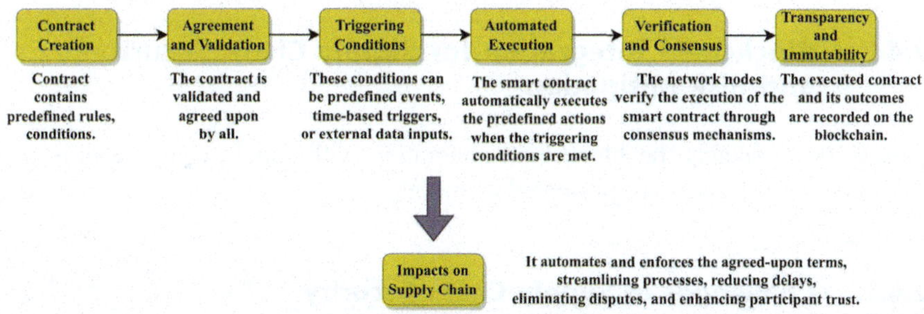

Fig. 2.6 The smart contract demonstrates the stages required in smart contract execution and the influence on supply chain operations

B. Cost savings: Smart contracts lower transaction costs in the supply chain by eliminating intermediaries and decreasing administrative overheads. Automation allows businesses to streamline their operations and save money.
C. Increased efficiency and accuracy: Smart contracts ensure that predefined terms and conditions are followed precisely. Process automation decreases human error, improves accuracy, and increases supply chain efficiency.
D. Enhanced trust and transparency: Smart contracts run on blockchain technology that is transparent and auditable. All stakeholders may see the set terms and conditions, ensuring transparency and fostering trust in supply chain interactions.

- Supply Chain Use Cases for Smart Contracts

Smart contracts can be used in various supply chain scenarios (Law 2017). Smart contracts, for example, can automate freight bookings, optimize routes, and trigger automatic payments upon successful delivery in logistics and transportation. Smart contracts can help with invoice factoring and supply chain financing by automating payment settlements based on established parameters. Smart contracts can also automate compliance activities, including validating certificates or regulatory requirements and triggering relevant actions or warnings when noncompliance is identified.

Finally, this section looks at how blockchain technology might increase supply chain transparency. It investigates the problems of traditional supply chain transparency and the potential for blockchain to fix them. The chapter goes deeper into the impact of blockchain on traceability and provenance, allowing stakeholders to validate product origin and authenticity. It also looks at smart contracts and how they might be used to automate supply chain activities, resulting in greater efficiency and accuracy. In addition, the chapter contains real-world examples of blockchain and smart contracts being utilized successfully in supply chain management.

2.4 Blockchain Integration for Supply Chain Security and Risk Mitigation

This section explains the blockchain integration with supply chain security and risk mitigation. Details are presented as follows.

2.4.1 Challenges in Supply Chain Security

This subsection explains the challenges in supply chain security with and without blockchain.

- Security Risks in Traditional Supply Chains

Traditional supply chains are exposed to various security risks, including theft, counterfeiting, unauthorized access, data breaches, and manipulation. The need for more transparency and visibility in supply chain procedures makes detecting and preventing security breaches harder. Furthermore, the engagement of various stakeholders and the reliance on manual processes make supply chains more vulnerable to security threats (Etemadi et al. 2021).

- Impact of Security Breaches on Supply Chains

Supply chain security breaches can have profound implications. They can result in financial losses, reputational damage, business disruption, and legal ramifications. Counterfeit products can reach consumers, posing health and safety dangers. Security breaches can potentially jeopardize sensitive information, like client data, trade secrets, and intellectual property, resulting in a loss of competitive advantage.

2.4.2 Blockchain's Role in Supply Chain Security

- Immutable and Tamper-Proof Nature of Blockchain

The immutability and tamper-proof character of data kept on the blockchain are ensured by blockchain technology. Once a transaction has been recorded on the blockchain, it cannot be changed or removed, ensuring excellent data integrity and security (Azzi et al. 2019). Figure 2.4 shows how this feature makes blockchain a perfect tool for minimizing security threats in supply chains.

- Authentication and Verification of Products

The blockchain provides product authentication and verification throughout the supply chain. Stakeholders can verify the authenticity and integrity of products by recording the movement and ownership of goods on the blockchain. This decreases the possibility of counterfeit items entering the supply chain and ensures that only authentic products reach the end user.

- Secure Sharing of Information

Blockchain enables the secure exchange of information among supply chain partners. Rather than depending on risky centralized databases, blockchain enables decentralized storage of information across several nodes. This decentralized design, combined with encryption and access controls, improves data security and prevents unauthorized access or breaches.

2.4.3 Supply Chain Risk Mitigation with Blockchain

- Enhanced Supply Chain Visibility and Transparency

Blockchain integration in supply chains improves visibility and transparency, allowing stakeholders to detect and manage any hazards ahead of time. The blockchain's real-time tracking and traceability of commodities allow for rapidly discovering bottlenecks, delays, or disruptions. This enables early responses and risk-mitigation measures to reduce the impact of risks (Wang et al. 2020).

- Improved Supplier Verification and Compliance

Blockchain technology can speed up vetting suppliers and ensure compliance with regulatory regulations and industry standards. Organizations may quickly access and verify the validity of suppliers by recording supplier information, certifications, and audit reports on the blockchain. Smart contracts can enforce compliance by triggering actions or penalties when noncompliance is found.

- Supply Chain Auditability and Accountability

The transparent and auditable feature of blockchain improves supply chain audibility and accountability. Every transaction and event on the blockchain leaves a permanent and traceable record. This allows auditors and regulatory organizations to evaluate supply chain process compliance and uncover any irregularities or fraudulent actions. Blockchain promotes accountability among supply chain players because their actions are visible and cannot be readily tampered with.

- Supply Chain Resilience and Risk Response

Blockchain technology can improve supply chain resilience by allowing swift responses to risks and interruptions. Blockchain provides real-time visibility into inventory levels, alternative suppliers, and logistics options during a supply chain disruption, such as a natural disaster or a product recall. This allows organizations to adapt and minimize the impact on operations, allowing for more efficient risk response. Table 2.3 will comprehensively summarize the application scenarios where blockchain integration can improve supply chain security, risk mitigation, and overall resilience.

Finally, this section delves into the significance of blockchain integration in improving supply chain security and risk mitigation. It explores how blockchain technology tackles the security concerns that traditional supply networks confront. The chapter emphasizes blockchain's immutable and tamper-proof aspect, which enables authentication, verification, and safe information sharing. It also examines how blockchain can help with supply chain visibility, supplier verification,

Table 2.3 Use cases of blockchain integration for supply chain security and risk mitigation

Use cases	Description
Product traceability	Enable end-to-end traceability of products across the supply chain, providing a transparent record of each transaction and movement, reducing the risk of counterfeiting, and ensuring the authenticity and quality of products
Supplier verification	Verify suppliers' authenticity and credibility by recording their information, certifications, and compliance records on the blockchain, ensuring the integrity of the supply chain and reducing the risk of working with unverified or non-compliant suppliers
Secure data sharing	Facilitate secure and encrypted sharing of sensitive supply chain information between authorized participants, ensuring data privacy and confidentiality and preventing unauthorized access or tampering
Supply chain auditing	Improve supply chain auditing and compliance by recording all transactions and events on the blockchain, creating an immutable audit trail for regulatory compliance, detecting anomalies, and ensuring transparency and accountability
Risk response and resilience	Enhance supply chain resilience by leveraging real-time visibility provided by the blockchain, enabling quick responses to disruptions, identifying alternative suppliers, and optimizing logistics routes to mitigate risks and maintain continuity

compliance, audibility, and risk response. Organizations can improve their supply chains' security, resilience, and integrity by adopting blockchain in supply chain management.

2.5 Blockchain Integration for Supply Chain Efficiency and Cost Savings

2.5.1 Inefficiencies in Traditional Supply Chains

- Complexity and Fragmentation

Traditional supply chains can suffer from complexity and fragmentation because of the involvement of various stakeholders, disparate systems, and manual processes. These issues result in supply chain delays, inaccuracies, and cost increases.

- Lack of Transparency and Visibility

To support effective decision-making and stakeholder participation, more openness in supply chain activities is essential. Transparency causes information asymmetry, delays in discovering and addressing problems, and higher costs from redundant or inefficient actions.

- Inefficient Documentation and Reconciliation

Manual reconciliations and paper-based documentation may be inefficient and error-prone. Inefficient documentation procedures cause supply chain delays, disputes, and higher administrative expenditures.

2.5.2 Blockchain's Impact on Supply Chain Efficiency

This subsection presents the blockchain's impact on supply chain efficiency. Details are as follows.

- Streamlined Documentation and Data Management

Blockchain technology enables the digitization and automation of supply chain documentation activities. Organizations may remove the need for human documentation, enhance processing time, and decrease mistakes by recording transactions on the blockchain. Digital documentation makes it easier for supply chain participants to share data more effectively, enhancing coordination and decreasing redundancy.

- Real-Time Visibility and Tracking

Blockchain technology enables real-time insight into the movement of commodities, transactions, and events across the supply chain. Because of this openness, stakeholders can keep track of inventory levels, manufacturing processes, and spot bottlenecks or disruptions quickly. Real-time tracking enables improved decision-making, more effective resource allocation, and more efficient logistics, all of which result in cost savings and enhanced efficiency.

- Supply Chain Optimization Through Data Analytics

The usage of blockchain allows for the collection and analysis of massive volumes of supply chain data. Organizations may obtain important insights into supply chain performance, demand patterns, and optimization possibilities by applying advanced analytics and machine learning algorithms to blockchain data. Data-driven decision-making provides process improvements, more accurate demand forecasts, and inventory optimization, leading to cost savings and increased efficiency.

2.5.3 Supply Chain Cost Savings with Blockchain

- Reduced Administrative Costs

Intermediaries, manual paperwork, and reconciliation processes are all eliminated by blockchain. Organizations can drastically cut administrative expenses associated with processing, verification, and record-keeping by automating supply chain transactions and paperwork. Smart contracts save money by automating payment settlement and eliminating the need for formal agreements or arbitration (Perboli et al. 2018).

- Improved Inventory Management

Organizations may optimize inventory management by having real-time visibility into inventory levels and movement on the blockchain. Organizations can cut inventory carrying costs, minimize stockouts, and increase supply chain responsiveness by properly managing stock levels, predicting demand patterns, and speeding up order fulfillment.

- Efficient Supplier Management

Integrating blockchain improves supplier management by offering transparency, traceability, and performance data. Based on verifiable data on the blockchain, organizations can discover dependable suppliers, track supplier performance, and negotiate better terms. This reduces costs due to better supplier selection, lower risks, and better pricing negotiations.

- Streamlined Logistics and Transportation

Blockchain enables optimal logistics and transportation management by providing real-time visibility into shipment status, location, and conditions. This allows for more accurate route planning, lower transportation costs, and improved delivery reliability. By automating freight reservations, optimizing carrier selection, and expediting customs operations, smart contracts can cut costs and boost operational efficiency (Perboli et al. 2018).

This section investigates how blockchain integration can improve supply chain efficiency and save costs by revealing inefficiencies in existing supply networks. It investigates how blockchain technology might aid in streamlined paperwork, real-time visibility, data analytics, and supply chain optimization. The chapter also emphasizes the cost-cutting benefits of blockchain, such as improved inventory management, effective supplier management, and optimized logistics and transportation. By embracing blockchain technology, organizations can achieve operational improvements, cost reductions, and competitive advantages in supply chain management.

2.6 Overcoming Challenges and Adoption Considerations for Blockchain Integration in Supply Chain Management

2.6.1 Challenges in Blockchain Integration

Technical Complexity

Implementing blockchain technology in supply chain management necessitates a deep understanding of the technological factors involved. Blockchain architecture, network scalability, interoperability, and data integration may present organizational issues (Kouhizadeh et al. 2021). Managing this technical complexity necessitates skill and meticulous planning.

Cost and Investment
Integrating blockchain into the supply chain incurs initial expenditures such as infrastructure setup, development, and continuous maintenance. Organizations must examine the return on investment (ROI) and long-term benefits of blockchain integration compared to older solutions. Financial issues are essential in the decision-making process.

Regulatory and Legal Frameworks
In many places, the regulatory landscape surrounding blockchain technology is still emerging. Organizations must manage many legal and regulatory obligations concerning data protection, intellectual property rights, smart contracts, and cross-border transactions. It is critical to ensure compliance with applicable rules and regulations to avoid legal issues (Sulkowski 2018).

Collaboration and Governance
Collaboration among numerous stakeholders, including suppliers, manufacturers, logistics providers, and customers, is required for blockchain integration. It is vital for successful implementation to establish a shared governance model, define roles and duties, and handle possible conflicts of interest. Collaboration issues can occur due to differences in organizational cultures, incentives, and technological adoption levels (Naef et al. 2022). Table 2.4 provides a visual representation of these obstacles, explaining each in detail and offering advice on overcoming them.

2.6.2 Adoption Considerations

Use Case Identification

Before using blockchain, businesses must find appropriate use cases where the technology may add significant value. Assessing pain spots, inefficiencies, and areas that require transparency, traceability, or automation is critical. Use cases that

Table 2.4 Challenges in blockchain integration

Challenges	Description
Technical complexity	Implementing and managing the technical infrastructure and blockchain nodes
Cost implications	Initial investment, maintenance costs, and scalability challenges
Regulatory framework	Navigating legal and regulatory complexities related to blockchain adoption
Governance and standards	Establishing industry-wide governance models and interoperability standards

correspond with an organization's strategic goals and deliver quantifiable benefits, like cost reductions, efficiency increases, or risk mitigation, should be prioritized.

Data Privacy and Security

Organizations must carefully examine data privacy and security measures since blockchain entails the sharing and keeping data over a decentralized network. Encryption techniques, access controls, and privacy-preserving mechanisms must be implemented to secure sensitive information. Data protection standards, such as the General Data Protection Regulation, must be followed to maintain customer trust.

Scalability and Interoperability

Blockchain networks must accommodate large volumes of transactions and data created by complicated supply chains. Organizations should evaluate the scalability of blockchain systems and consider solutions capable of handling the expected transaction volume. Interoperability with existing systems and technologies, such as ERP (Enterprise Resource Planning) and IoT devices, is also critical for seamless integration (Qin and Gervais 2018).

Change Management and Training

Blockchain technology necessitates a paradigm shift in attitude, methods, and workflows. Organizations should engage in change management techniques to educate staff and stakeholders on the advantages and consequences of blockchain integration. Training programs and workshops can aid in developing the skills and knowledge needed to operate with blockchain systems.

Proof of Concept and Pilots

Organizations can perform proof of concept (POC) initiatives and pilots before using a wide-scale blockchain to validate the technology's applicability and viability. POCs enable organizations to test blockchain integration's technical and business components in a controlled environment, uncover potential issues, and optimize their deployment strategy (George et al. 2019).

Collaboration and Ecosystem Building
Collaboration with technology suppliers, industry consortia, and regulatory authorities is frequently required for successful blockchain integration. Organizations should actively participate in industry forums and engage with relevant stakeholders to share knowledge, standards, and best practices. Building a solid ecosystem of blockchain-enabled partners, suppliers, and customers can generate additional value and create network effects (Agrawal et al. 2022).

Finally, Sect. 2.7 covers the problems and implications of adopting blockchain in supply chain management. It addresses technical obstacles, cost considerations, regulatory frameworks, collaborative challenges, and governance concerns. Key adoption issues include use case identification, data privacy and security, scalability, change management, proof of concept, and ecosystem building. Organizations may maximize the benefits of blockchain integration and drive change in their supply chain operations by recognizing and efficiently addressing these difficulties and considerations.

2.7 Future Trends and Challenges in Blockchain Integration

2.7.1 Emerging Trends in Blockchain Integration

Integration of Blockchain with Emerging Technologies like IoT and AI

Integrating blockchain with other new technologies, such as the IoT and Artificial Intelligence (AI), is an exciting trend in supply chain management. Organizations can record real-time data about items by connecting IoT devices and sensors with blockchain, enabling transparency and traceability across the supply chain. AI can use this data to generate meaningful insights and automate decision-making processes, improving supply chain efficiency even more.

Decentralized Supply Chain Networks and Consortium Blockchains
The concept of blockchain-enabled decentralized supply chain networks is gaining traction. Organizations are investigating the use of consortium blockchains for supply chain operations, in which various stakeholders interact and share a distributed ledger. These networks increase trust, transparency, and efficiency by eliminating the need for intermediaries and promoting direct peer-to-peer interactions. Consortium blockchains can improve supply chain performance by fostering collaboration, data exchange, and standardization across industries.

The Role of Cryptocurrencies and Tokenization in Supply Chain Finance
Tokenization and cryptocurrency have the potential to transform supply chain finance. Blockchain-based cryptocurrencies like Bitcoin and Ethereum enable secure and efficient financial transactions inside supply chains. Tokenization allows physical or digital assets to be represented as digital tokens on a blockchain, allowing for fractional ownership, greater liquidity, and simplified cross-border

payments. These technologies can improve payment processes, lower transaction costs, and broaden financial inclusion in supply chain financing.

2.7.2 Regulatory and Legal Considerations

Addressing Regulatory Challenges and Compliance Issues

As blockchain usage grows, authorities face the challenge of developing transparent frameworks and laws to oversee blockchain-based supply chains. Organizations must traverse changing regulatory landscapes while adhering to regulations governing data privacy, consumer protection, anti-money laundering, and intellectual property rights. Collaboration among industry stakeholders and regulatory agencies is critical for building appropriate frameworks that support innovation while addressing legal concerns. Here are a few examples of laws and acts in various countries that have incorporated blockchain:

1. **United States:**
   - **Digital Asset Regulatory Reform Act (DARRA)**: This proposed law proposes to create a clear regulatory framework for digital assets in the United States, including cryptocurrencies and blockchain technology.
2. **Switzerland:**
   - **Blockchain Act**: Switzerland enacted a comprehensive blockchain law to govern distributed ledger technology. It establishes the legal requirements for tokenization, smart contracts, and trading platforms.
3. **Singapore:**
   - **Payment Services Act**: This legislation was passed in Singapore, and it provides provisions for cryptocurrency payment services and digital payment token services. It creates a regulatory framework for bitcoin service providers.
4. **Japan:**
   - **Payment Services Act**: The Payment Services Act of Japan recognizes cryptocurrency as lawful payment methods. It also mandated licensing for bitcoin exchanges operating in the nation.
5. **United Arab Emirates (UAE):**
   - **Smart Dubai Blockchain Strategy**: The UAE's Smart Dubai Blockchain Strategy, while not a law, intends to make Dubai the first blockchain-powered government by 2020. Its primary goal is to deploy blockchain technology in government services.

Intellectual Property Rights and Data Privacy in Blockchain-Based Supply Chains

Blockchain's immutability and transparency can pose intellectual property rights and data privacy issues. Organizations must implement safeguards to protect sensitive information and guarantee that only authorized parties can access specific blockchain data. Smart contracts can be programmed to enforce data privacy standards and limit access to sensitive information. Furthermore, legal frameworks must be established to manage intellectual property challenges and preserve innovation in blockchain-based supply chains. Here are a few examples of IPR:

1. **Patents:**
   - **Example**: Bank of America's Cryptocurrency Risk Detection System (https://patents.google.com/patent/US20180071283A1/en).
2. **Copyrights:**
   - **Example**: Bitcoin Whitepaper by Nakamoto (2008).
3. **Trademarks:**
   - **Example**: Ethereum Trademark (https://ethereum.org/en/).

Legal Frameworks and Initiatives for Blockchain Adoption in Supply Chain Management
Governments and international organizations are rapidly recognizing blockchain's promise in supply chain management and launching initiatives to encourage its implementation. For example, pilot programs and regulatory sandboxes are being built to test blockchain technology and uncover regulatory gaps. Industry consortiums and organizations such as the International Organisation for Standardisation (ISO) are working to create interoperability and consistency in blockchain implementations through standardization efforts (König et al. 2020). These activities help to create an atmosphere conducive to blockchain adoption in supply chain management.

2.7.3 Overcoming Adoption Challenges and Future Outlook

Scaling Blockchain Solutions for Large-Scale Supply Chains

Scaling the technology to accommodate the complexities and transaction volumes of large-scale supply chains is one of the problems of blockchain integration. Organizations must handle scalability challenges like network throughput, transaction speed, and storage capacity as use grows (Bose et al. 2018). Blockchain protocol advancements, such as sharding and layer-two solutions, are being developed to boost scalability and ensure that blockchain can meet the needs of global supply chains.

Education and Training for Blockchain Integration
Organizations must engage in education and training programs to fully realize the promise of blockchain in supply chain management (Steiu 2020). Building internal blockchain understanding and competence is critical for effective adoption. Employees and stakeholders can benefit from training efforts that assist them

Table 2.5 Future trends and challenges in blockchain integration

Trends and challenges	Description
Integration with IoT	Exploring the integration of blockchain with the IoT to enable secure and automated data exchange between devices and the supply chain
Consortium blockchains	The emergence of consortium blockchain allows multiple stakeholders to collaborate and share data securely while maintaining privacy and control
Supply chain finance	Utilizing cryptocurrencies and tokenization in supply chain finance enables faster and more efficient cross-border transactions and settlements

in grasping the benefits, challenges, and best practices of blockchain integration. Collaboration with academic institutions and industry specialists helps spread information and develop a trained workforce in blockchain integration.

Predictions for the Future of Blockchain Integration in Supply Chain Management

Blockchain integration in supply chain management has a bright future. By enabling unprecedented transparency, traceability, and efficiency, the technology has the potential to revolutionize supply networks. We should expect more blockchain usage across industries, resulting in more simplified and resilient supply chains. Blockchain networks will become more integrated as interoperability improves and standards are defined, enabling smooth collaboration and data exchange across supply chain partners. Table 2.5 summarizes the anticipated trends and challenges in using blockchain technology.

2.8 Conclusion

Integrating blockchain in supply chain management necessitates careful consideration of technological, financial, legal, and cultural considerations, as well as teamwork to achieve standards and adoption. Blockchain in supply chains will grow, combining with IoT and AI, extending decentralized networks, and utilizing crypto and tokens in finance, emphasizing the dynamic nature of the technology. Increased scalability, interoperability, and user-friendliness will underlie the predicted boom in adoption and realization of blockchain's disruptive potential in supply chain management. Organizations may achieve a competitive advantage in the dynamic environment of supply chain management by utilizing the potential of blockchain integration to improve transparency, efficiency, data security, and cooperation. Adopting blockchain technology is a strategic choice that has the ability to inspire innovation, optimize processes, and produce value for organizations and their stakeholders. Organizations must stay agile as the blockchain environment changes in order to capitalize on the benefits it brings, paving the way for a more transparent, efficient, and sustainable future in supply chain management.

References

Agrawal TK, Angelis J, Khilji WA, Kalaiarasan R, Wiktorsson M (2022) Demonstration of a blockchain-based framework using smart contracts for supply chain collaboration. Int J Prod Res 1–20

Azzi R, Chamoun RK, Sokhn M (2019) The power of a blockchain-based supply chain. Comput Ind Eng 135:582–592

Bose S, Raikwar M, Mukhopadhyay D, Chattopadhyay A, Lam KY (2018) BLIC: a blockchain protocol for manufacturing and supply chain management of ICS. In: 2018 IEEE international conference on Internet of Things (iThings) and IEEE green computing and communications (GreenCom) and IEEE cyber, physical and social computing (CPSCom) and IEEE smart data (SmartData), pp 1326–1335. IEEE

Copacino WC (1997) Supply chain management: the basics and beyond, vol 1. CRC Press

Etemadi N, Borbon-Galvez Y, Strozzi F, Etemadi T (2021) Supply chain disruption risk management with blockchain: a dynamic literature review. Information 12(2):70

George RP, Peterson BL, Yaros O, Beam DL, Dibbell JM, Moore RC (2019) Blockchain for business. J Invest Compliance 20(1):17–21

Hamledari H, Fischer M (2021) Construction payment automation using blockchain-enabled smart contracts and robotic reality capture technologies. Autom Constr 132:103926

Hellani H, Sliman L, Samhat AE, Exposito E (2021) On blockchain integration with supply chain: overview on data transparency. Logistics 5(3):46

https://ethereum.org/en/

https://patents.google.com/patent/US20180071283A1/en

König L, Korobeinikova Y, Tjoa S, Kieseberg P (2020) Comparing blockchain standards and recommendations. Future Internet 12(12):222

Kouhizadeh M, Saberi S, Sarkis J (2021) Blockchain technology and the sustainable supply chain: theoretically exploring adoption barriers. Int J Prod Econ 231:107831

Law A (2017) Smart contracts and their application in supply chain management. Doctoral dissertation, Massachusetts Institute of Technology

Mann S, Potdar V, Gajavilli RS, Chandan A (2018) Blockchain technology for supply chain traceability, transparency and data provenance. In: Proceedings of the 2018 international conference on blockchain technology and application, pp 22–26

Mohanta BK, Panda SS, Jena D (2018) An overview of smart contract and use cases in blockchain technology. In: 2018 9th international conference on computing, communication and networking technologies (ICCCNT), pp 1–4. IEEE

Naef S, Wagner SM, Saur C (2022) Blockchain and network governance: learning from applications in the supply chain sector. Prod Plan & Control 1–15

Nakamoto S (2008) Bitcoin: a peer-to-peer electronic cash system, p 21260. Decentralized Business Review

Perboli G, Musso S, Rosano M (2018) Blockchain in logistics and supply chain: a lean approach for designing real-world use cases. IEEE Access 6:62018–62028

Qin K, Gervais A (2018) An overview of blockchain scalability, interoperability and sustainability. Hochschule Luzern Imperial College London Liquidity Network

Roy T, Garza-Reyes JA, Kumar V, Kumar A, Agrawal R (2022) Redesigning traditional linear supply chains into circular supply chains–a study into its challenges. Sustain Prod Consum 31:113–126

Steiu MF (2020) Blockchain in education: opportunities, applications, and challenges. First Monday

Sulkowski AJ (2018) Blockchain, law, and business supply chains: the need for governance and legal frameworks to achieve sustainability. Sulkowski AJ (2019) Blockchain, business supply chains, sustainability, and law: the future of governance, legal frameworks, and lawyers, pp 303–345

Wang S, Ouyang L, Yuan Y, Ni X, Han X, Wang FY (2019) Blockchain-enabled smart contracts: architecture, applications, and future trends. IEEE Trans Syst, Man, Cybern Syst 49(11):2266–2277

Wang M, Wu Y, Chen B, Evans M (2020) Blockchain and supply chain management: a new paradigm for supply chain integration and collaboration. Oper Supply Chain Manag Int J 14(1):111–122

Zheng Z, Xie S, Dai H, Chen X, Wang H (2017) An overview of blockchain technology: architecture, consensus, and future trends. In: 2017 IEEE international congress on big data (BigData congress), pp 557–564. IEEE

Zero-Knowledge Proofs in Blockchain-Enabled Supply Chain Management

3

Sameeksha Prasad, Namita Tiwari, Meenu Chawla, and Deepak Singh Tomar

Abstract

The global supply chain has been influenced by recent technology advancements and manufacturing growth. To enhance transparency, traceability, and trust in supply chain management, innovative solutions are required. One promising tool is blockchain technology, although it presents challenges in preserving privacy and confidentiality of sensitive information. Zero-knowledge proofs (ZKPs) are a powerful cryptographic tool that can validate the authenticity of goods, certifications, and transaction records in supply chain management systems without exposing sensitive business information. By utilizing ZKPs, participants can maintain data integrity and prove provenance while keeping proprietary details confidential. In this way, supply chain transparency is maintained for authorized parties only. ZKPs can enhance privacy in supply chain transactions by validating data integrity without disclosing the actual information. This enables confidentiality while ensuring proof of compliance with specific standards or regulations. Additionally, ZKPs help prevent and detect fraudulent activities by identifying anomalies in supply chain data through advanced cryptographic techniques. The use of ZKPs is a critical tool in ensuring the security and reliability of supply chains. This article provides

S. Prasad (✉) · N. Tiwari · M. Chawla · D. S. Tomar
Department of Computer Science and Engineering, Maulana Azad National Institute of Technology, Bhopal 462003, MP, India
e-mail: sameekshaprasad.27@gmail.com

N. Tiwari
e-mail: namita_tiwari21@rediffmail.com

M. Chawla
e-mail: chawlam@manit.ac.in

D. S. Tomar
e-mail: deepaktomar@manit.ac.in

A. Kumar et al. (eds.), *Sustainable Security Practices Using Blockchain, Quantum and Post-quantum Technologies for Real Time Applications*, Contributions to Environmental Sciences & Innovative Business Technology,
https://doi.org/10.1007/978-981-97-0088-2_3

a comprehensive analysis of the research on blockchain-based management of the supply chain specifically focusing on ZKPs application. Furthermore, this article identifies technical considerations, scalability problems, and interoperability challenges that may arise while incorporating such mechanisms. Zero-knowledge proofs hold great promise for transforming supply chain management through increased data integrity, improved privacy protection, and fraud prevention. By incorporating ZKPs into their processes, businesses can establish trust with partners and consumers alike while boosting efficiency and mitigating risks across the entire supply chain ecosystem. This article is a valuable resource for those interested in supply chain data privacy and security. It highlights the significant role of ZKPs in such areas and supports further advancements in blockchain-based management.

Keywords

Blockchain • Privacy • Supply-chain management • Security • Zero-knowledge proof

3.1 Introduction

Supply chain management plays a critical role in today's globalized and interconnected business landscape, facilitating the efficient flow of goods, services, and information across various entities (Hamisi 2011; Tien et al. 2019; Hassija et al. 2020). However, traditional supply chain systems have long been plagued by challenges such as lack of transparency, data fragmentation, counterfeit products, and trust deficits among stakeholders (Hamisi 2011; Tien et al. 2019). Blockchain technology has emerged as a viable remedy to get around these limitations, offering a decentralized, immutable, and transparent platform for managing different operations performed in supply chain.

The blockchain and supply chain management together has immense potential in resolving crucial pain points (Tien et al. 2019; Hassija et al. 2020) while opening new doors for improving efficiency, traceability, and trust (Liu 2020; Waller et al. 2019). By utilizing the distributed ledger technology offered by blockchain, the supply chain participants can record and verify transactions securely and transparently, leading to an accurate tamper-proof record of the entire supply chain journey.

Blockchain-enabled supply chain management emerged as a response to the limitations of traditional supply chains (Waller et al. 2019). The decentralized and transparent platform provided by blockchain technology revolutionized supply chain processes, addressing critical issues such as inadequate transparency, data silos, and trust deficits. With the rise of cryptocurrencies came the prominence of this technology's benefits in enhancing transparency, real-time tracking, traceability, reducing fraud, and streamlining processes (Schmidt and Wagner 2019).

Blockchain-enabled supply chain management optimizes process operations and boosts efficiency, transparency, and trust in the industry.

The management of supply chains through blockchain technology has enormous benefits (Liu 2020; Prabadevi et al. 2021). These include reducing fraud and errors, minimizing delays from paperwork, improving inventory management, identifying issues promptly, reducing courier costs, and enhancing consumer and partner trust. Although maintaining data privacy is challenging in a public and transparent blockchain environment while ensuring trustworthiness (Prabadevi et al. 2021). ZKPs offer an intriguing approach that enables information verification without revealing underlying data; this is where ZKPs come into play and help in enhancing data integrity, confidentiality, and selective disclosure in the context of blockchain-enabled supply chain management (Kilian 1992; Ling et al., 2013). This study explores the possible benefits of using ZKPs for managing supply chains that are based on blockchains efficiently while addressing their associated challenges and discussing their future prospects.

In supply chain management, Zero-Knowledge Proofs offer a ground-breaking solution to data security, privacy, and credibility issues (Hasan 2019). This cryptographic technique facilitates the verification of product authenticity, ensuring transparent visibility of supply chain and enabling secure sharing of data in supply chain finance. Organizations can use them to enhance cooperation, establish trust, and build robust supply chains in various industries. However, coping with scalability challenges while implementing ZKPs practically into complex supply chain ecosystems requires further research and development.

ZKPs can be used by supply chain management for various reasons (Spekman et al. 1998; Bamberger et al. 2022; Gabay 2019).

1. Verification of Product Authenticity

Ensuring the authenticity of products in supply chains is a crucial matter that demands attention (Marucheck et al. 2011). Counterfeit goods pose enormous risks to consumer safety and negatively impact brand identities. Fortunately, ZKPs offer a possible answer to this problem. ZKPs enable the verification of product authenticity without revealing sensitive manufacturing processes or trade secrets.

Manufacturers can prove the authenticity of their products without revealing sensitive information using a cryptographic method called ZKPs (Santis and Persiano 1992). In this process, a prover creates proof to validate the claim's correctness while an independent verifier verifies it without accessing any classified data. This unique feature of ZKP ensures confidentiality and enables stakeholders to verify product authenticity and origin with confidence.

When verifying product authenticity, ZKPs also come in handy for supplying the item's delivery history (Prabadevi et al. 2021; Islam 2023). With cryptographic methods at play, suppliers can provide a valid proof of the journey while ensuring compliance with quality standards and confirming that the product is genuine. This evidence gives stakeholders such as consumers, retailers, and regulators an opportunity to verify its veracity independently rather than relying solely on supply chain actors' trustworthiness.

2. Supply Chain Transparency

Modern businesses require supply chain transparency to provide stakeholders with insights into product movements and compliance auditing (Marucheck et al. 2011). However, maintaining transparent supply chains while safeguarding confidential data can be challenging. That's where ZKPs come in handy (Santis and Persiano 1992). ZKPs offer a cryptographic solution that enhances transparency without compromising confidentiality.

ZKPs allow supply chain claims to be verified without compromising important details such as a customer can prove they satisfy minimum age criteria without disclosing their actual date of birth. They provide proof of compliance, traceability, and authenticity while keeping proprietary processes and confidential information safe. This technology enables suppliers to demonstrate adherence to sustainability or fair-trade standards through cryptographic evidence (Bellés-Muñoz et al. 2021). By using ZKPs, stakeholders can gain verifiable evidence that builds trust in the supply chain ecosystem.

3. Secure Data Sharing and Collaboration

Supply chain stakeholders need to share confidential information securely and collaborate on insights and resources (Marucheck et al. 2011). This process involves safeguarding privacy, integrity, and accuracy of the shared data, ensuring secure transmission, storage, and access control mechanisms. When executed efficiently within the supply chain ecosystem, it enables effective coordination for decision-making purposes while maintaining optimal security standards.

ZKPs provide a secure cryptographic framework for exchanging and collaborating on data in the supply chain (Santis and Persiano 1992). With ZKPs, participants can confidentially share insights and work together on computations and tasks related to the supply chain without revealing sensitive information or needing trust beyond what is necessary. This cultivates transparency and maintains integrity of data shared among collaborators in the ecosystem.

4. Privacy-Preserving Supply Chain Finance

Supply chain finance transactions require the secure sharing of financial data among participants while preserving confidentiality (Marucheck et al. 2011). Privacy-preserving supply chain finance is an approach that applies privacy-enhancing solutions like ZKPs for protecting sensitive information and maintain privacy within the supply chain ecosystem (Islam 2023). This approach enables secure transactional collaboration while protecting confidential data in financial transactions.

ZKPs can be utilized by participants to prove the validity of financial statements or claims while keeping confidential data safe (Santis and Persiano 1992). These cryptographic proofs enable truthful assertions such as payment history, transaction amounts, supply chain performance metrics, and creditworthiness to be demonstrated. Other parties can easily verify these proofs without accessing sensitive underlying financial information, thus protecting its privacy.

ZKPs can improve blockchain-based supply chain management by enhancing data privacy and confidentiality. This allows participants to selectively share information while keeping the rest hidden, which is especially useful for competing companies within the same supply chain. ZKPs also increase trust among supply chain actors by enabling verifiable and auditable processes without revealing sensitive information. This transparency fosters collaboration, reduces fraud, and enhances overall transaction integrity. Additionally, these proofs help ensure compliance with regulatory requirements for securely managing sensitive data in accordance with privacy and protection laws. This way, supply chain management can benefit greatly from ZKPs and blockchain technology working together.

This paper aims to explore and analyze the role of ZKPs in blockchain-enabled supply chain management. It will delve into the technical aspects of ZKPs, their integration with blockchain technology, and their specific applications and benefits within supply chain operations. Furthermore, it will discuss the challenges and potential solutions for implementing ZKPs in supply chain management, considering scalability, efficiency, and the evolving landscape of privacy and data protection regulations.

The rest of the paper is organized as follows. In Sect. 3.2, description of preliminary concepts of blockchain-based supply chains and ZKPs are presented. Further, research work related to ZKP implementations in supply chain management is presented in Sect. 3.3. Section 3.4 concludes the paper.

3.2 Preliminaries

3.2.1 Blockchain Technology

Blockchain is innovative digital ledger that allows for decentralization and distribution of recorded transactions across multiple computers or nodes (Crosby 2016; Gupta et al. 2020; Abeyratne and Monfared 2016). It operates by forming a secure chain of blocks, where the transaction records are contained in each block. Every block of a blockchain also contains a reference to the preceding block. Blockchains are transparent, tamper-resistant, and facilitate secure communication among diverse users worldwide.

The origins of blockchain can be traced back to Bitcoin's introduction, which is recognized as the first and most widely recognized cryptocurrency (Crosby 2016). In 2008, Satoshi Nakamoto published a whitepaper describing a decentralized digital currency system that was underpinned by a blockchain framework (Abeyratne and Monfared 2016; Taskinsoy 2019). The decentralization, transparency, and security properties of blockchain became the basis for much more blockchain-based development in various industries today (Liu 2020; Gupta et al. 2020).

Research has evolved to encompass the creation of alternate blockchain platforms that boast unique capabilities and features. Ethereum, a platform introduced in 2015, pioneered the concept of smart contracts—programmable and self-executing agreements on the blockchain (Gupta et al. 2020; Meshcheryakov and Ivanov 2020).

As blockchain technology gained popularity, scholars and industry professionals delved into the various challenges associated with it (Islam 2023; Taskinsoy 2019). This includes scalability, privacy, and interoperability; issues which were explored in depth. To tackle these matters, researchers examined several consensus algorithms like Proof of Stake and Byzantine Fault Tolerance to enhance efficiency and energy consumption (Taskinsoy 2019). In order to improve data secrecy in blockchain systems, additional techniques like zero-knowledge proofs, homomorphic encryption, and secure multi-party computing were implemented (Bellés-Muñoz et al. 2021; Taskinsoy 2019).

The study of blockchain technology involved analyzing its potential benefits and limitations in various industries like supply chain management, healthcare, finance, intellectual property rights, and more (Bellés-Muñoz et al. 2021; Gupta et al. 2020). The use cases were examined to understand how this revolutionary technology can transform different sectors.

Blockchain serves several purposes as illustrated in Fig. 3.1. Detailed description of every purpose of blockchain is as follows (Gupta et al. 2020; Abeyratne and Monfared 2016; Taskinsoy 2019; Meshcheryakov and Ivanov 2020; Komalavalli et al. 2020):

1. **Decentralization**: Blockchains are decentralized systems that operate without a central authority or intermediary, giving participants complete autonomy over transactions. Instead, the network of individuals or nodes validates and records each transaction to maintain distributed consensus.
2. **Distributed ledger**: A blockchain serves as a distributed ledger where every participating node stores its copy. This ensures redundancy which enhances the system's security and resilience since there is not a single failure point.
3. **Blocks and transactions**: Blocks are created from transactions and appended to the blockchain in chronological order. Each block usually includes a collection

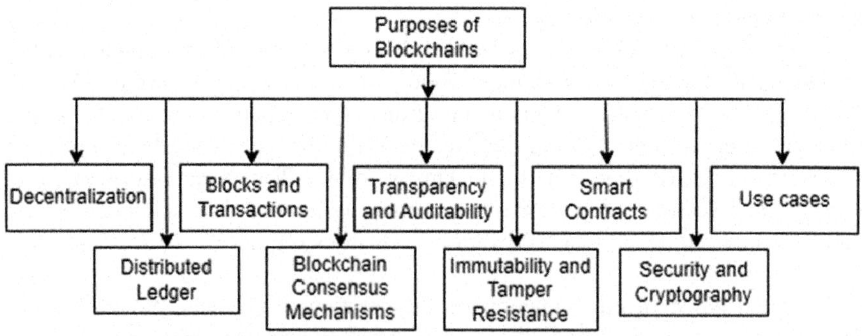

Fig. 3.1 Purposes served by blockchain

of transactions along with a timestamp and a unique identifier known as a cryptographic hash (Taskinsoy 2019; Aste et al. 2017). Each block's hash is generated using its content and the previous block's hash, resulting in a chain-like structure.

4. **Blockchain consensus mechanisms**: This mechanism ensures that all nodes concur on the state of the blockchain. The most popular mechanisms include Proof of Work, Proof of Stake, and Practical Byzantine Fault Tolerance (Feng 2020). They enable trust and prevent malicious activities like double-spending or tampering with transaction history.
5. **Immutability and tamper resistance**: Following the addition of a block to the blockchain, modifying or deleting it becomes incredibly difficult. Since each block's hash is linked to the previous one's hash, creating an interconnected chain that would be disrupted if any block were changed (Aniello et al. 2017). This makes blockchains perfect for applications where data integrity and audit trails are critical factors.
6. **Transparency and auditability**: Blockchains promote accountability and trust through their transparent design (Farooq et al. 2020). Anyone can view the entire transaction history, allowing for verification and auditing of transactions to ensure system integrity. This transparency fosters public confidence in the blockchain network.
7. **Smart contracts**: Smart Contracts are a feature supported by many blockchains. These contracts, also known as smart agreements, are self-executing and enforce predefined rules and conditions (Mohanta et al. 2018). Their automation capabilities allow for the enforcement of agreements while enabling complex business logic within blockchain ecosystems.
8. **Security and cryptography**: Blockchains use cryptography to protect transactions and data (Ferrag and Shu 2021). Public key cryptography safeguards participants' identities and keeps their transactions private, authentic, and integral. Cryptographic hashes ensure that data is tamper-proof, while digital signatures verify transaction authenticity.
9. **Use cases**: Blockchains have found applications in various domains beyond crypto currencies (Abeyratne and Monfared 2016; Taskinsoy 2019; Zou et al. 2019). They are used for supply chain management, healthcare record-keeping, voting systems, identity management, intellectual property rights, and decentralized finance (DeFi). By offering transparency, security, and decentralization, blockchains bring potential benefits and disruptions to these industries.

Blockchains offer a decentralized and reliable method for recording and verifying transactions, allowing for secure data sharing. Cryptography combined with consensus mechanisms ensures trustless operation and efficiency. This infrastructure creates opportunities for new decentralized applications and business processes.

3.2.2 Blockchain Interoperability

Blockchain interoperability involves the seamless communication and interaction between different blockchain networks or platforms (Belchior 2021; Lafourcade and Lombard-Platet 2020). It aims to address the challenge of exchanging data or assets across separate blockchain systems, which may have varying protocols, consensus mechanisms, or smart contract languages.

Blockchain technology has immense potential, but to unlock it fully, interoperability is key (Lafourcade and Lombard-Platet 2020). When different blockchain networks can connect and exchange data seamlessly, participants gain the freedom to transact and share information without limitations. To discuss blockchain interoperability effectively some basic features of blockchain interoperability as illustrated in Fig. 3.2 are considered, these crucial points are:

1. **Interoperability standards**: Developing common standards and protocols is essential for achieving blockchain interoperability (Pillai et al. 2022). These standards define how different blockchain networks can communicate, verify transactions, and exchange data. Examples of interoperability standards include Interledger Protocol (ILP) (Hope-Bailie and Thomas 2016), Blockchain Interoperability Alliance (BIA) (Han et al. 2023), and Enterprise Ethereum Alliance (EEA) (Gleim et al. 2019).
2. **Cross-chain communication**: Interoperability solutions can enable different blockchain networks to communicate with each other, allowing for the secure transfer of transactions and data (Han et al. 2023). Techniques like atomic swaps, sidechains, and pegged assets help in transferring value seamlessly across multiple blockchains.

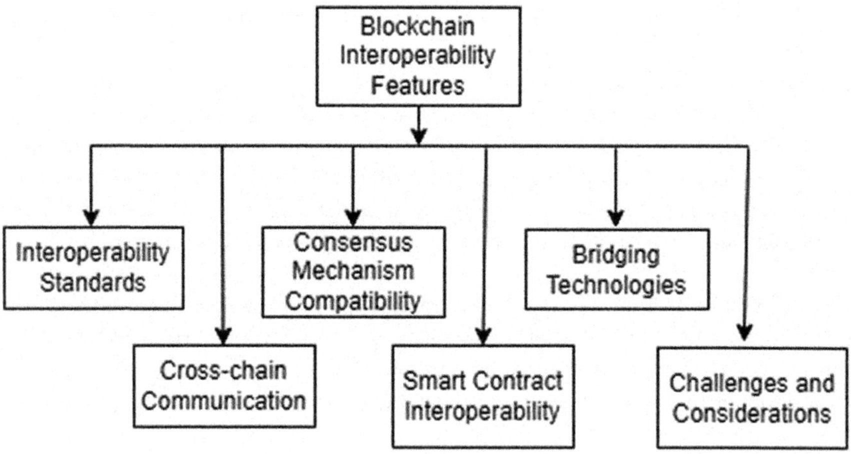

Fig. 3.2 Features of blockchain interoperability

3. **Consensus mechanism compatibility**: When it comes to blockchain networks, there are various consensus mechanisms in use such as PoW, PoS, or DPoS (Pan et al. 2021). However, to enable secure cross-chain transactions, interoperability solutions need to ensure compatibility between these mechanisms despite their differences. This requires careful consideration and planning to maintain consistency and security across different chains.
4. **Smart contract interoperability**: Smart contracts are self-executing contracts that facilitate transactions based on predefined rules (Khan et al. 2021). Achieving smart contract interoperability involves ensuring that smart contracts written in different programming languages or platforms can interact with each other. Projects like Polkadot (Burdges et al. 2020), Cosmos (Kwon and Buchman 2019), and Inter Blockchain Communication (IBC) (Kan et al. 2018) aim to enable smart contract interoperability.
5. **Bridging technologies**: To achieve interoperability between different blockchains, bridging technologies serve as intermediaries (Dannen and Dannen 2017). These tools transfer assets and data while ensuring their integrity and security. Some examples are blockchain oracles, middleware layers, and interoperability-focused blockchain platforms.
6. **Challenges and considerations**: Achieving blockchain interoperability poses various challenges in terms of technology, governance, and scalability (Gaur 2020; Zheng et al. 2017). Achievers must meticulously address concerns about security, privacy, scalability, and regulatory compliance to establish reliable cross-chain interactions.

The field of blockchain interoperability is rapidly evolving and boasts ongoing research and development efforts (Gad 2022). This promising technology has the potential to foster collaboration, enhance efficiency, and unlock new use cases by seamlessly integrating diverse blockchain networks.

3.2.3 Smart Contracts

On a blockchain network, smart contracts are agreements or transactions that can be enforced by themselves and have rules and conditions pre-programmed (Mohanta et al. 2018; Zou et al. 2019; Khan et al. 2021). They operate without the need for intermediaries, allowing for automation and streamlined processes.

Smart contracts operate in a blockchain, and their function can be comprehensively explained by breaking down the process into the following steps (Khan et al. 2021; Grishchenko et al. 2018; Alharby and Moorsel 2017) as shown in Fig. 3.3. These steps are as follows:

1. **Contract creation**: A participant in the blockchain network creates a smart contract. The contract includes the terms, conditions, and rules that will govern the agreement or transaction.

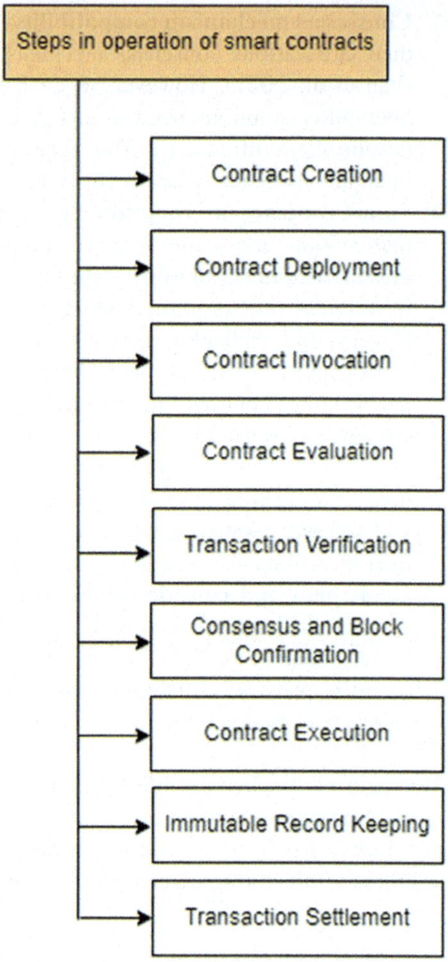

Fig. 3.3 Smart contracts workflow

2. **Contract deployment**: The distributed ledger securely adds the smart contract as an unchangeable component of its ecosystem when it has been deployed to the blockchain network. Its presence is ensured across multiple network nodes for maximum redundancy and resilience.
3. **Contract invocation**: Participants in the blockchain network can communicate with the smart contract once it has been deployed by sending particular transactions to its address on the blockchain. This process of invoking the contract is essential for executing its predefined operations and functions accurately.
4. **Condition evaluation**: Smart contracts are designed with pre-set conditions or triggers that activate specific actions when particular events occur or criteria are met. Upon meeting these specified conditions, the contract automatically executes the corresponding code or logic. Condition evaluation is a critical aspect of smart contracts, enabling seamless and secure transaction execution.

5. **Transaction Verification**: The blockchain network's decentralized nodes validate and verify the transactions associated with the smart contract. This process ensures that the transaction meets the required conditions and that it complies with the consensus rules of the blockchain network.
6. **Consensus and block confirmation**: After a transaction is verified, it gets grouped with other confirmed transactions and forms a block. This block then undergoes the consensus process of the network, such as PoW or PoS (Pan et al. 2021), before being added to the blockchain.
7. **Contract execution**: Nodes in the network execute the smart contract's code after block confirmation. The contract automatically enforces agreed-upon rules and conditions, triggering desired actions on the blockchain.
8. **Immutable record keeping**: Record-keeping in smart contracts is immutable. The blockchain's distributed ledger records the contract's execution and outcome, rendering it transparent and tamper-proof. Such an unalterable record grants all participants access to the history of the contract, ensuring its credibility can be verified with ease.
9. **Transaction settlement**: Smart contracts are capable of facilitating the transfer of digital assets or cryptocurrencies between involved parties upon execution. The specified assets are transferred based on predetermined rules and conditions agreed upon in advance.

The working of smart contracts in blockchain offers numerous benefits, including automation, transparency, immutability, and elimination of intermediaries (Khan et al. 2021). They enable trustless and efficient execution of agreements and transactions, revolutionizing a number of sectors including finance, supply chain management, and decentralized applications (DApps).

3.2.4 Supply Chain Management

Supply chain management is a process that involves several interconnected steps or activities (Ferguson 2000; Varma et al. 2006). These stages act together to ensure the smooth and efficient movement of goods or services from their point of origin to their consumption destination. The stages involved in supply chain management are depicted in Fig. 3.4.

The major stages in the supply chain management include:

1. **Planning stage**: In the planning stage, a strategic approach is developed to meet customer demands while making efficient use of available resources. This involves predicting demand, determining production levels, and deciding on inventory levels.
2. **Sourcing stage**: In the sourcing stage, the third-party suppliers providing raw materials, components, or services are identified. It involves selecting suppliers through negotiation and contract management to ensure timely delivery while maintaining cost-effectiveness.

Fig. 3.4 Stages of supply chain management

3. **Production stage**: During production, raw materials undergo manufacturing or assembly processes to become finished products. Optimizing production efficiency involves a range of activities including production scheduling and quality control.
4. **Inventory management stage**: The process of managing inventory entails maintaining the suitable amount of stock to cater to customers' needs while reducing expenses incurred from storing excess items. It involves closely monitoring stock levels, making predictions about customer demands, and implementing measures to control the inventory.
5. **Logistics stage**: Logistics involves managing the movement and storage of products across the supply chain. It includes transportation, warehousing, order fulfillment, and distribution. The foremost aim of logistics is to guarantee accurate and timely delivery of goods to customers.
6. **Customer service stage**: In managing customer relationships, the step of customer service is crucial for ensuring satisfaction. This involves several aspects like order processing, handling inquiries or complaints, and providing after-sales support to customers.
7. **Demand management stage**: Demand Management impacts the production and inventory levels of an organization by defining customer demand patterns through careful analysis and targeted influence. This practice includes demand forecasting, promotional activities, and effectively managing customer expectations to achieve optimal results.
8. **Returns management stage**: Returns management involves handling product returns and reverse logistics. This step includes managing product recalls, processing returns, and implementing efficient processes for product repairs, replacements, or recycling.
9. **Sustainability and risk management stage**: The crucial step in supply chain management involves sustainability and risk management. This involves addressing environmental, social, and ethical concerns throughout the supply chain. To ensure this goal is met, sustainable practices must be implemented while also managing risks associated with the supply chain.

In the overall supply chain management process, each step is crucial. Achieving operational efficiency and customer satisfaction requires effective coordination and collaboration among stakeholders.

3.2.5 Zero-Knowledge Proof Techniques

Various cryptographic techniques can be employed to implement ZKPs in blockchain systems (Kassaras and Maglaras 2020; Gong et al. 2022) as illustrated in Fig. 3.5.

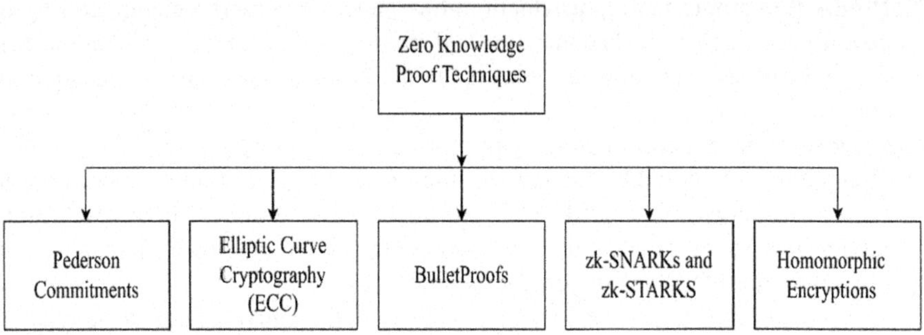

Fig. 3.5 Commonly known ZKP techniques

From Fig. 3.5, some commonly used ZKP methods include:

1. **Pedersen Commitments**: Using Pedersen commitments, a participant can agree to a value without disclosing it and then subsequently demonstrate knowledge of the committed value (Metere and Dong 2017). They are used in ZKPs to demonstrate possession of certain information without revealing the actual data.
2. **Elliptic Curve Cryptography (ECC)**: ECC is a form of public-key cryptography that uses elliptic curves' mathematical properties (Chatzigiannakis et al. 2011). ECC plays a crucial role in blockchain systems, particularly for generating and verifying cryptographic proofs like ZKPs. Efficient and secure operations are essential for implementing ZKPs, which ECC provides with precision.
3. **Bulletproofs**: Bulletproofs are a specific kind of non-interactive zero-knowledge proof, allowing for simpler and more efficient range proofs (Ma et al. 2023). These protocols make it possible to validate that certain values lie within defined parameters or ranges. By demonstrating the presence of positive transaction amounts in a detailed but streamlined manner, Bulletproofs reduce resource consumption compared to other types of identification methods.
4. **zk-SNARKs** (Zero-Knowledge Succinct Non-interactive Argument of Knowledge) **and zk-STARKs** (Zero-Knowledge Scalable Transparent Argument of Knowledge): zk-SNARKs (Mayer 2016) and zk-STARKs (Berentsen et al. 2022) are types of advanced ZKP constructions. They offer efficient and concise ways to prove complex computations or statements, verifying their accuracy or fulfilling certain requirements. Such techniques empower the creation of highly secure and confidential blockchain applications.
5. **Homomorphic encryption**: A cryptographic method known as homomorphic encryption permits computations on masked data while keeping their anonymity (Solomon et al. 2021). Hence, sensitive data computation can be performed in private and securely. Homomorphic encryption authorizes calculations without decryption, while ZKPs validate encrypted data's correctness without disclosing inputs or outputs information.

ZKP methods are crucial in managing the supply chain, especially in blockchain technology (Wu et al. 2019). They ensure sensitive data remains private, bolstering trust between participants and enabling efficient verification of critical statements. These techniques enhance the security, transparency, and integrity of supply chain operations which improve overall efficiency and effectiveness in managing supply chain processes.

3.3 Related Work

Chang and Chen (2020) represented the survey of the blockchain technology's current state, possible uses, and directions for future developments in supply chain management. To conduct a comparative analysis, the researchers took into account a number of fundamental characteristics, including traceability and transparency, stakeholder involvement and collaboration, supply chain integration and digitalization, as well as common frameworks on platforms based on blockchain.

Esmaeilian et al. (2020) gave a thoughtful review of how Industry 4.0 and blockchain technologies might help move supply chains toward sustainability. Innovative business models, smart logistics and transportation, and IoT-enabled energy management are some of the major capabilities that were taken into account for promoting sustainability in Industry 4.0. Similarly, the researchers highlighted blockchain's potential to increase sustainability through incentive mechanisms and tokenization to encourage environmentally friendly consumer behavior, product lifecycle visibility improvement, as well as increased productivity efficiency while decreasing both development and operational costs. However, the study also revealed some adverse impacts resulting from blockchain implementation on sustainable practices throughout supply chain networks.

Herbe et al. (2023) introduced distributed ledger technology as a tool for effective supply chain management. They conducted empirical research applying the 'theory of affordance actualisation' to Distributed Ledger Technology (DLT) (Li and Kassem 2021) affordances. The process included tracking logistics and product trails, confirming the product's source and history, exchanging data on electronic goods models, streamlining supply chain finance, and automating payments. The results demonstrate successful actualization of these affordances.

Malik et al. (2021) established a framework called TradeChain, which uses blockchain technology for separating identification and traceability in supply chains. The main challenge of balancing transparency and privacy was addressed by using ZKPs to verify multiple identities on the trade ledger using the merchants' private credentials. While the traceability information is stored on public blockchain, and participant identities are kept separately in a private blockchain. This approach ensures that traceability information remains immutable and transparent while participant identities remain confidential, providing a secure and efficient solution for supply chain management without compromising the confidentiality of participant information.

Anita et al. (2022) presented a Blockchain-based Anti-Counterfeit framework for supply chain management that ensures product authentication and detects RFID tag cloning. The uPort Decentralised Identity (DID) is utilized in the fully integrated privacy-protecting blockchain network, and Zero-Knowledge Succinct Non-Interactive Argument of Knowledge (zk-SNARK) (Mayer 2016) is employed to give users anonymity and preserve their privacy. The BA2C model employs PoA (Pan et al. 2021) consensus which provides better latency and throughput outcomes than traditional consensus models such as Proof of Work (PoW) (Pan et al. 2021) and Proof of Stake (PoS) (Pan et al. 2021). The experimental results showed that the presented approach was highly efficient.

Zhang et al. (2023) focused on improving the coordination and processing of the grain food supply chain through a blockchain-based multi-chain technology using ZKP. They created a multi-chain network mode for grain food by examining important supply chain linkages. They also offered a hierarchical encryption method for private data storage, a relay cross-chain communication mode, and a full consensus process under the multi-chain design. The model's results showed decreased storage redundancy, which increased data security.

Subashini and Hemavathi (2023) suggested using a permissioned blockchain system to track the origin of shelf-stable agricultural products. Their proposed architecture featured four tiers, parallel side chains, ZKPs, and Interplanetary File Systems (IPFS) (Huang et al. 2020) to enhance confidentiality and scalability for single or multiple items. To test their approach, they conducted simulations using Ethereum Rinkeby and Polygon which showed improved execution time, latency, and throughput without added overheads.

Moosa et al. (2023) proposed a unified model for data exchange on the blockchain. The proposed solution employs a decentralized network of peers to mitigate supply chain risks such as Man-in-The-Middle (MiTM) attacks (Conti et al. 2016), weak cryptography (Schneier 1998), repudiable transactions (Treiblmaier 2018), single points of failure (SPOF) (Lynch 2009), and poor access controls. The authors intend to connect all isolated steps within the supply chain together to streamline processes in a secure manner. The model can be used as an additional layer of security for data exchange between services and smart devices.

Gabay et al. (2019) introduced a framework to protect the privacy of electric vehicles (EVs) while charging. They combined ZKPs and blockchain smart contracts for authentication, scheduling, and charging without third-party involvement. The framework's efficiency improved by utilizing one smart contract that serves all EVs and ensures data confidentiality for involved parties such as electric vehicle service providers (EVSPs), public, and EV charging stations. Using Zokrates (Eberhardt and Tai 2018) to implement the proposed framework proved affordable in terms of time and Ethereum cost for real-life applications. This solution could be effective in managing transportation nodes in supply chain management.

Malik et al. (2022) developed PrivChain, a new framework aimed at ensuring both provenance and privacy preservation in blockchain-enabled supply chains. The framework responds to the challenge of maintaining data privacy without sacrificing transparency and traceability during supply chain transactions. Various cryptographic techniques, such as zero-knowledge proofs and ring signatures, were considered to allow for selective disclosure of information while preserving the confidentiality of sensitive data. The PrivChain was carefully designed to provide a secure and privacy-preserving solution for managing supply chains that safeguard against data manipulation while also protecting sensitive information's confidentiality.

The review of the literature is done for a variety of other works as well. To maintain traceability and privacy across the supply chain, the authors put up a number of techniques based on how ZKPs might be connected with supply networks built on blockchain technology. The several methods of integrating ZKPs with blockchain-based supply chain management are discussed in Table 3.1 in terms of their description and outcomes. The techniques used for implementation are described in the description. The findings of the methodology are described in the outcomes.

3.4 Conclusion

The main aim of this research was to study the security and reliability weakness of blockchain-based supply chain management, and mitigating these risks, with ZKPs integrated supply chain solutions. Integrating ZKPs in supply chain management using blockchain technology proposes remarkable benefits such as improved privacy, security, and trust. With ZKPs, verification processes have become more efficient, private, and secure than ever before. Further research is necessary to address potential vulnerabilities and attack vectors linked with ZKPs implementations in supply chain management. To improve supply chain safety, we recommend for ZKP solutions to be continuously upgraded. Also, interoperability between different blockchain platforms and networks must be tested and inspected thoroughly for seamlessly integrating ZKP-based supply chain solutions.

The risk associated with ZKP integrated supply chains is that it is not yet resilient to quantum computers. Ensuring the resistance of quantum computing attacks in ZKPs is paramount to safeguarding supply chain data in blockchain-based networks. By integrating post-quantum cryptographic algorithms, ZKPs can withstand powerful quantum computers threat, guaranteeing long-lasting security and privacy protection. In future, more stress can be laid on developing standardized post-24 quantum cryptographic algorithms that establish a strong foundation for quantum-resistant ZKPs implementation.

Table 3.1 Comparison table

Author	Year	Description	Outcome
Uesugi, Takio, Yoshinobu Shijo, and Masayuki Murata	2020	Suggested for protecting sensitive data in supply chain systems, using homomorphic encryption and secure multi-party computation techniques. The proposed framework utilizes ethereum smart contracts, zero-knowledge proofs, and a public blockchain to establish a secure shared database for all participants in the supply chain	The encryption approach aimed to secure distribution information while zero-knowledge proofs authenticated trusted supply chain parties anonymously in the blockchain. This Ethereum smart contract implementation performed cost-effectively at a maximum fee of 2.6 USD per party
Lam, Ching, Michelle Helena van Velthoven, and Edward Meinert	2020	Established a protocol that uses blockchain technology to build a connected platform for managing manufacturing and supply chain logistics. The project focused on cell and gene therapy as an application. Safety measures including zero-knowledge proofs and product release by exception techniques were implemented for supply chain safety	The analysis showed that the overall safety and reliability of advanced therapies, particularly in areas like cell and gene therapy, is enhanced. And using zero-knowledge proofs and product release by exception, monitoring costs are reduced to a great extent
Hrušovský, Martin, Gerald Reiner, and Alfred Taudes	2022	Focused on how to address information sharing risks in supply chains. They discussed the potential solution of using blockchain technology, and also explored key blockchain concepts such as smart contracts, zero-knowledge proofs, homomorphic encryption, and secure multi-party computation for tasks like quality management, planning, and tracking and tracing	The technology presented had positive outcomes. It enhanced supply chain resilience by improving transparency, visibility, and collaboration among actors involved in the supply chain

(continued)

Table 3.1 (continued)

Author	Year	Description	Outcome
H. Al-Aswad, H. Hasan, W. Elmedany, M. Ali and C. Balakrishna	2019	Projected a new model to share sensitive data securely between services and devices without compromising confidentiality or privacy. By leveraging zero-knowledge proofs, the model facilitates data exchange while keeping actual information concealed. Blockchain-based security is incorporated into the model to add a layer of trustlessness and enhance data sharing security. The paper focuses on two specific case studies in which this secure data sharing approach can be applied: the Bahraini government's National Healthcare Insurance Scheme and VAT collection	The given model used in these situations prioritizes patient privacy and bolsters overall security for VAT reporting and B2B transactions in the healthcare industry
Hellani, Houssein, Hellani, Houssein, et al.	2021	Presented various aspects of blockchain integration with supply chains have been presented, highlighting their potential benefits such as data verification, immutability, and smart contracts. Furthermore, the utilization of ZKPs as a cryptographic technique within the blockchain is being explored and given emphasis	The secure and tamper-resistant platform used for supply chain data verification ensures both privacy and confidentiality while maintaining the integrity of the outcomes
Ravi, Deebthik, et al.	2022	Introduced a new system that offers secure data sharing among supply chain participants. This technique uses encryption, ZKPs, and smart contracts to safeguard sensitive information. The Hyperledger Fabric platform was used for implementation and evaluation purposes	The Hyperledger Fabric platform's implementation of the system proves its ability to maintain transparency and privacy in managing supply chains with great effectiveness

(continued)

Table 3.1 (continued)

Author	Year	Description	Outcome
Katti, Jayashree, et al.	2022	Discussed an innovative approach that involves using smart contracts and cryptographic techniques to secure and verify food-related data at every stage of the supply chain. The technique employs digital signatures and hash functions for added security	The techniques attained for enhanced efficiency, reduced fraud, and improved food safety in the supply chain
Maouchi, Mourad El, OğuzhanErsoy, and ZekeriyaErkin	2019	Introduced DECOUPLES, a decentralized and privacy-preserving system that ensures traceability at every stage of supply chains. The system employed cryptographic techniques, including zero-knowledge proofs and ring signatures, to achieve unlinkability between transactions and preserve privacy	The outcomes of the analysis suggested that the integrity of supply chain transactions can be verified by stakeholders without revealing sensitive information. And the traceability of the supply chain is secured
Xu, Shiwei, et al.	2022	Provided zkrpChain, a data auditing system for consortium blockchains that ensures confidentiality. This system uses a range of ZKPs to enable multi-party authentication while maintaining the privacy of data stored on the blockchain	The analysis demonstrated that the multiple auditors can verify data correctness and integrity without disclosing specific values. The privacy in consortium blockchains is enhanced and the security and trustworthiness of data auditing is improved
Powell, Warwick, et al.	2022	Presented the combination of distributed ledger technology (DLT) and fully homomorphic encryption (FHE) to improve information sharing and enhance supply chain efficiency. This approach employs zero knowledge information sharing to address the limitations in hyperconnected supply chains without compromising data control	The evidence presented aimed at showcasing better collaboration, improved data integrity, and operational efficiency in supply chain management

References

Abeyratne SA, Monfared RP (2016) Blockchain ready manufacturing supply chain using distributed ledger. Int J Res Eng Technol 5(9):1–10

Al-Aswad H, Hasan H, Elmedany W, Ali M, Balakrishna C (2019) Towards a blockchain-based zero-knowledge model for secure data sharing and access. In: 2019 7th international conference on future internet of things and cloud workshops (FiCloudW), Istanbul, Turkey, pp 76–81. https://doi.org/10.1109/FiCloudW.2019.00027

Alharby M, Van Moorsel A (2017) Blockchain-based smart contracts: a systematic mapping study. arXiv:1710.06372

Aniello L et al (2017) A prototype evaluation of a tamper-resistant high performance blockchain-based transaction log for a distributed database. In: 2017 13th European dependable computing conference (EDCC). IEEE

Anita N, Vijayalakshmi M, Shalinie SM (2022) Blockchain-based anonymous anti-counterfeit supply chain framework. Sādhanā 47:208. https://doi.org/10.1007/s12046-022-01984-2

Aste T, Tasca P, Matteo TD (2017) Blockchain technologies: the foreseeable impact on society and industry

Bamberger KA et al (2022) Verification dilemmas in law and the promise of zero-knowledge proofs. Berkeley Technol Law J 37(1)

Belchior R et al (2021) A survey on blockchain interoperability: past, present, and future trends. ACM Comput Surv (CSUR) 54(8):1–41

Bellés-Muñoz M et al (2021) New privacy practices for blockchain software. IEEE Softw 39(3):43–49

Berentsen A, Lenzi J, Nyffenegger R (2022) A walk-through of a simple Zk-STARK proof. SSRN 4308637

Burdges J et al (2020) Overview of polkadot and its design considerations. arXiv:2005.13456

Chang SE, Chen Y (2020) When blockchain meets supply chain: a systematic literature review on current development and potential applications. IEEE Access 8:62478–62494

Chatzigiannakis I et al (2011) Elliptic curve based zero knowledge proofs and their applicability on resource constrained devices. In: 2011 IEEE eighth international conference on mobile ad-hoc and sensor systems. IEEE

Conti M, Dragoni N, Lesyk V (2016) A survey of man in the middle attacks. IEEE Commun Surv Tutor 18(3):2027–2051

Crosby M et al (2016) Blockchain technology: beyond bitcoin. Appl Innov 2(6–10):71

Dannen C, Dannen C (2017) Bridging the blockchain knowledge gap. In: Introducing Ethereum and solidity: foundations of cryptocurrency and blockchain programming for beginners, pp 1–20

De Santis A, Persiano G (1992) Zero-knowledge proofs of knowledge without interaction. In: Proceedings of the 33rd annual symposium on foundations of computer science. IEEE Computer Society

Eberhardt J, Tai S (2018) Zokrates-scalable privacy-preserving off-chain computations. In: 2018 IEEE international conference on internet of things (iThings) and IEEE green computing and communications (GreenCom) and IEEE cyber, physical and social computing (CPSCom) and IEEE smart data (SmartData). IEEE

Esmaeilian B, Sarkis J, Lewis K, Behdad S (2020) Blockchain for the future of sustainable supply chain management in Industry 4.0. Resour Conserv Recycl 163:105064. ISSN 0921-3449.https://doi.org/10.1016/j.resconrec.2020.105064

Farooq MS, Khan M, Abid A (2020) A framework to make charity collection transparent and auditable using blockchain technology. Comput Electr Eng 83:106588

Feng J et al (2020) Towards random-honest miners selection and multi-blocks creation: proof-of-negotiation consensus mechanism in blockchain networks. Future Gener Comput Syst 105:248–258

Ferguson BR (2000) Implementing supply chain management. Prod Invent Manag J 41(2):64

Ferrag MA, Shu L (2021) The performance evaluation of blockchain-based security and privacy systems for the internet of things: a tutorial. IEEE Internet Things J 8(24):17236–17260

Gabay D (2019) A privacy framework for decentralized applications using blockchains and zero knowledge proofs

Gabay D, Akkaya K, Cebe M (2019)A privacy framework for charging connected electric vehicles using blockchain and zero knowledge proofs. In: 2019 IEEE 44th LCN symposium on emerging topics in networking (LCN symposium), Osnabrueck, Germany, pp 66–73. https://doi.org/10.1109/LCNSymposium47956.2019.9000682

Gad AG et al (2022) Emerging trends in blockchain technology and applications: a review and outlook. J King Saud Univ-Comput Inf Sci 34(9):6719–6742

Gaur N (2020) Blockchain challenges in adoption. Manag Financ 46(6):849–858

Gleim B et al (2019) Enterprise Ethereum alliance off-chain trusted compute specification V1. 0. Enterp. Ethereum Alliance

Gong Y et al (2022) Analysis and comparison of the main zero-knowledge proof scheme. In: 2022 international conference on big data, information and computer network (BDICN). IEEE

Grishchenko I, Maffei M, Schneidewind C (2018) A semantic framework for the security analysis of Ethereum smart contracts. In: Principles of security and trust: 7th international conference, POST 2018, held as part of the European joint conferences on theory and practice of software, ETAPS 2018, Thessaloniki, Greece, April 14–20, Proceedings 7. Springer International Publishing

Gupta S, Sinha S, Bhushan B (2020) Emergence of blockchain technology: fundamentals, working and its various implementations. In: Proceedings of the international conference on innovative computing and communications (ICICC)

Hamisi S (2011) Challenges and opportunities of Tanzanian SMEs in adapting supply chain management. Afr J Bus Manag 5(4):1266

Han P et al (2023) A survey on cross-chain technologies. In: Distributed ledger technologies: research and practice

Hasan J (2019) Overview and applications of zero knowledge proof (ZKP). Nanjing University of Posts and Telecommunications, Nanjing

Hassija V et al (2020) A survey on supply chain security: application areas, security threats, and solution architectures. IEEE Internet Things J 8(8):6222–6246

Hellani H et al (2021) On blockchain integration with supply chain: overview on data transparency. Logistics 5(3):46

Herbe A et al (2023) How to effectively use distributed ledger technology in supply chain management? Int J Prod Res 1–26

Hope-Bailie A, Thomas S (2016) Interledger: Creating a standard for payments. In: Proceedings of the 25th international conference companion on world wide web

Hrušovský M, Reiner G, Taudes A (2022) Applying blockchain technologies for increasing supply chain resilience. In: Supply chain resilience: insights from theory and practice. Springer International Publishing, Cham, pp 109–117

Huang H et al (2020) When blockchain meets distributed file systems: an overview, challenges, and open issues. IEEE Access 8:50574–50586

Islam MD (2023) A survey on the use of blockchains to achieve supply chain security. Inf Syst 102232

Kan L et al (2018) A multiple blockchains architecture on inter-blockchain communication. In: 2018 IEEE international conference on software quality, reliability and security companion (QRS-C). IEEE

Kassaras S, Maglaras L (2020) Zkps: does this make the cut? Recent advances and success of zero-knowledge security protocols. arXiv:2006.09990

Katti J et al (2022) A block-chain based approach for food supply chain management. In: 2022 13th international conference on computing communication and networking technologies (ICCCNT). IEEE

Khan SN et al (2021) Blockchain smart contracts: applications, challenges, and future trends. Peer-to-peer Netw Appl 14:2901–2925

Khan S et al (2021) Towards interoperable blockchains: a survey on the role of smart contracts in blockchain interoperability. IEEE Access 9:116672–116691

Kilian J (1992) A note on efficient zero-knowledge proofs and arguments. In: Proceedings of the twenty-fourth annual ACM symposium on theory of computing

Komalavalli C, Saxena D, Laroiya C (2020) Overview of blockchain technology concepts. In: Handbook of research on blockchain technology. Academic Press, pp 349–371

Kwon J, Buchman E (2019) Cosmos whitepaper. Netw Distrib Ledgers 27

Lafourcade P, Lombard-Platet M (2020) About blockchain interoperability. Inf Process Lett 161:105976

Lam C, van Velthoven MH, Meinert E (2020) Developing a blockchain-based supply chain system for advanced therapies: protocol for a feasibility study. JMIR Res Protocols 9(12):e17005

Li J, Kassem M (2021) Applications of distributed ledger technology (DLT) and blockchain-enabled smart contracts in construction. Autom Constr 132:103955

Ling S et al (2013) Improved zero-knowledge proofs of knowledge for the ISIS problem, and applications. In: Public-key cryptography–PKC 2013: 16th international conference on practice and theory in public-key cryptography, Nara, Japan, February 26–March 1, 2013. Proceedings 16. Springer Berlin Heidelberg

Liu Z (2020) Literature review of supply chain finance based on blockchain perspective. Open J Bus Manag 9(1):419–429

Lynch GS (2009) Single point of failure: the 10 essential laws of supply chain risk management. Wiley

Ma C et al (2023) Data transaction privacy protection mechanism based on bulletproof in multi-channel blockchain. In: International conference on cyber security, artificial intelligence, and digital economy (CSAIDE 2023), vol 12718. SPIE

Malik S, Gupta N, Dedeoglu V, Kanhere SS, Jurdak R (2021) TradeChain: decoupling traceability and identity in blockchain enabled supply chains. In: 2021 IEEE 20th international conference on trust, security and privacy in computing and communications (TrustCom), Shenyang, China, pp 1141–1152. https://doi.org/10.1109/TrustCom53373.2021.00155

Malik S, Dedeoglu V, Kanhere SS, Jurdak R (2022) PrivChain: provenance and privacy preservation in blockchain enabled supply chains. In: 2022 IEEE international conference on blockchain (blockchain), Espoo, Finland, pp 157–166. https://doi.org/10.1109/Blockchain55522.2022.00030

Maouchi ME, Ersoy O, Erkin Z (2019) DECOUPLES: a decentralized, unlinkable and privacy-preserving traceability system for the supply chain. In: Proceedings of the 34th ACM/SIGAPP symposium on applied computing

Marucheck A et al (2011) Product safety and security in the global supply chain: issues, challenges and research opportunities. J Oper Manag 29(7–8):707–720

Mayer H (2016) zk-SNARK explained: basic principles. https://blog.coinfabrik.com/wp-content/uploads/2017/03/zkSNARK-explained_basic_principles.pdf

Meshcheryakov A, Ivanov S (2020) Ethereum as a hedge: the intraday analysis. Econ Bull 40(1):101

Metere R, Dong C (2017) Automated cryptographic analysis of the Pedersen commitment scheme. In: Computer network security: 7th international conference on mathematical methods, models, and architectures for computer network security, MMM-ACNS 2017, Warsaw, Poland, August 28–30, 2017, Proceedings 7. Springer International Publishing

Mohanta BK, Panda SS, Jena D (2018) An overview of smart contract and use cases in blockchain technology. In: 2018 9th international conference on computing, communication and networking technologies (ICCCNT). IEEE

Moosa H et al (2023) A combined blockchain and zero-knowledge model for healthcare B2B and B2C data sharing. Arab J Basic Appl Sci 30(1):179–196

Pan J, Song Z, Hao W (2021) Development in consensus protocols: from PoW to PoS to DPoS. In: 2021 2nd international conference on computer communication and network security (CCNS). IEEE

Pillai B et al (2022) Level of conceptual interoperability model for blockchain based systems. In: 2022 IEEE crosschain workshop (ICBC-CROSS). IEEE

Powell W et al (2022) Garbage in garbage out: The precarious link between IoT and blockchain in food supply chains. J Ind Inf Integr 25:100261

Prabadevi B et al (2021) Toward blockchain for edge-of-things: a new paradigm, opportunities, and future directions. IEEE Internet Things Mag 4(2):102–108

Ravi D et al (2022) Privacy preserving transparent supply chain management through Hyperledger fabric. Blockchain: Res Appl 3(2):100072

Schmidt CG, Wagner SM (2019) Blockchain and supply chain relations: a transaction cost theory perspective. J Purch Supply Manag 25(4):100552

Schneier B (1998) Security pitfalls in cryptography. Edi Forum-Oak Park, vol 11. The EDI Group, Ltd

Solomon R, Weber R, Almashaqbeh G (2021) smartFHE: privacy-preserving smart contracts from fully homomorphic encryption. Cryptology ePrint Archive

Spekman RE, Kamauff JW, Myhr N (1998) An empirical investigation into supply chain management: a perspective on partnerships. Supply Chain Manag: Int J 3(2):53–67

Subashini B, Hemavathi D (2023) Scalable blockchain technology for tracking the provenance of the agri-food. Comput Mater Continua 75(2)

Taskinsoy J (2019) Blockchain: a misunderstood digital revolution. In: Things you need to know about blockchain, 8 October 2019

Tien NH, Anh DBH, Thuc TD (2019) Global supply chain and logistics management. Academic Publications, Delhi

Treiblmaier H (2018) The impact of the blockchain on the supply chain: a theory-based research framework and a call for action. Supply Chain Manag Int J 23(6):545–559

Uesugi T, Shijo Y, Murata M (2020) Short paper: design and evaluation of privacy-preserved supply chain system based on public blockchain. arXiv:2004.07606

Varma S, Wadhwa S, Deshmukh SG (2006) Implementing supply chain management in a firm: issues and remedies. Asia Pac J Mark Log 18(3):223–243

Waller MA et al (2019) Integrating blockchain into supply chain management: a toolkit for practical implementation. Kogan Page Publishers

Wu H et al (2019) Data management in supply chain using blockchain: challenges and a case study. In: 2019 28th International conference on computer communication and networks (ICCCN). IEEE

Xu S et al (2022) zkrpChain: towards multi-party privacy-preserving data auditing for consortium blockchains based on zero-knowledge range proofs. Future Gener Comput Syst 128:490–504

Zhang B, Xu J, Wang X, Zhao Z, Chen S, Zhang X (2023) Research on the construction of grain food multi-chain blockchain based on zero-knowledge proof. Foods 12:1600. https://doi.org/10.3390/foods12081600

Zheng Z et al (2017) Blockchain challenges and opportunities: a survey. Int J Web Grid Serv 14(4)

Zou W et al (2019) Smart contract development: challenges and opportunities. IEEE Trans Softw Eng 47(10):2084–2106

A Sustainable Blockchain and Asymmetric Broadcast Encryption-Based Secure E-Healthcare System

4

Snehlata Yadav and Vivek Yadav

Abstract

The value of healthcare records has skyrocketed due to the emergence of many new diseases and the ongoing pandemic situations. With these healthcare records spread across numerous healthcare provider entities in various locations, it has become time-consuming to manage them in physical form. Consequently, there is a growing need to transition to electronic healthcare records (EHRs). However, the current approach of managing EHRs at the centralized level creates a single point of failure and exposes users to numerous security risks which include amateur-level attacks and data breaches orchestrated by adversaries. Handling electronic health records from the Internet of Medical Things(IoMT) poses significant challenges due to the sensitive information involved, which makes it a prime target for attackers. Furthermore, managing modern healthcare systems is complex and expensive, demanding highly secure storage space. The recent COVID-19 pandemic has inflicted unprecedented costs and human suffering on a scale never seen before in the modern world. As long as diseases continue to resurface, we must strive to slow down pandemics by leveraging information systems empowered by new IT technologies. Blockchain technology emerges as one potential solution for achieving this goal. By enhancing health record management, blockchain can address these problems. Its rising popularity has sparked extensive research into

S. Yadav (✉)
Department of Computer Science and Engineering, Maulana Azad National Institute of Technology, Link Road 3, Bhopal 462003, MP, India
e-mail: yadavsnehlata@gmail.com

Department of Computer Science and Engineering, Government Women's Polytechnic College, shivaji Nagar, Bhopal 462016, MP, India

V. Yadav
Department of IT, Khoj Information Technology Inc., Gurugram 122001, Haryana, India

A. Kumar et al. (eds.), *Sustainable Security Practices Using Blockchain, Quantum and Post-quantum Technologies for Real Time Applications*, Contributions to Environmental Sciences & Innovative Business Technology,
https://doi.org/10.1007/978-981-97-0088-2_4

different transaction schemes that specialize in privacy preservation across various fields. However, the public nature of blockchain, which exposes transaction information, introduces significant privacy risks. Therefore, it is crucial to construct an efficient scheme that is flexible in nature and ensures privacy preservation of transaction contents and reliable auditability, aspects that previous works have failed to adequately address. We propose a framework of broadcast encryption and a specific instance to provide conditional data access control. It relies on a specific dedicated asymmetric-key-based broadcast encryption scheme, which stores user encrypted credentials for the blockchain conditional data access control. Simultaneously, conditional access of the encrypted data that is stored in an off-chain source of the data is applicable to encrypted data. The blockchain environment facilitates interactive communication between users of data and data providers, providing the necessary data access control information. Our proposed solution is based on a blockchain enabled privacy-preserving transaction construction that utilizes an efficient asymmetric-key broadcast encryption method. By implementing this approach, we aim to overcome the limitations of previous works and ensure secure and private handling of healthcare transactions within the blockchain ecosystem.

Keyword

Blockchain Technology · Broadcast encryption · Healthcare · Sustainable blockchain · Security · Privacy

4.1 Introduction

Every new technological advancement attracts a group of enthusiastic pioneers who eagerly delve into its potential societal and economic advantages. They strive to harness the technology for the betterment of society. However, it is important to acknowledge that a few adversaries may also attempt to exploit these innovations for malicious purposes.

The field of cryptography has experienced significant changes in recent decades due to the rise and widespread use of the Internet and web-based interfaces. Within this domain, different specialized areas have emerged to cater to specific needs such as key management protocols, non-repudiation, encryption mechanism (Yadav and Tiwari 2020), and authentication protocols. Primitives, which are fundamental and trusted cryptographic standards, play a crucial role in enabling other cryptographic protocols to secure IT-based systems in today's interconnected ecosystem (Yadav 2020).

Blockchain technology (Jain et al. 2021) has garnered significant interest as a novel computing technology due to its consensus confirmation mechanism and distributed data storage. Its applications include various sectors, such as finance, healthcare industry, and the Internet of Things, and its popularity continues to grow. While blockchain offers increased transparency and eliminates the need for a trusted inter-

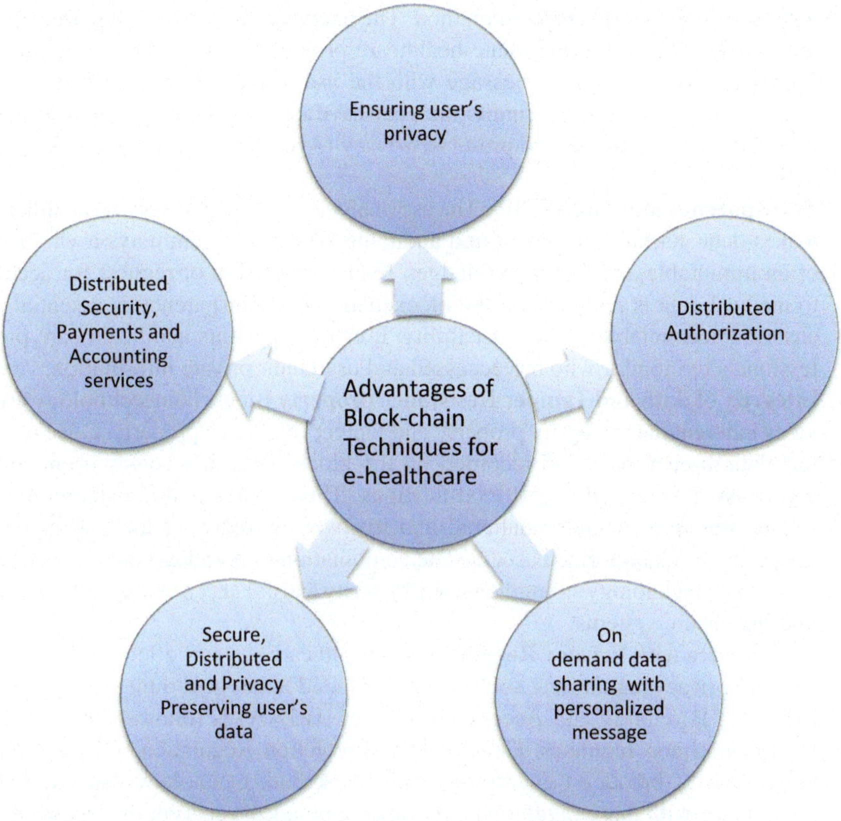

Fig. 4.1 Advantages of blockchain in e-healthcare (EHR)

mediary in certain transactions, it also raises concerns regarding privacy preservation. Evaluating privacy risks and implementing privacy protection measures are crucial aspects that deserve attention. The advantages of blockchain are depicted in Fig. 4.1.

Broadcast encryption can be effectively integrated with blockchain technology in e-healthcare (Fan et al. 2021; Chen et al. 2021) to enhance privacy and security in data sharing and communication. Therefore, broadcast encryption ensures that sensitive medical information is securely transmitted and distributed to the authorized recipients while preserving the privacy of the data. The blockchain provides a platform for verifying transactions, ensuring transparency of data among users and trust in the e-healthcare scenario.

1. **Privacy-Preserving Data Sharing in e-healthcare** Broadcast encryption (Boneh et al. 2005; Phuong et al. 2015; Acharya 2020; Delerablée 2007; Abdalla et al. 2005; Lubicz and Sirvent 2008) enables efficient and secure distribution (Kumar et al. 2016) of e-healthcare data to a group of authorized recipients. In the e-healthcare scenario, this property can be achieved by encrypting health records

with an asymmetric encryption method. The encrypted message is then transmitted to a specific set of legitimate healthcare providers, researchers, or patients. They can retrieve original message with the use of their decryption key. Thus ensuring that only legitimate individuals can have access and only they can decrypt the shared healthcare data to preserve privacy (Azees et al. 2017) and confidentiality.

2. **Transparency and Auditability** The use of blockchain technology in e-healthcare makes data sharing transparent and auditable. The blockchain plays a vital role of an immutable and distributed ledger. Every transaction or request for access to medical data is recorded on the blockchain. This transparency is essential in ensuring accountability and traceability. It enables patients and healthcare professionals to monitor who has accessed and used their private information.
3. **Integrity of data and Tamper Resistance property** Blockchain technology provides inherent data integrity property and tamper resistance property. Once medical data is encrypted and recorded on the blockchain, it becomes immutable and resistant to unauthorized modifications. This ensures that the shared medical information remains unaltered and trustworthy, reducing the risk of data tampering or unauthorized access. The combination of broadcast encryption and blockchain technology strengthens data integrity and enhances the overall security of e-healthcare systems.
4. **Compliance with Privacy Regulations** E-healthcare systems often need to comply with privacy regulations, such as for the United States government, the Health Insurance Portability and Accountability Act (HIPAA) is followed and for the European Union regulates, General Data Protection Regulation (GDPR). The integration of broadcast encryption with blockchain technology can facilitate compliance with these regulations. By design, broadcast encryption ensures that sensitive medical data is securely shared only with authorized recipients, reducing the risk of data breaches. The blockchain's transparency and auditability features further support regulatory compliance by providing a verifiable record of data access and usage.

Privacy issues associated with blockchain-based transactions are broadly categorized into identity and data privacy. Specifically, data privacy in blockchain refers to the concept that observers should have limited access to transaction details, with only essential information being made publicly available. Safeguarding the privacy of blockchain transactions is of utmost importance, given the inherent nature of the technology. This manuscript aims to examine the advancements in various healthcare applications, particularly within the realms of IoT and blockchain technology. This research paper aims to enable privacy-preserving data sharing through the use of blockchain and broadcast encryption cryptographic primitive. To tackle this challenge, we employ a comprehensive approach that combines broadcast encryption, blockchain technology, and smart contracts.

4.2 Motivation

In the recent decade, there has been a notable surge in research papers focusing on blockchain technology, with researchers contributing to its continuous advancement and development. The growing interest in blockchain technology, coupled with the increasing significance of healthcare and information technology globally, indicates a promising future for blockchain as a prominent and evolving technology. Given the ongoing advancements in blockchain technology, it is imperative to revisit the current state of research and the strengths of this emerging technology. In particular, researchers have shown a keen interest in exploring the integration of blockchain in the healthcare sector. The secure data sharing along with broadcast encryption and blockchain technology converges two distinct fields. Heathcare sector requires a secure and distributed data sharing to only legitimate users in a way that revoked users learn nothing from the encrypted data.

4.2.1 Challenges in e-Healthcare Data Sharing and Integration of Blockchain with Broadcast Encryption

Some of the challenges in e-healthcare data sharing and integration of blockchain with broadcast encryption are identified as follows:

1. **Data Privacy and Confidentiality** One of the primary challenges in clinical data sharing and integration is maintaining the privacy and confidentiality of sensitive patient information. Blockchain and broadcast encryption in the cloud can address this challenge, but there are still concerns regarding the exposure of patient identities and the potential for unauthorized access to data.
2. **Scalability and Performance** Blockchain technology, while providing transparency and immutability, can face scalability issues when it comes to handling large volumes of clinical data. Similarly, broadcast encryption in the cloud may experience performance limitations when encrypting and decrypting data for a large number of authorized recipients. Ensuring efficient data sharing and integration without compromising performance is crucial.
3. **Interoperability and Data Standardization** Clinical data comes from various sources and systems that may employ different data formats, standards, and terminologies. Achieving interoperability and seamless integration of heterogeneous data across different blockchain networks and cloud environments (Chen et al. 2019; Sohal et al. 2022; Du et al. 2018; Al-Qerem et al. 2020; Wang et al. 2019) can be complex. It requires consensus on data standards, mapping mechanisms, and semantic interoperability to ensure meaningful data exchange.
4. **Consent Management and Governance** Obtaining proper patient consent for data sharing and defining governance models for data access and usage are critical challenges. Blockchain and broadcast encryption should incorporate robust consent management mechanisms to handle patient preferences, data ownership,

and consent revocation. Additionally, defining governance frameworks to ensure responsible data handling and mitigate potential ethical concerns is essential.

5. **Security and Trust** Blockchain technology and broadcast encryption provide inherent security features, but vulnerabilities and risks still exist. Secure key management, encryption algorithms, and protection against attacks or malicious actors are crucial for maintaining the integrity and trustworthiness of clinical data. Ensuring the security of blockchain networks and cloud infrastructure is paramount to prevent data breaches or unauthorized modifications.
6. **Regulatory Compliance** The healthcare industry is subject to strict regulations and compliance requirements, such as HIPAA in the United States and GDPR in the European Union. Adhering to these regulations while leveraging blockchain and broadcast encryption technologies can be challenging. Compliance with data protection laws, patient consent management, and auditability of data access becomes even more complex in distributed and decentralized environments.
7. **Education and Adoption** Blockchain technology and broadcast encryption in the cloud are relatively new concepts in the healthcare domain. There is a need for education and awareness among healthcare professionals, policymakers, and patients regarding the benefits, challenges, and risks associated with these technologies. Promoting adoption and building trust in blockchain-based clinical data sharing and integration require concerted efforts to overcome resistance and address misconceptions.

Organization of Manuscript

The rest of the manuscript is organized as follows: Sect. 4.3 covers literature survey for the integration of blockchain network with broadcast encryption. Section 4.4 presents the introduction of broadcast encryption and blockchain technology. In Sect. 4.5, the framework of the integration of Broadcast Encryption and blockchain technology-based secure e-healthcare system. Section 4.6 concludes the paper.

4.3 Related Work

In recent times, there have been notable advancements in the utilization of broadcast encryption (BE) and its integration with blockchain technology (Jin et al. 2020; Chen et al. 2021; Gupta et al. 2021). Several studies and reports have highlighted the application of these combined technologies in various domains. One such example is presented in Du et al. (2018), where a medical information service platform was created using a hierarchical identity-based BE architecture and blockchain. According to Jin et al. (2020), the combination of blockchain and BE was also used for secure healthcare data sharing. Additionally, Lin et al. (2021) introduced an approach for jointly employing BE and blockchain in e-healthcare systems. In the context of fog computing, Zhang et al. (2019) proposed a dynamic and contributory broadcast encryption scheme for establishing secure channels. The main focus of Sohal et al. (2022) was on securing consumers' precious data from cloud service providers(which are not trusted) using a broadcast encryption-based keys management system. Another study (Chen et al. 2019) explored the integration of identity-based broadcast encryption system for providing data access control over a cloud enabled storage services. Furthermore, Chen et al. (2021) presented a combined approach using broadcast encryption and blockchain for enabling secure key management and the establishment of secure transmission of data over the Internet of Things (IoT) environment (Al-Qerem et al. 2020; Chugh et al. 2016; Kumar et al. 2022). It is worth noting that the above-mentioned references primarily employed public-key-based BE methods. These studies demonstrate the growing interest and wide-ranging applications of BE combined with blockchain technology (Mihaljević et al. 2023; Yao and Zhang 2023). The integration of these approaches enables enhanced security, privacy, and access control in various domains, fostering advancements in medical information services, clinical data sharing, telecare systems, fog computing, cloud storage services, and IoT environments. A Comparison of Blockchain-based broadcast encryption techniques is shown in Table 4.1.

4.4 An Overview of Blockchain and Broadcast Encryption

4.4.1 Background of Blockchain Technology

In 2008, Satoshi Nakamoto introduced Bitcoin (Nakamoto 2009), a groundbreaking decentralized digital cryptocurrency. Bitcoin utilized blockchain technology to address the double-spending issue. The introduction of blockchain (Chugh et al. 2020; Kaushik and Kumar 2023) had a significant impact across various industries, demonstrating its versatility and adaptability in sectors such as banking, law, governance, prediction business, identity management, supply chain sectors, and healthcare industry worldwide. Fundamentally, blockchain is a type of peer-to-peer (P2P) network-based distributed ledger technology. The hash of the previous block, combined with the current block's own hash and block data, serve as references for each block in the blockchain. Each block is built on a set of transactions that are added during block formation.

There are several types of blockchain (Fig. 4.2), each with its own characteristics and purposes. Here are some of the main types:

1. **Public Blockchain** Public blockchains are open and decentralized networks accessible to anyone. They allow for transparent and permissionless participation, where everyone is allowed to join the network, validate transactions, and contribute to the consensus process. Examples include Bitcoin and Ethereum.
2. **Private Blockchain** Private blockchains are permissioned networks where access and participation are restricted to specific entities or participants. Organizations use these blockchains to maintain control over data and network governance. Private blockchains enable more privacy and scalability compared to public blockchains. Permissioned Blockchains are similar to private blockchains, where users must be granted permission to join the network. However, they may differ in

Table 4.1 Comparison of Blockchain-based broadcast encryption techniques

Scheme	Contribution	Advantages	Limitations
Yao et al. (2023)	BPRT scheme combines blockchain technology and personalized broadcast encryption for privacy-preserving transactions in decentralized networks	Enhanced Privacy, Decentralization, efficiency, and Tamper-Proof Transactions	Integrating blockchain and broadcast encryption could increase system complexity. Volume and encryption efficiency might impact transaction processing speed
Mihaljević et al. (2023)	This scheme leverages blockchain and symmetric-keys broadcast encryption to enhance access control in IoT environments and combines blockchain's security and transparency with broadcast encryption for access management	Strengthened Access Control, security, and transparency	Symmetric-Key Management, scalability, and adoption complexity
Jin et al. (2020)	This paper focuses on utilizing blockchain technology to enhance the security, privacy, and integration of clinical data sharing	Data Integrity, privacy Preservation, interoperability, and auditable transactions	Integration complexity, regulatory considerations, scalability, and user Adoption
Li et al. (2022)	Introduces a novel hierarchical searchable encryption scheme that employs blockchain-based indexing for secure and efficient data retrieval	Secure data retrieval, efficient indexing, privacy preservation, and tamper-proof indexing	Implementation complexity, scalability, and adoption barriers
Lin et al. (2018)	This paper presents the BSeIn system, which utilizes blockchain to establish secure mutual authentication and fine-grained access control in Industry 4.0	Secure Mutual Authentication, Fine-Grained Access ControlTrust and Transparency Industry 4.0 Readiness	Resource OverheadScalability
Lin et al. (2021)	This scheme introduces a secure online treatment service using blockchain technology, ensuring data integrity, authentication, and transparency in healthcare	Data integrity, authenticity, trust, transparency, privacy-preserving	Complex integration, Usability Scalability

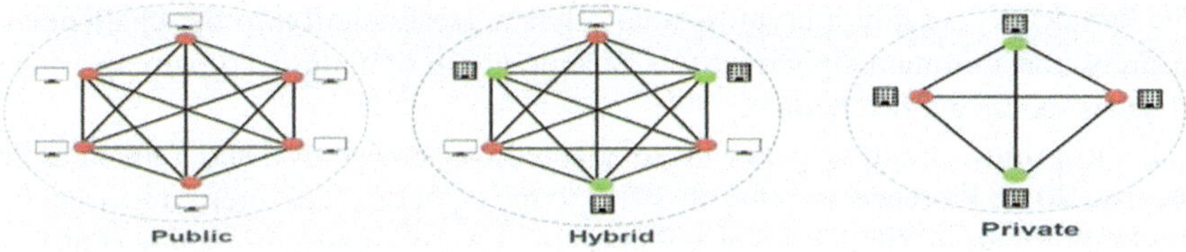

Fig. 4.2 Types of blockchain Technology in e-healthcare (EHR)

their level of decentralization, consensus mechanisms, and governance models. These blockchains are broadly used in enterprise settings, enabling access control and scalability mechanism.

3. **Hybrid Blockchain** It combines elements of both public and private blockchains. Hence it allows for both public and private transactions, enabling organizations to choose the level of transparency and privacy (Vijayakumar et al. 2017, 2018) required for specific use cases. It is often used in situations where some transactions or data need to be publicly accessible, while others require privacy and restricted access.
4. **Consortium Blockchain** A consortium blockchain, also known as a federated blockchain, is similar to a hybrid blockchain in that it combines private and public blockchain functionality. However, it differs in that numerous organizational members operate together on a decentralized network. It is essentially a private blockchain with limited access to a specific group, minimizing the hazards associated with only one entity controlling the network on a private blockchain. It is more secure, scalable, and efficient than a public blockchain. It, like private and hybrid blockchain, provides access controls.
   The consensus mechanisms in a consortium blockchain are controlled by predefined nodes. A validator node initiates, receives, and validates transactions. Transactions can be received or initiated by member nodes.
5. **Sidechain** It is a kind of independent blockchain that is interoperable with the main blockchain, also known as the parent chain. It usually allows for the execution of specific applications or smart contracts. Thus providing scalability and customization while leveraging the security and consensus of the main blockchain.
6. **Cross-Chain Blockchain** Cross-chain blockchains aim to connect and facilitate interoperability between multiple independent blockchains. They enable the transfer of assets or data between different blockchains, enhancing scalability and expanding the functionality of blockchain networks.

4.4.2 Broadcast Encryption

Broadcast encryption (Fiat and Naor 1993) cryptographic primitive is a valuable cryptographic primitive that has been extensively studied and is essential for many real-life applications. Its initial introduction by Fiat and Naor (1993) gained significant popularity after the concrete realization of the scheme of Naor et al. (2001).

The broadcast encryption cryptosystem offers a great solution to the challenge of securely transmitting a ciphertext to a specific group of authorized participants via an under secure public channel.

In this primitive, subscribers who are granted access (Agrawal and Yamada 2020; Du et al. 2018; Florence and Suresh 2019) to the ciphertext are referred to as authorized subscribers and they become members of the set *S*, while non-members of *S* are considered revoked subscribers. The broadcast algorithm works by separating legitimate users from those whose access has been revoked, which can vary depending on the broadcast message. Importantly, despite the fact that numerous revoked members conspire, they are unable to learn anything about the encrypted message, which demonstrates the collusion resistance property of broadcast encryption (Dubois et al. 2013; Acharya and Dutta 2018; Liu et al. 2016).

Due to its desirable properties, broadcast encryption (Ak 2012) has found potential enormous applications in various fields, including pay TV, satellite TV, secure e-mailing systems, DRM system, and encrypted file systems employed in cloud computing environments (Mandal and Nuida 2020; Galbraith et al. 2008; Sohal et al. 2022; Wu et al. 2011; Sabitha and Rajasree 2020; Agrawal and Yamada 2020; Sahai and Waters 2005). Based on the keys employed, the deployment of broadcast encryption (Yadav and Tiwari 2022, 2023) can be categorized into two settings: symmetric-key framework and asymmetric-key framework.

While leveraging symmetric keys, a key generation facility is in charge of providing each authorized member with the secret decryption key in advance, even before the ciphertext is communicated. Only the broadcasting company is in charge of data transmission in this circumstance. A session key is used by the data provider to encrypt the plaintext, and the public encryption keys of the set *S*'s valid members are used to further encrypt the session key. However, with this configuration, the secret key needs to be updated each time a new user logs in or a departing user logs out.

The public-key (asymmetric-key) setting efficiently addresses the "one affects all" problem in broadcast encryption. In this setting, all users in set *S* possess both encryption key as well as decryption key. While the broadcaster entity and its supporting entities may act as data providers, the message's encryption prevents unauthorized users from accessing the actual data. Furthermore, the public-key setting mitigates the issue of having to update secret keys every time a new user wants to sign up for the system. However, it introduces the challenge of preventing key compromise (Fukushima et al. 2009) by members of set *S* during the protected communication of secret keys to all system users. Figure 4.3 depicts a typical scenario of broadcast encryption.

4.5 System Model

This section will go over the details of the proposed approach, which combines broadcast encryption (Acharya and Dutta 2017; Acharya 2020; Smart and Vercauteren 2007; Chen et al. 2020; Boneh et al. 2005) and blockchain-based access control.

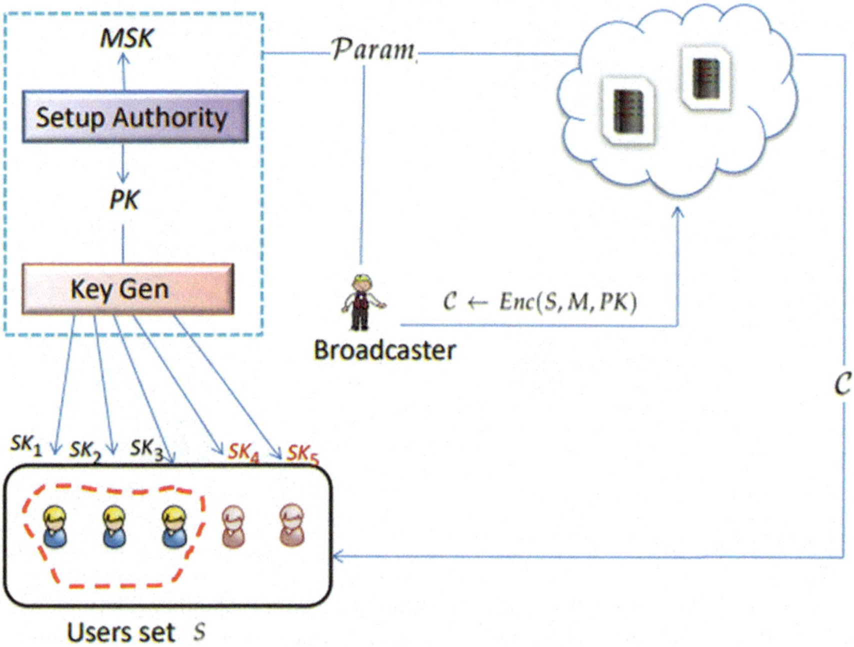

Fig. 4.3 Broadcast Encryption work flow (Yadav and Tiwari 2023)

4.5.1 Proposed Framework

In the context of an online e-healthcare system (Kaur et al. 2021; Yadav and Tiwari 2023, 2022) (see Fig. 4.4), patient data, including COVID-19 related data and other healthcare data, are collected from vivid data producers, such as doctors from different hospital authorities. To ensure the security of the data, it is securely uploaded to a centralized cloud server, which acts as a storage manager, in an encoded form.

A trustworthy authority, whether a hospital consortium or another, provides the key used in this encryption. This approach safeguards the data during its transit from the data owners to the storage server, protecting it from unauthorized access. Failure to address this issue of secure data transmission (Verma et al. 2009) can have severe consequences, as patient information may be at risk of being compromised. We have witnessed numerous incidents of large-scale data breaches, where millions of user data have been exposed. In the scenario of COVID time, making sure the COVID-19 data is secure and private is extremely important, in compliance with government guidelines. By employing encryption process during data transit, the e-healthcare system aims to protect sensitive patient information against unauthorized access. Thus, it helps maintain the confidentiality and integrity of the data and reduces the potential risks associated with information leaks.

Ohtake et al. (2010) first presented a cryptographic primitive in the discipline of cryptography at the ProvSec 2010 conference. The primary objective of this BEPM (Lü et al. 2017; Boneh et al. 2014) is to make it easier for a broadcaster entity and

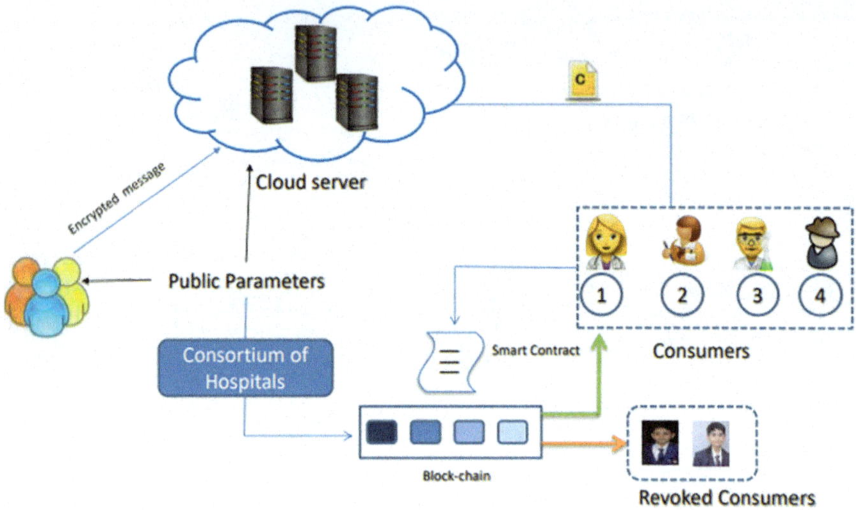

Fig. 4.4 System model

a legitimate group of users to securely communicate with each other. Thus allowing the broadcaster entity to send a public broadcast header while also supporting the transmission of customized encrypted messages to authorized individuals. However, in this scheme, the broadcaster entity first establishes a session key and unique keys for system users. These keys are then used to encrypt both a common confidential broadcast message and personalized individual messages through symmetric-key encryption. To recover the received messages, subscribers require a decryption key which is issued by a trusted third party. This decryption key allows them to retrieve the session key and their personalized key, enabling them to access and decrypt both the common broadcast message and their individualized messages. BEPM (Mandal and Nuida 2020; Chen et al. 2020; Han et al. 2013; Mandal and Nuida 2020; Mihaljević et al. 2023) has demonstrated its utility in various real-world applications where maintaining anonymity of not only the broadcast message shared by a certain group but also to the customized messages for users. By providing a mechanism for secure and customized communication, BEPM contributes to enhancing the confidentiality and privacy of sensitive information in different contexts. To develop a solution that establishes a trustworthy e-healthcare system online, such a system must prioritize the patients' privacy and security while simultaneously enabling authorized users to access this data, thereby enhancing patient care and improving public health outcomes. The proposed system model comprises five components.

1. The proposed approach involves the creation of personalized data by the data owner for each authorized user within a designated class. Subsequently, a unified set of encrypted information is transmitted to the entire group in a manner that conceals the identities of the legitimate users from external parties. This ensures the privacy and confidentiality of the user set. Only authorized users possess the

capability to retrieve their personalized data, and only authorized users have the power to retrieve their personalized data;

2. Hospital authorities act as data creators;
3. Through a consortium of hospitals, the cloud server is fed with data generated and gathered by networked devices, and data consumers are able to access it pursuant to access control rules;
4. Owners of the data are responsible for information encryption, information publication on storage systems, and the hospital consortium is in charge of defining access control rights;
5. In accordance with the permissions granted to them, data users may obtain access and retrieve data from data storage;
6. Data encryption is feasible to store and access through cloud servers;
7. The blockchain network facilitates communication between data providers and users and also records access control data.

The integration of IoT (Internet of Things) (Singh et al. 2018) and cloud technology with Blockchain Technology plays a vital role in achieving these objectives. By leveraging IoT devices and cloud infrastructure, remote access to patient information becomes possible, improving the accessibility and convenience of healthcare services for both patients and healthcare practitioners. This integration enhances the efficiency of data management and enables seamless and secure data exchange while adhering to privacy and security standards.

4.6 Conclusion

This paper contributes by demonstrating the substantial benefits of integrating broadcast encryption with blockchain technology within e-healthcare systems. This integration serves to elevate the levels of privacy, security, and regulatory adherence. It establishes a framework for secure data sharing, precision-driven access control, traceable auditability, and robust resistance against tampering. The combined utilization of these technologies empowers e-healthcare systems to uphold the sanctity of sensitive medical data, fostering trust, and facilitate a secure environment for collaborative efforts among healthcare providers, researchers, and patients. Integration of blockchain with quantum-safe broadcast encryption is left as an open problem.

References

Abdalla M, Fouque P-A, Pointcheval D (2005) Password-based authenticated key exchange in the three-party setting. In: Proceedings of the 8th international conference on theory and practice in public key cryptography. PKC'05. Springer, Berlin, Heidelberg, pp 65–84

Acharya K (2020) Secure and efficient public key multi-channel broadcast encryption schemes. J Inform Secur Appl 51:102436

Acharya K, Dutta R (2017) Provable secure constructions for broadcast encryption with personalized messages. In: International conference on provable security. Springer, pp 329–348
Acharya K, Dutta R (2018) Constructions of secure multi-channel broadcast encryption schemes in public key framework. In: Camenisch J, Papadimitratos P (eds) Cryptology and network security. Springer, Cham, pp 495–515
Agrawal S, Yamada S (2020) Optimal broadcast encryption from pairings and lwe. Lecture Notes in Computer Science (including subseries. Lecture notes in artificial intelligence and Lecture notes in bioinformatics), vol 12105, LNCS, pp 13–43. https://doi.org/10.1007/978-3-030-45721-1_2
Ak M (2012) Optimization techniques and new methods for broadcast encryption and traitor tracing
Al-Qerem A, Alauthman M, Almomani A, Gupta BB (2020) IOT transaction processing through cooperative concurrency control on fog-cloud computing environment. Soft Comput 24(8):5695–5711
Azees M, Vijayakumar P, Deboarh LJ (2017) Eaap: efficient anonymous authentication with conditional privacy-preserving scheme for vehicular ad hoc networks. IEEE Transa Intell Transp Syst 18(9):2467–2476
Boneh D, Boyen X, Goh E-J (2005) Hierarchical identity based encryption with constant size ciphertext. In: Proceedings of the 24th annual international conference on theory and applications of cryptographic techniques. EUROCRYPT'05. Springer, Berlin, Heidelberg, pp 440–456 (2005). https://doi.org/10.1007/11426639_26
Boneh D, Gentry C, Waters B (2005) Collusion resistant broadcast encryption with short ciphertexts and private keys. In: Shoup V (ed) Advances in cryptology—CRYPTO 2005. Springer, Berlin, Heidelberg, pp 258–275
Boneh D, Waters B, Zhandry M (2014) Low overhead broadcast encryption from multilinear maps. In: Annual cryptology conference. Springer, pp 206–223
Chen L, Li J, Zhang Y (2019) Adaptively secure anonymous identity-based broadcast encryption for data access control in cloud storage service. KSII Transa Internet Inform Syst (TIIS) 13(3):1523–1545
Chen L, Li J, Zhang Y (2020) Anonymous certificate-based broadcast encryption with personalized messages. IEEE Trans Broadcast 66(4):867–881
Chen T, Zhang L, Choo K-KR, Zhang R, Meng X (2021) Blockchain-based key management scheme in fog-enabled IoT systems. IEEE Internet of Things J 8(13):10766–10778
Chugh N, Kumar A, Aggarwal A (2016) Security aspects of a RFID-sensor integrated low-powered devices for internet-of-things. In: 2016 fourth international conference on parallel, distributed and grid computing (PDGC). IEEE, pp 759–763
Chugh N, Sharma DK, Singhal R, Jain S, Srikanth P, Kumar A, Aggarwal A (2020) Blockchain-based decentralized application (DApp) design, implementation, and analysis with healthcare 4.0 trends. In: Basic & clinical pharmacology & toxicology, vol 126, pp 139–140. WILEY 111 RIVER ST, HOBOKEN 07030-5774, NJ USA
Delerablée C (2007) Identity-based broadcast encryption with constant size ciphertexts and private keys. In: Proceedings of the advances in crypotology 13th international conference on theory and application of cryptology and information security. ASIACRYPT'07. Springer, Berlin, Heidelberg, pp 200–215. http://dl.acm.org/citation.cfm?id=1781454.1781471
Dubois R, Guillevic A, Sengelin Le Breton M (2013) Improved broadcast encryption scheme with constant-size ciphertext. In: Abdalla M, Lange T (eds) Pairing-based cryptography—Pairing 2012. Springer, Berlin, Heidelberg, pp 196–202
Du Y, Liu J, Guan Z, Feng H (2018) A medical information service platform based on distributed cloud and blockchain. In: 2018 IEEE international conference on smart cloud (SmartCloud). IEEE, pp 34–39
Fan Q, Chen J, Deborah LJ, Luo M (2021) A secure and efficient authentication and data sharing scheme for internet of things based on blockchain. J Syst Archit 117:102112
Fiat A, Naor M (1993) Broadcast encryption. In: Annual international cryptology conference. Springer, pp 480–491

Florence ML, Suresh D (2019) Enhanced secure sharing of PHR's in cloud using user usage based attribute based encryption and signature with keyword search. Cluster Computing 22:13119–13130

Fukushima K, Kiyomoto S, Tanaka T, Sakurai K (2009) Ternary subset difference method and its quantitative analysis. In: Chung K-I, Sohn K, Yung M (eds) Information security applications. Springer, Berlin, Heidelberg, pp 225–239

Galbraith SD, Paterson KG, Smart NP (2008) Pairings for cryptographers. Discrete Appl Math 156(16):3113–3121. https://doi.org/10.1016/j.dam.2007.12.010. Applications of Algebra to Cryptography

Gupta BB, Li K-C, Leung VC, Psannis KE, Yamaguchi S (2021) Blockchain-assisted secure fine-grained searchable encryption for a cloud-based healthcare cyber-physical system. IEEE/CAA J Automatica Sinica 8(12):1877–1890

Han J, Park J, Lee D (2013) Transmission-efficient broadcast encryption scheme with personalized messages. IEICE Trans Fundam Electron Commun Comput Sci E96.A:796–806. https://doi.org/10.1587/transfun.E96.A.796

Jain S, Ahuja NJ, Srikanth P, Bhadane KV, Nagaiah B, Kumar A, Konstantinou C (2021) Blockchain and autonomous vehicles: Recent advances and future directions. IEEE Access 9:130264–130328

Jin H, Xu C, Luo Y, Li P (2020) Blockchain-based secure and privacy-preserving clinical data sharing and integration. In: Algorithms and architectures for parallel processing: 20th international conference, ICA3PP 2020, New York City, NY, USA, October 2–4, 2020, Proceedings, Part III 20. Springer, pp 93–109

Kaur M, Singh D, Kumar V, Gupta BB, Abd El-Latif AA (2021) Secure and energy efficient-based e-health care framework for green internet of things. IEEE Trans Green Commun Network 5(3):1223–1231

Kaushik K, Kumar A (2023) Demystifying quantum blockchain for healthcare. Secur Privacy 6(3):284

Kumar A, Gopal K, Aggarwal A (2016) Design and analysis of lightweight trust mechanism for secret data using lightweight cryptographic primitives in manets. Int J Netw Secur 18(1):1–18

Kumar A, Ottaviani C, Gill SS, Buyya R (2022) Securing the future internet of things with post-quantum cryptography. Secur Privacy 5(2):200

Lin H, Zhang H, Yan H, Wang H, Shi Y, Gao F, Wen Q (2021) A secure online treatment blockchain service. Wirel Perso Commun 117:1773–1795

Lin C, He D, Huang X, Choo K-KR, Vasilakos AV (2018) Bsein: a blockchain-based secure mutual authentication with fine-grained access control system for industry 4.0. J Netw Comput Appl 116:42–52

Liu W, Liu J, Wu Q, Qin B, Li Y (2016) Practical chosen-ciphertext secure hierarchical identity-based broadcast encryption. Int J Inform Secur 15(1):35–50

Li Y, Zhou F, Ji D, Xu Z (2022) A hierarchical searchable encryption scheme using blockchain-based indexing. Electronics 11(22):3832

Lü L-Q, Yang X-Y, Wang J-J, Cheng L (2017) Low overhead broadcast encryption with personalized message based on multilinear maps. Yingyong Kexue Xuebao/J Appl Sci 35:626–633 (2017). https://doi.org/10.3969/j.issn.0255-8297.2017.05.008

Lubicz D, Sirvent T (2008) Attribute-based broadcast encryption scheme made efficient, pp 325–342

Mandal M, Nuida K (2020) Identity-based outsider anonymous broadcast encryption with simultaneous individual messaging. In: International conference on network and system security. Springer, pp 167–186

Mandal M, Nuida K (2020) Identity-based outsider anonymous broadcast encryption with simultaneous individual messaging. In: Network and system security: 14th international conference, NSS 2020, Melbourne, VIC, Australia, November 25–27, 2020, Proceedings 14. Springer, pp 167–186

Mihaljević MJ, Knežević M, Urošević D, Wang L, Xu S (2023) An approach for blockchain and symmetric keys broadcast encryption based access control in IoT. Symmetry 15(2):299

Nakamoto S (2009) Bitcoin: a peer-to-peer electronic cash system. Cryptography Mailing list at https://metzdowd.com

Naor D, Naor M, Lotspiech J (2001) Revocation and tracing schemes for stateless receivers. In: Kilian J (ed) Advances in cryptology CRYPTO 2001. Springer, Berlin, Heidelberg, pp 41–62
Ohtake G, Hanaoka G, Ogawa K (2010) Efficient broadcast encryption with personalized messages. In: International conference on provable security. Springer, pp 214–228
Phuong TVX, Yang G, Susilo W, Chen X (2015) Attribute based broadcast encryption with short ciphertext and decryption key, pp 252–269
Sabitha S, Rajasree MS (2020) Multi-level on-demand access control for flexible data sharing in cloud. Clust Comput 0123456789
Sahai A, Waters B (2005) Fuzzy identity-based encryption. In: Cramer R (ed) Advances in cryptology—EUROCRYPT 2005. Springer, Berlin, Heidelberg, pp 457–473
Singh D, Ranvijay YR (2018) A state-of-art approach to misbehaviour detection and revocation in vanet: survey. Int J Ad Hoc and Ubiquit Comput 28(2):77–93
Smart N, Vercauteren F (2007) On computable isomorphisms in efficient asymmetric pairing-based systems. Discrete Appl Math 155:538–547 (2007). https://doi.org/10.1016/j.dam.2006.07.004
Sohal M, Bharany S, Sharma S, Maashi MS, Aljebreen M (2022) A hybrid multi-cloud framework using the IBBE key management system for securing data storage. Sustainability 14(20):13561
Verma E, Abraham J, Yadav S (2009) Design of energy efficient scheme for conducting secure e-class application on manet. In: 2009 first international conference on computational intelligence, communication systems and networks. IEEE, pp 464–469
Vijayakumar P, Chang V, Deborah LJ, Balusamy B, Shynu P (2018) Computationally efficient privacy preserving anonymous mutual and batch authentication schemes for vehicular ad hoc networks. Fut Gen Comput Syst 78:943–955
Vijayakumar P, Azees M, Chang V, Deborah J, Balusamy B (2017) Computationally efficient privacy preserving authentication and key distribution techniques for vehicular ad hoc networks. Clust Computi 20(3):2439–2450
Wang H, Zhang Y, Chen K, Sui G, Zhao Y, Huang X (2019) Functional broadcast encryption with applications to data sharing for cloud storage. Inform Sci 502:109–124
Wu Q, Qin B, Zhang L, Domingo-Ferrer J (2011) Fully distributed broadcast encryption. In: Proceedings of the 5th international conference on provable security. ProvSec'11. Springer, Berlin, Heidelberg, pp 102–119. http://dl.acm.org/citation.cfm?id=2042756.2042768
Yadav S (2020) An efficient affinity propagation clustering technique. Int J Adv Sci Technol 29(04):9555–9573
Yadav S, Tiwari N (2020) Recent advancements in chaos-based image encryption techniques: a review. In: Social networking and computational intelligence: proceedings of SCI-2018, pp 639–647
Yadav S, Tiwari N (2022) An efficient and secure data sharing method using asymmetric pairing with shorter ciphertext to enable rapid learning in healthcare. Comput Intell Neurosci 2022 (2022)
Yadav S, Tiwari N (2023) Privacy preserving data sharing method for social media platforms. PloS One (181):0280182
Yadav S, Tiwari N (2023) Secure and efficient online data sharing scheme using broadcast encryption with personalised message system
Yao S, Zhang D (2023) Bprt: a blockchain-based privacy-preserving transaction scheme based on an efficient broadcast encryption with personalized messages. Trans Emerg Telecommun Technol 34(1):4675
Zhang L (2019) Key management scheme for secure channel establishment in fog computing. IEEE Trans Cloud Comput 9(3):1117–1128

5 Computational Analysis of Interference for Sustainable Operation of Wireless Networks

Ram Krishan

Abstract

In wireless networks, data transmission is affected by interference from radio waves, and some data packets get discarded during transmission due to weak signal strength. As a result, data transmission is hazardous, and network operations are unreliable. This work addresses the investigation of co-channel and adjacent channel interference for sustainable operation of IEEE 802.11 networks. The goal of this work is to measure the level of signal interference experienced by access points and its impact on network performance. The Riverbed modeler tool was used to monitor and compute interference in the wireless network. According to the study results, the major cause of excessive interference is improper wireless channel distribution to access points.

Keywords

Interference • Wireless network • Analysis • Frequency • Riverbed modeler

5.1 Introduction

An IEEE 802.11 wireless local area network (WLAN or Wi-Fi) provides wireless access between computing devices using a radio signal instead of a wired medium (Pahlavan and Krishnamurthy 2021). Wireless LANs are implemented (Krishan and Laxmi 2015a) to provide a link between the existing wired network and wireless users, giving wireless access to these users (Oni and Blostein 2015). In 1987

R. Krishan (✉)
Department of Computer Science, Mata Sundri University Girls College, Mansa, India
e-mail: ramkrishan@pbi.ac.in

A. Kumar et al. (eds.), *Sustainable Security Practices Using Blockchain, Quantum and Post-quantum Technologies for Real Time Applications*, Contributions to Environmental Sciences & Innovative Business Technology,
https://doi.org/10.1007/978-981-97-0088-2_5

the IEEE committee started to work on WLANs and work was to develop wireless communication for 2.4 and 5 GHz ISM (industrial, scientific, and medical) frequency bands (Krishan and Laxmi 2015b). The 802.11 standards focus was on developing Medium Access Control (MAC) Protocol and Physical (PHY) Layer specifications. In 1997 the IEEE introduced 802.11 the first standard of WLAN (Sharma and Singh 2016). This standard of 802.11 provides a data transfer rate of only 2 Mbps at a 2.4 GHz frequency (Haider et al. 2018). In 1999, the 802.11 standards were divided into two new standards 802.11a and 802.11b. According to the standard, 802.11a makes use of orthogonal frequency division multiplexing (OFDM), which results in the maximum rate at which data may be sent on the 5 GHz frequency being 54 Mbps (Haider et al. 2018). The direct-sequence spread spectrum (DSSS) media access technique is used by the 802.11b standard, which provides a data rate of 11 Mbps at the 2.4 GHz band and delivers the same data rate as the original standard. The 802.11 g protocol was first released the same year, in 2003. The 802.11 g protocol works on the 2.4 GHz band and employs the same OFDM technology as its predecessor, the 802.11a standard. This standard provides a data transmission rate that is equal to 54 Mbps. The group of (BSS) Basic Service Set and Extended Service Set constitutes a WLAN structure (ESS). Every BSS simulates an AP's coverage, but an ESS participates in numerous BSSs and promotes mobility inside the WLAN (Sarkar et al. 2021). The service area of an access point is the coverage area where wireless clients are permitted to travel and stay connected (AP). The 802.11 WLAN standards define two operating modes:

5.1.1 Infrastructure Mode

In infrastructure mode, as shown in Fig. 5.1, wireless stations (STA) are connected to the network through an access point. In this mode, each station (STA) connects to an access point in order to have wireless access. The access point (AP) is connected to the wired link. To broaden the network's coverage and provide more wireless station access, a number of APs can be added. A basic service set, or BSS, was created by an access point (AP) and the stations within its region of coverage (Obelovska et al. 2021). The BSS allows communication between every station. Every BSS is identified with a BSSID. BSSID is 6-byte (48 bits) and it is the MAC address of the corresponding access point (AP). An ESS is created when many BSSs are linked to one another (Extended Service Set). The distribution system (DS), which constitutes an extended service set, provides the link between various access points in the ESS (ESS). Each ESS in the network has a unique identification called an Extended Service Set Identification (ESSID), which is a string of 32 bytes. ESSID is acting as the name of the network. In a nutshell, ESSID is the SSID used as the network name. A station attempting to join the network must be aware of its SSID. A roaming user's (Oni and Blostein 2015) ability to switch between access points within the ESS allows them to do so in order to take advantage of different access points' stronger signal reception. The distribution system provides communication between access points to exchange

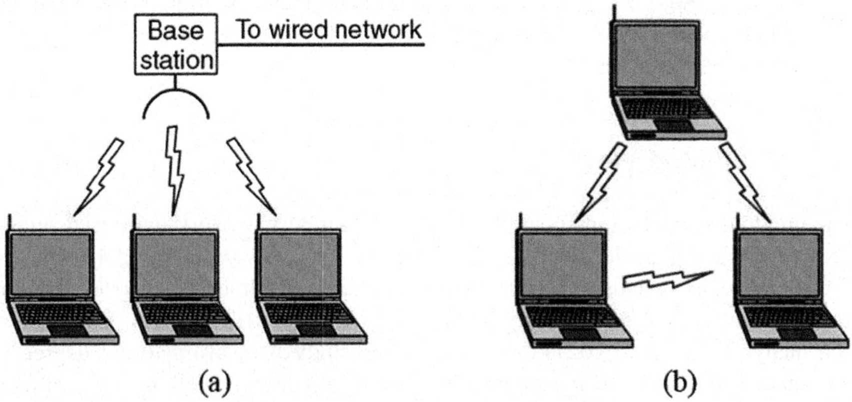

Fig. 5.1 Wireless LANs **a** Infrastructure Mode **b** Ad-hoc Mode

station information and data transmission between them (2011). The functionality of stations to move between access points is called roaming.

5.1.2 Ad-Hoc Mode

In ad-hoc mode, wireless stations can connect only to a peer-to-peer network. In this mode, each station acts as both station and an access point (AP). This kind of station setup is known as the independent basic service set (IBSS). In IBSS, there are two stations that can communicate with each other without any access point. An IBSS is a wireless network that can be recognized by SSID. In an ad-hoc network if a station is in the range of any other station, then they are able to communicate otherwise not. The range of each station is also known as the range of IBSS. In ad-hoc network mode there no distribution system is present to transfer data frames between the stations. As the setup, IBSS is known as a constrained wireless network.

The following are the objectives and contributions of this work:

1. To understand the interference in the wireless network and the design of the simulation model using the Riverbed modeler simulator.
2. The second objective is to calculate and compare the total network interference of the designed model.
3. Finally, the effects of interference on the performance of the wireless network were analyzed.

This chapter is organized into six sections: Sect. 5.1 introduces the wireless local area network and how it functions. Section 5.2 presents the earlier works that dealt with the interference-related issue of wireless networks. The idea of interference and its causes are presented in Sect. 5.3. The performance of wireless networks was studied in Sect. 5.4 in relation to several parameters. Section 5.5 presents the riverbed simulation-based interference analysis of wireless networks. The chapter concludes with Sect. 5.6.

5.2 Related Work

The previous contributions made to the area of wireless network interference control are discussed in this section. IEEE 802.11 networks using the 2.4 GHz and 5 GHz frequency bands experience co-channel and neighboring channel interference, as well as the performance of access points (APs), which are investigated and analyzed (Krishan 2021). The goal of the study is to evaluate the degree of frequency-specific signal interference that is encountered as well as how it affects the wireless network. In a study of (Kakar and Sezgin 2017), the two primary research viewpoints of how to reliably address interference for situations where CSIT is imprecise were examined, namely broadcast and X-channel on instances for non-distributed and distributed networks. As the pre-log factor of feasible rates, they utilize the well-known (generalized) degrees of freedom (GDoF) measure to evaluate the performance of different systems. The author of (Amzucu et al. 2014) presents measurement findings on indoor propagation, microwave oven (MWO) interference, and their effects on the effectiveness of the WSN. The tests of the propagation show that the multipath has a considerable impact. For example, altering the communication route or moving a node a few centimeters can result in a 30dB change in received power. They provide a method for cluster-based cooperative interference management (Kakar and Sezgin 2017) that is based on cognitive network resource allocation and interference pattern formation. This technology prevents base stations from interfering with one another on the same layer and decreases the amount of cross-layer interference that primary users experience by reducing the number of base station cells that cause it via a cooperative channel and power allocation.

5.3 Concept of Wireless Network Interference

Interference with Wi-Fi networks is a common and challenging problem. WLAN's portability is its greatest strength; however, many common electrical devices may interfere with wireless connections. WLAN networks are vulnerable to interference, which might cause unexpected behavior even if no configuration changes have been made.

5.3.1 What Leads to Interference in Wireless Radio Frequency?

IEEE 802.11 wireless networks are susceptible to interference because they employ unlicensed frequencies that are shared by other technology including mobile phones, microwaves, and surveillance cameras. Additionally, interference from other wireless access points using the same channel is possible. The transmission power and the proximity of the interfering device have a significant impact on the effect of interference. Interference will have a greater impact on access points that are nearby and sharing the same channel than on those that are further away.

5.3.2 Reasons for Wireless Interference

This sub-section explains the major reasons for wireless interference:

- With so many radios using the same frequency, interference is possible if the radios are left on their factory-default channels.
- In a wireless network, "hidden nodes" are nodes that are unreachable to other nodes.
- Overlapping channels may result in co-channel interference and neighboring channel interference. Another possibility is that your network isn't utilizing all of the available channels due to interference from adjacent business communications.
- Some non-network gadgets, such as wireless video cameras, auto-alarms, cordless phones, and microwaves can obstruct wireless channels. These devices are typically operating at 2.4 GHz frequency.
- Broad RF spectrum emissions can result from poor electrical connections and RF jamming is an intentional attempt to interfere with the network by sending out a strong signal.

5.4 Factors Affecting WLAN Performance

Factors that affect the performance of WLAN are discussed as under.

5.4.1 Interference

In Wireless LANs, the medium is shared by all devices which causes Radio Frequency interference (Amzucu et al. 2014). When 802.11 stations and devices transmit data on the same frequency, then interference starts to occur between all stations and devices. In a shared medium when a station is sending data to an access point or access point to one of the stations, then if any other station wants to transmit the data he will wait until the medium is free. Since 802.11 wireless networks use the unlicensed ISM frequency band, which is also used by microwave ovens, security cameras, and mobile phones, this can cause interference.

5.4.2 Scattering and Multipath Fading

Obstructions build transmission mistakes in view of diminished power at specific frequencies, the time distinction between various segments (delay spread) makes the received signal spread in the time–space. In most wireless channels the transmitted signal lands at the collector from different directions in a variety of ways. Correspondence between the transmitter and the receiver can be demonstrated in a few ways to represent a line-of-sight (LOS) microwave radio connection, as is broadly utilized as a part of across-the-country networks.

5.4.3 Antenna Type

The originating foci of the transmitting and receiving antennas (Meng and Liu 2019) are assumed, and the route is shown as a beam. A beam is an ideal projectile that travels in a straight line and reflects from objects in accordance with the rules of geometric optics. In a wireless setting, the signal can be received via one of the following two paths:

1. Direct line-of-sight (LOS) relationship between the transmitter and recipient.
2. Following the ground's reflection, the way appears.

5.4.4 Antenna Gain

The amount of radiation emitted by the antenna in one direction divided by the amount emitted in the other direction that would have been produced if the antenna had recognized its power isotropically is known as the antenna gain (Dolińska et al. 2013). The gain of an antenna is measured in decibels (dB). Antenna Directivity shows the maximum antenna gain in a given direction.

Antenna Diversity—Antenna diversity qualities are critical in averting multipath fading particularly to increase antenna range.

Antenna Cable Length—Antenna loss in the link relies on the length of the cable. The losses that occur due to the cable length degrade SNR significantly and ought to be minimized.

5.4.5 System Design

There is a minimum value of the threshold signal to be received by the receiver to accomplish a certain bit rate. In the event that the accepted signal is lower than the minimum value, the bit rate could be diminished and affect the performance.

Transmit Power—Higher output of transmitted power gives better performance however is generally constrained by administrative necessities (e.g., FCC).

Turn Around Time (TAT)—TAT is defined as the time it takes to switch between transmitter and receiver modes.

5.4.6 Network Load

This sub-section will talk about the things that are responsible for network load.

Number of Users—Wireless medium is shared by multiple users so as the number of users increases they affect the network performance.

Traffic Mix—A traffic mix is a way to describe how users act. There are many methods, from simple to complicated, which can be used to explain how users act in networks.

5.4.7 Location Factors

This sub-section will go through how location affects network performance.

AP and User distance—If the distance between the AP and the user expands, the data rate will drop.

Direction—Network performance could be affected by the direction of the user and depends on the location of an access point (AP) and antenna.

5.5 Interference Analysis

In this section, a design of the network model is presented for interference analysis of WLAN. This section will cover the detailed simulation analysis and the effect of interference on the performance of wireless networks.

5.5.1 Network Model Design

The proposed simulation model of WLAN is constructed and analyzed using the Riverbed Modeler software (Krishan 2021). In this network model, each access point's interference is computed for two alternative network scenarios. In scenarios 1 and 2, simulations were performed with 4 APs and 8 APs, respectively. The simulation network setup depicted in Fig. 5.2 shows four different areas, with each red circle having an access point and its associated users in the network. There are

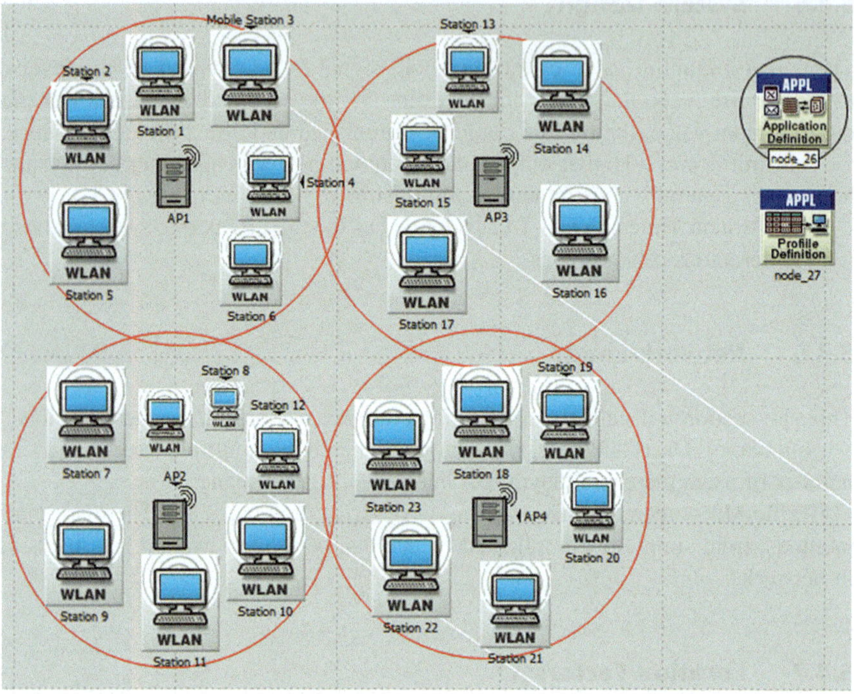

Fig. 5.2 Wireless LAN riverbed scenario

a total of 24 users on the network, some of whom are roaming and some who are static. In this model, an application profile for HTTP data transfer is used to investigate the simulation results.

5.5.2 Simulation Results

The simulation results in findings include the outcome for each access point as well as the total network interference in watts. Table 5.1, for the network, displays the statistics of interference that affected access points AP1 to AP4. The X and Y axis of Fig. 5.3 present the network access points and interference, respectively. The values of interference calculated for each access point can be seen visually in this graph. The total interference in the network was recorded as 0.00522 watts.

Similarly, Table 5.2 and Fig. 5.4 show statistical and graphical representations of the interference that access points AP1 to AP8 in the network experienced, respectively. The overall interference of the network with 4 APs and 8 APs was recorded as 0.00522 and 0.00754 watts, respectively.

Table 5.1 Interference on access points AP1 to AP4

Name of AP	Interference
AP1	0.00129
AP2	0.00102
AP3	0.00132
AP4	0.00159
Total interference (Watts)	0.00522

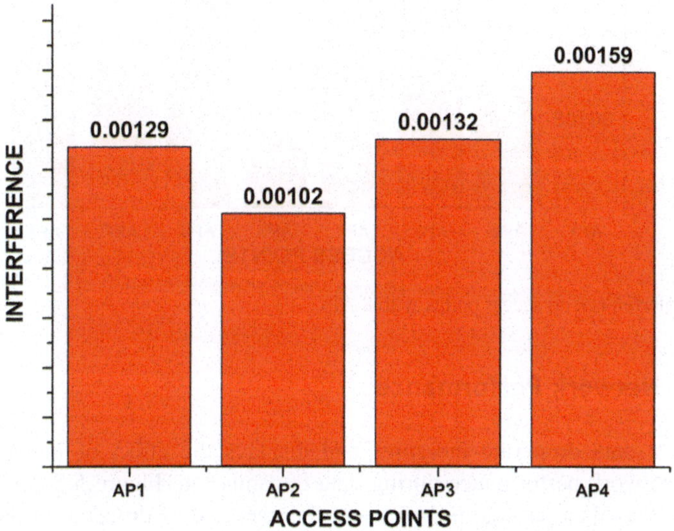

Fig. 5.3 Interference on AP1 to AP4

Table 5.2 Interference on access points AP1 to AP8

Name of AP	Proposed scheme
AP1	0.00106
AP2	0.00066
AP3	0.00089
AP4	0.00089
AP5	0.00107
AP6	0.00106
AP7	0.00066
AP8	0.00125
Total interference (Watts)	0.00704

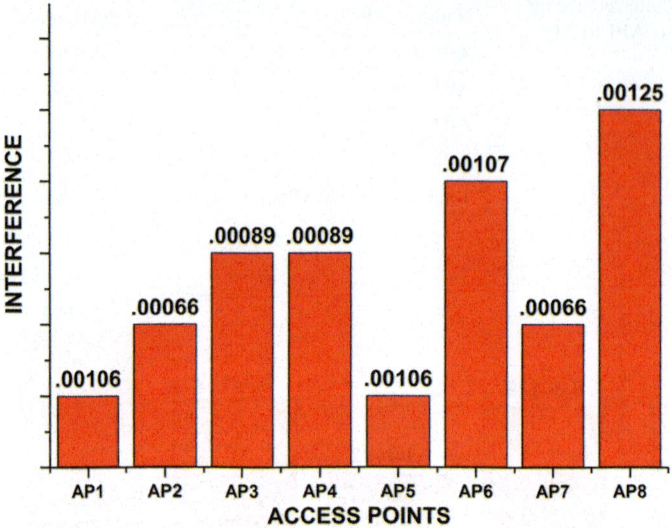

Fig. 5.4 Interference on access points with 8 APs

5.5.3 Network Performance

The packet data drop rate measures and displays the effect of interference on wireless network performance during data transmission. Figure 5.5 depicts the X and Y axes as data packets in bytes and the packet data drop rate in percentage (%), respectively, to record the network data drop rate. With 8 APs in the network, the average packet data drop rate was 0.466, while with 4 APs, it was 0.386. It has been discovered that when there are four APs instead of eight, the data drop rate is reduced.

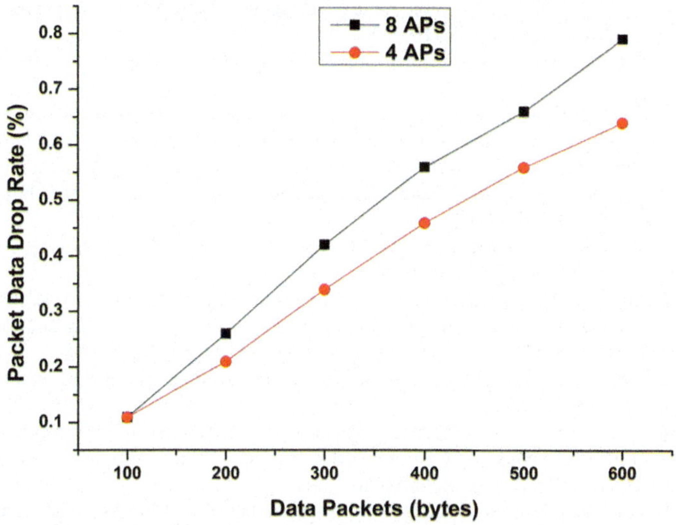

Fig. 5.5 Packet data drop rate of wireless network

5.6 Conclusion

This chapter describes research that was conducted to determine the degree of signal interference encountered by access points and its influence on wireless network performance. The Riverbed modeler tool was used to detect and compute wireless network interference. The simulation results of the interference for 4 APs and 8 APs are measured as 0.00522 and 0.00754 watts. The degree of interference in the network with 4 APs is lower than the interference in the network with 8 APs, according to the simulation findings shown on the graph. It has been observed that the packet data loss rate of the network would rise as the number of access points (APs) increased. In the future, this research may be expanded to include more complicated and dense network situations in order to determine the sustainable operation of wireless networks.

References

[10]Cisco (2011) WLAN radio frequency design considerations. Enterp Mobil 7.3 Des Guid, pp 1–36

Amzucu DM, Li H, Fledderus E (2014) Indoor radio propagation and interference in 2.4 GHz wireless sensor networks: Measurements and analysis. Wireless Pers Commun 76(2):245–269

Dolińska I, Masiukiewicz A, Rządkowski G (2013) The mathematical model for interference simulation and optimization in 802.11n networks. CEUR Workshop Proc. 1032:99–110

Haider Z, Saleem M, Jamal T (2018) Analysis of interference in wireless networks, no. August. http://arxiv.org/abs/1810.13164

Kakar J, Sezgin A (2017) A survey on robust interference management in wireless networks. Entropy 19(7):362. https://doi.org/10.3390/e19070362

Krishan R (2019) Performance characterization of dual-band Microstrip fractal antenna. Intern J Inform Technol Elect Eng 8(5)

Krishan R (2021) Performance characterization of load balancing algorithm 13(3):915–922

Krishan R (2022) Terahertz band for wireless communication—a review, in *Terahertz Wireless Communication Components and System Technologies*, Ist., A. B. y Mohammed El Ghzaoui, Sudipta Das, Trupti Ranjan Lenka, Ed. Springer Singapore, pp 153–161. https://doi.org/10.1007/978-981-16-9182-9_10

Krishan R, Laxmi V (2015) IEEE 802.11 WLAN Load Balancing for Network Performance Enhancement. https://doi.org/10.1016/j.procs.2015.07.371

Krishan R, Laxmi V (2015) Channel management algorithm for modeling network throughput in IEEE 802.11 WLAN, in International Conference on Advanced Computing and Communication Technologies, ACCT, vol. 2015. https://doi.org/10.1109/ACCT.2015.15

Krishan R, Laxmi V (2016) Algorithm for optimized load balancing of WLAN. Int J Comput Sci Inf Secur 14(11):872–878

Meng Y, Liu X (2019) Resource allocation and interference management for multi-layer wireless networks in heterogeneous cognitive networks. Eurasip J Wirel Commun Network 2019(1). https://doi.org/10.1186/s13638-019-1514-1

Obelovska K, Panova O, Karovič V (2021) Performance analysis of wireless local area network for a high-/low-priority traffic ratio at different numbers of access categories. Symmetry 13(4):693. https://doi.org/10.3390/sym13040693

Oni PB, Blostein SD (2015) AP association optimization and CCA threshold adjustment in dense WLANs. *2015 IEEE Globecom Work. GC Wkshps 2015 - Proc.*, pp 0–5. https://doi.org/10.1109/GLOCOMW.2015.7414115

Pahlavan K, Krishnamurthy P (2021) Evolution and impact of Wi-Fi technology and applications: a historical perspective. Int J Wirel Inf Networks 28(1):3–19. https://doi.org/10.1007/s10776-020-00501-8

Sarkar NI, Mussa O, Gul S (2021) Impact of people's movement on wi-fi link throughput in indoor propagation environments: an empirical study. Electron 10(7). https://doi.org/10.3390/electronics10070856

Sharma P, Singh G (2016) Comparison of Wi-Fi IEEE 802.11 standards relating to media access control protocols. Int J Comput Sci Inf Secur 14(10):856–862. https://www.researchgate.net/publication/332174122_Comparison_of_Wi-Fi_IEEE_80211_Standards_Relating_to_Media_Access_Control_Protocols

Shirley Andjamba T, Singh Jat D, Lusilao Zodi G-A (2016) Interference analysis of IEEE 802.11 wireless networks: a case study of Namibia University of Science and Technology. https://ieeexplore.ieee.org/document/7892726/

Advanced Techniques for Digital Evidence Preservation: The Power of Blockchain and Machine Learning

6

Rizwan Ur Rahman, Deepak Singh Tomar, Gaurav Pramod Kacharea, Meeraj Mahendra Gawde, Tenzin Tsundue, Pavan Kumar, and Hamiden Abd El Wahed Khalifa

Abstract

Cybercrime is one of the fastest-growing crimes worldwide, and they are increasing in volume, sophistication, and cost. According to numerous reports such as Cybersecurity Ventures and others it is estimated that every seven seconds, cyber attackers penetrated into Cyber Systems. As a result, one of the essential parts of any system for storing and handling all the events is the log system. However, the system is not robust, and detecting an anomaly in logs has

R. U. Rahman (✉) · G. P. Kacharea · M. M. Gawde · T. Tsundue
School of Computing Science and Engineering, VIT Bhopal University, Bhopal 466114, India
e-mail: rizwan.rahman12@gmail.com; rizwan.ur@vitbhopal.ac.in

G. P. Kacharea
e-mail: gaurav.pramod2018@vitbhopal.ac.in

M. M. Gawde
e-mail: meeraj.gawde@gmail.com; mahendra.gawde2018@vitbhopal.ac.in

T. Tsundue
e-mail: tenzin.tsundue2018@vitbhopal.ac.in

D. S. Tomar
Department of Computer Science and Engineering, Maulana Azad National Institute of Technology, Bhopal 462003, India
e-mail: deepaksinghtomar@manit.ac.in

P. Kumar
School of Advanced Sciences and Language, VIT Bhopal University, Bhopal 466114, India
e-mail: pavankmathsor@gmail.com

H. A. E. W. Khalifa
Deapartment of Operations and Management Research, Faculty of Graduate Studies for Statistical Research, Cairo University, Giza 12613, Egypt
e-mail: hamiden@cu.edu.eg

A. Kumar et al. (eds.), *Sustainable Security Practices Using Blockchain, Quantum and Post-quantum Technologies for Real Time Applications*, Contributions to Environmental Sciences & Innovative Business Technology,
https://doi.org/10.1007/978-981-97-0088-2_6

been challenging because of the continuous and ever-changing log events and their mutability property. Attackers attempt to modify the logs in order to avoid being discovered, which extends the time between detection and triage. In this work, we propose a novel model using Blockchain to problem of log analysis by suggesting two modules, anomaly detection using different machine learning models and Distributed Immutable storage system for securely storing the logs. We also present descriptive and user-friendly Web Application by integrating all modules using HTML, CSS, and Flask Framework on the Heroku cloud environment. Using proposed Hybrid Machine Learning Model, we are able to achieve 99.7% accuracy for detecting network anomalies.

Keywords

Blockchain • Distributed system • Immutable storage • Logs • Isolation forest • KNN • Machine learning

6.1 Introduction

As people and technology both advance, many requests are being made to the server, which leads to more generation of Log data than ever before. However, on the flip side, more cyber-attacks are occurring where log analysis and anomaly detection are shaping into vital roles (Simoes et al. 2022; Oliner et al. 2012). With security aspects, log entries are an essential part where developers must be aware of users injecting malicious content into logs called log injection. Security Logging and monitoring Failures are so at risk that they made Open Web Application Security Project (OWASP) Top 10 2022 (Albahar et al. 2022; Candel et al. 2023).

Typically, organizations that routinely check and analyze logs are better able to spot mistakes. The organization may even be able to identify issues with a sophisticated log analysis tool, considerably reducing the time and expense of remedy. The log also assists the security engineers in reviewing the events that preceded the error, which may make it simpler to troubleshoot and fix the problem going forward (Zhu et al. 2019).

Furthermore, for large companies such as Meta and google, the amount of data is so large that even with experts, just manual presence has not been enough, which machine learning has come to the rescue with automatic anomaly detection with less or no human assistance. However, with no proper standard up-to-date actual log labeled dataset as reference and log messaging, changing every few years has been challenging (Chen and Chang 2022; Rout and Natarajan 2022).

According to security analysts, there will be a cyber security attack occurring every seven seconds around the globe by the end of 2023 (Ghiasi et al. 2023). The focal point of this work is to develop suitable solution that will act as both a proactive and reactive approach to tackle the cyber security related risks and issues occurring in organizations. This solution processes the most minor yet crucial component of data generated from any activity performed within an organization's systems Logs. Numerous projects and research has been done for anomaly detection in a different range of Log details. Furthermore, for research reference purposes, building varieties of test beds for generating random log files has been a significant advancement in the cyber domain. From Clustering to Neural Network, all kinds of Machine Learning models have been tested and compared across already existing data set available (Abdullahi et al. 2022). However, most standard datasets are outdated, and the realness of the dataset has been a challenge.

This chapter proposes Blockchain for log storage and data sets from the most realistic test beds with suitable anomaly detection algorithms. This work proposes a model for detecting threats and vulnerabilities at different levels in systems. These are server logs and network devices logs using multiple machine learning methods for instance KNN and Isolation Forest to analyze logs and prevent any forgery in a log of the system. In order to maintain the indignity of logs we also propose distributed immutable storage.

The remainder of this text is organized as follows: Sect. 6.2 discusses the background and overview of Blockchain and related terms in different headings such as Blockchain, Distributed System, Merkle Tree, and Logs. Section 6.3 discusses the related work of log analysis and presents the state-of-the-art comparison of solutions for Log Analysis. We introduce the proposed model in Sect. 6.4. The evaluation of the proposed model is presented in detailed in four different headings: Apache Logs Analysis, Network Log Analysis, Blockchain, and Demonstration of Web App. Section 6.5 gives detailed result analysis. The final portion concludes with ideas for future work in Sect. 6.6.

6.2 Related Work

Zhang et al. (2019) proposed LogRobust detection of anomaly based on log analysis to mitigate the problem of anomaly detection on unstable logs by finding the critical semantics information and processing through Bi-LSTM model collecting from the Hadoop and Microsoft System. Wang et al. (2020a) proposed anomaly log detection with automatic neighbor selection for supervised KNN algorithm mitigating the problem of proper k neighbors selection for large-scale data with different distribution. A similar logs min hash algorithm is used for grouping, and an MVP tree is used for each bucket. Wang et al. (2020b) have reviewed the robustness of the LSTM network for the detection and classification of anomalous events in sensor files. In comparison, RNN could only compete in foreseen anomalies events with LSTM. The authors conclude that LSTM performs far better than RNN in log-based anomaly detection an anomaly.

Wang et al. (2022) considered log files as a standard plain text file to apply NLP to extract information about the log events, including the most popular embedding technique, which is based upon Google's word2vec algorithm. This could perform online detection of any anomaly. Their future work includes the collation of their work with an unsupervised algorithm. Catillo et al. (2022) have proposed ADA which acronym of Adaptive Deep Log Anomaly Detector. It is unsupervised deep neural network framework that makes use of LSTM networks with the dynamic threshold. This framework has proven to increase the f1 score significantly compared to existing methods with decreased storage.

Pourmajidi and Miranskyy (2018) have proposed cloud blockchain-based log system to avoid log tampering and provided tamper detection solution named Logchain. Pourmajidi et al. (2019) explained about providing blockchain-based log storage as service. Validation of the blocks and integrating existing public and private blockchain services using the Ethereum public blockchain network. Oprea and Bâra (2021) worked upon Identity and Access Management (IAM) logs with the use of blockchain for Enterprise usage. Facilitates the temper proof IAM and also maintains the solubility for users and transactions, most effective for security auditing and monitoring in Enterprises. Huang (2019) proposed Framework for a Secure Log Storage system using blockchain technology. It uses temper proof and decentralization properties of blockchain by utilizing the Ethereum public blockchain for indexing and hashing log files. Medusa (Wang et al. 2018) is a blockchain-powered immutable storage system for storing logs. This work mainly focuses on storing weblogs in a blockchain, and it yields much higher throughput and significantly lower latency. Friedberg et al. (2015) proposed a novel system events and also monitors dependencies and event occurrences and therefore learns the natural system behavior.

Ambre and Shekokar (2015) proposed a probabilistic method that shows the percentage likelihood of a given event while still taking into account a false alarm rate that is acceptable. Reguieg et al. (2015) proposed a scalable and effective process and event analysis technique that includes parallel algorithms to enable quick event correlation for massive amounts of process data. Over large event datasets, the MapReduce framework suggests a two-stage process for identifying probable event connections and verifying them.

Bračevac et al. (2018) proposed a unique algorithm based on the principles of filtering and verification, which is also called RF-GraP, which provides a very effective correlation in a distributed systems environment. Kotenko et al. (2016) provide a detailed explanation of methods for correlating different security events and integrating diverse data sources for defense against cyber-physical threats. Landauer et al. (2020) set test beds with actual realistic generating log directory files for six days for evaluation purposes when testing for Intrusion Detection System (IDS). We have done systematic comparison on the following parameters: PR1—Anomaly Detection, PR2—Server Log Analysis, PR3—Network Log Analysis, PR4—Resource Optimization, PR5—Data Confidentiality Consideration, PR6—Distributed File System. Table 6.1 provides the most recent comparison of Log Analysis solutions.

Table 6.1 Most recent survey and comparison of solutions for log analysis

Author	Key contributions	PR1	PR2	PR3	PR4	PR5	PR6
Zhang et al. (2019)	LogRobust is able to identify and handle unstable logs	Yes	Yes	No	No	No	No
Wang et al. (2020a)	Efficient KNN neighbour search for anomaly detection	Yes	Yes	Yes	Yes	No	No
Wang et al. (2020b)	Evaluated simple LSTM model for anomaly detection	Yes	Yes	No	No	Yes	No
Wang et al. (2022)	Mined log as normal text using NLP	Yes	Yes	Yes	Yes	No	No
Catillo et al. (2022)	framework for lightweight anomaly detection	Yes	No	Yes	Yes	No	No
Friedberg et al. (2015)	White listing based anomaly detection approach with event correlation	No	Yes	No	Yes	No	Yes
Ambre and Shekokar (2015)	Frequency based event correlation system for insider threat detection	Yes	No	Yes	No	No	No
Reguieg et al. (2015)	Scalable event correlation system with MapReduce framework	Yes	Yes	No	No	Yes	Yes
Bračevac et al. (2018)	Filtering-and-Validation based event correlation over distributed system	Yes	No	Yes	No	No	No
Kotenko et al. (2016)	Heterogeneous data integration approach for event correlation	Yes	Yes	No	Yes	No	No
Landauer et al. (2020)	Cloud based logging system using blockchain	No	Yes	Yes	No	No	Yes
Oprea and Bâra et al. (2021)	Enterprise level security audit system for IAM logs	No	Yes	Yes	No	Yes	No
Huang et al. (2019)	Tamper-proof decentralized log storage system	No	Yes	No	No	Yes	Yes
Medusa et al. (2018)	Immutable storage system for web log data	Yes	No	No	Yes	No	No
Proposed work	Layered log analysis using ML & distributed immutable storage	Yes	Yes	Yes	Yes	Yes	Yes

6.3 Background and Related Terms

To better comprehend log analysis using distributed immutable storage, the terminologies below are introduced in this section. These terms are as follows: Blockchain, Distributed System, Proof-of-Work, Merkle Tree, Public Key Cryptography, Logs, Server Logs, Network Devices Log, and KNN Classifiers.

6.3.1 Blockchain

The word "Blockchain" describes a network of linked records that are highly resistant to alteration and are protected by encryption techniques. Blockchain technology was introduced in the year 2008. A complicated data structure using an asymmetric encryption algorithm and hash functions was detailed in the well-known research paper on the Bitcoin cryptocurrency "Bitcoin: a peer-to-peer electronic cash system" inscribed by Nakamoto in 2008 (Nakamoto 2008).

6.3.2 Distributed System

A distributed system is a dynamic system, a collection of software or hardware components called nodes linked together in a network. They work coherently and communicate to achieve a single end goal through message passing. These nodes may or may not be structured, depending on system requirements, and the complexities of the network are hidden to the end-user, making the system appear as a single computer to its users. Distributed Computing is a collection of independent systems or nodes interconnected in a network and can share idle resources currently not utilized by any other method. There are several disadvantages to this architecture. There is an increased amount of computational power involved (Rahman et al. 2017). It gets increased whenever a new node or server is added; there is a greater demand for storage, and the maintenance of this vast system is not easy. Inter-planetary File System (IPFS) is a peer to peer protocol for a distributed cloud platform. It uses Blockchain to ensure decentralized architecture and no sole point of failure. IPFS offers a separate naming system called IPNS for storing and accessing information quickly.

6.3.3 Proof-Of-Work

Proof of Work (abbreviated as PoW) is the first agreement algorithm that was implemented in the blockchain network. Currently, it is used by several blockchain technologies and add blocks. It has a decentralized ledger that holds information related to blocks and transactions. PoW to verify a transaction and add a fresh block to the Blockchain, consensus on the network is employed. With PoW, miners compete to validate a transaction and get rewarded (Schinckus 2021).

The probability of being selected to build the next block is linked to the individual computation power. A crucial part of the mining process is called PoW, when miners locate a legitimate block by resolving a mathematical conundrum. Miners must continuously alter the nonce value until the generated hash is higher than a predetermined threshold since the block header contains a 32-bit nonce field (Kiayias and Zindros 2020).

6.3.4 Merkle Tree

Blockchain data is securely encoded and hashed using the hash tree, often known as the Merkle tree. It permits effective mapping of large amounts of data, quicker blockchain data validation, and quicker transfer of large amounts of data across nodes in a peer to peer blockchain. We can quickly check the data consistency or where the change in data has occurred by crosschecking the root hash value without traversing the whole tree repeatedly. The root hash is used as the fingerprint for the entire data (Mohan and Gladston 2020).

In a Merkle tree, the data stored in the leaf nodes is hashed, then the hash values of both leaf nodes are combined and hashed again. This process goes on till the root of the tree is reached. Merkle trees can be used in distributed system applications where replica or similar data is stored in multiple places. It is currently used in bitcoin and blockchain (Jain et al. 2021).

6.3.5 Public Key Cryptography

Data is exchanged using two separate keys, one for encryption and the other for decryption, in asymmetric encryption known as public key cryptography. Here, the first two keys are generated: the public and private keys. The receiver's public key is used for data encryption, and then the receiver can use his private key to decrypt and hence receive the confidential data. There are five steps in the entire procedure: Key generation, Key exchange, encryption, data transfer, and decryption (Gill et al. 2022; Public-Key Cryptography–PKC 2023).

6.3.6 Logs

Logs can be defined as automated computer-generated data files that track activities performed by systems like operating systems, application servers, database servers, network devices, and many more during interaction with any user or another system. These activities could be logging, retrieving or inserting data, deleting or modifying data, etc. The primary purpose of these logs is to keep all the records of such activities that are or could be responsible for detecting any security issues in the systems in the future like its source, time, or medium (Jansen 2022).

6.3.6.1 Server Logs

It is simply a text file automatically generated within the system server that contains the records of all activities performed within it by users or another system like logging, input requests, and its response by who, whom, and when in a given period. Types of servers could be an application server, database server, network server, etc. (Majd et al. 2022).

6.3.6.2 Network Devices Log

This category includes logs from network devices such as routers, switches, load balancers, intrusion detection systems, and more. These logs offer in-depth details on what is occurring. It keeps track of activities including program startup and shutdown, commands run, information about logging in and out, connections made, and more. This extensive history of operations is a perfect example of why log data should be referred to as a network's footprints (Li et al. 2018).

6.3.7 Machine Learning for Log Analysis

American IBM employee Arthur Samuel first used the phrase "machine learning" in 1959 (Samuel 1967). He was a pioneer in the fields of artificial intelligence (AI) and video gaming. via the use of machine learning, systems may automate the process of learning and grow via experience without the need for separate coding. The goal of machine learning is to create software that can access, analyze, and use data for itself.

The learning process begins with gathering large amounts of data or observations, such as precise instructions or information, in order to uncover and create patterns in the data frame and improve future decisions based on the examples we supply. The main goal is to make it possible for computers to learn and develop autonomously without any help from humans and to adjust their behavior accordingly.

We used the following methods in our work for log analysis and anomaly detection. Isolation Forest is an unsupervised machine learning method that emphasizes anomaly detection over normal data points. This algorithm, like other ensemble tree approaches, is based on decision trees. Outliers or anomalies are data points that significantly deviate from the other observations. In comparison to frequent observations and value discrepancies, outliers are less common. Isolation forests have a variety of uses, including the detection of network anomalies, fraud, and unforeseen changes in sales. By choosing a number at random from the range between the highest and minimum values of the characteristic, outliers are isolated.

KNN algorithm for classification is a supervised learning algorithm that predicts the values based on the nearest already trained points with similar features. The algorithm has relatively high accuracy, and with more data points, the classifier constantly evolves but becomes computationally expensive due to high storage requirements.

6.3.8 Flask

Flask is a framework for developing web applications using python, which comes with a debugger and provided server. Due to the framework's lack of external libraries and inclusion of only necessary functionality, it is extremely lightweight. The static files in the template folder are in charge of passing dynamic data from the back-end code. Moreover, the flask does send the essential data by rendering templates using the jinja package (Lathkar 2021). With full completion of the flask web app, there are lots of options to operate in the cloud, and one of the reliable options is the Heroku app, which is a platform-as-a-service (PaaS) company letting programmers run their applications in the cloud for free with some restrictions (Yasrab 2018).

6.4 Proposed Hybrid Model for Log Analysis

To tackle the increasing problems raised due to attackers clearing their tracks and making the system and servers seem normal after every successful cyber-attack, we propose a model with an immutable log storage system and root cause analysis.

In this proposed model, Blockchain-based log monitoring system with the features of detecting anomalies on logs which will keep the logs tamper-proof and detect any anomalies in the present and future unseen logs is presented. The architectural diagram is shown in Fig. 6.1.

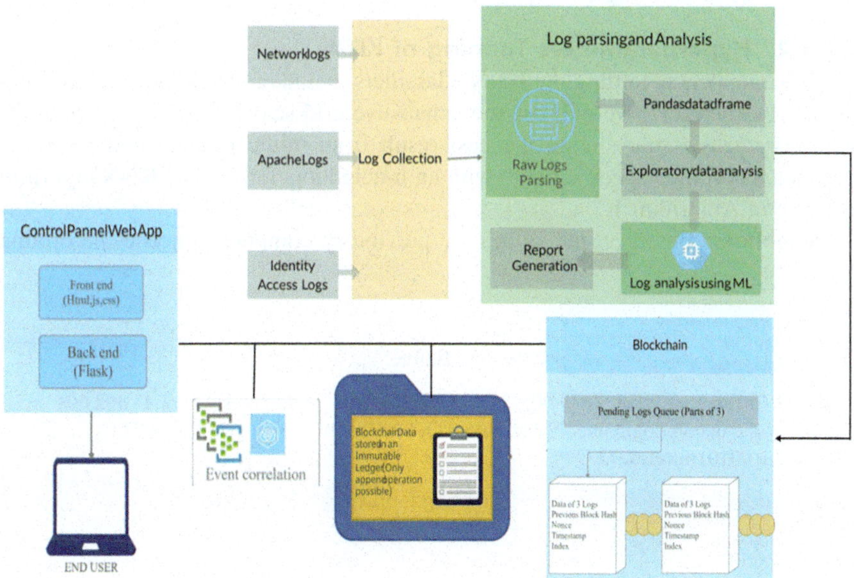

Fig. 6.1 Architectural diagram for proposed model

The proposed model consists of the following subcomponents such as Server Log Analysis, Network Log Analysis, Blockchain, and Distributed File Storage System. Description of each component is as follows.

6.4.1 Server Log Analysis

Apache Logs are basically information regarding events happening in the particular Apache Web servers. Apache Logs come in two kinds such as Access and Error Logs. The data is classified only to server administrators, and Log data includes IP, timestamp, the request-method, status-code, referrer, Byte size, User-Agent, etc. With a large number of Server Logs, the administrator gets a grasp of an incoming request and user behaviors which helps in the analysis of possible upcoming attacks, digging for the root cause of past problems, and so on. The Apache Access Logs formatting is shown in Fig. 6.2.

We have used KNN classifiers and the Isolation Forest algorithm to detect certain anomaly logs which deviate from the base distribution. The algorithms are chosen based on our research and understanding and those algorithms that have performed great results in existing articles.

6.4.1.1 KNN Classifiers

K-nearest Neighbors algorithm is the most well-known classification algorithm out there. It is simple to implement, and the accuracy drastically increases with more training data sets depending on the value of K. KNN algorithm does not work with missing values, but the data set used here is all labeled and preprocessed.

6.4.1.2 Hyperparameters Tunning of KNN

Hyperparameters regarding the KNN Classifiers include n-neighbors, weights, and metric. Furthermore, we will use the exhaustive grid search technique to optimize hyperparameters after slicing the best result from initial training and test score between a specified specific range with an n-neighbors range. The KNN algorithm is given in Algorithm 1.

Neighbors: Based on the values we previously computed, choose the optimal k.

Before

192.168.10.190 - - [29/Feb/2020:00:00:12 +0000] "POST /login.php HTTP/1.1" 302 601 "http://mail.cup.com/login.php" "Mozilla/5.0 (X11; Ubuntu; Linux x86_64; rv:73.0) Gecko/20100101 Firefox/73.0"

After Log formatting

ip_addr	ident_check	connecting_user	time_stamp	time_zone	request_method	status_code	byte_size	referer	user_agent
192.168.10.190	-	-	29/Feb/2020:00:00:12	0000	POST	302	601	http://mail.cup.com/login.php	Mozilla/5.0

Fig. 6.2 Apache access logs formatting

Weights: Examine whether giving the data points weights will help the model. While "distance" weighs points according to the inverse of their distances, which means that closer points will have greater weight than further points, "uniform" applies no weight.

Metric: The comparison will be calculated using the distance measure.

Algorithm 1: KNN Classifier Algorithm

```
Input: AIT Log Data Set V1.0
Output: Alert to user through web app
Collect Apache logs from the two servers mail.cup.com and mail.spiral.com
Processing the data according to features and label data
Taking mail.cup.com server logs as Training set and other as Test set
Process the data into selective features with columns and convert to dataframe
Make a column with the field name, then load the dataset using it
if Object type Dataset then
Make additional columns 0 to n_classes 1 with feature names as prefix using one hot encoding
remove the original feature
remove one of the generated column to avoid multilinear
reiterate process for all required labels
end if
Select KNN classifier
Tune the k Neighbors parameter by selecting certain radius of values
Choose the k values with best results
Initialize the grid parameters for each parameters including k neighbors values
Perform Exhaustive Grid Search Technique with grid parameters
Get the best parameter values from Grid Search
Train and Test the data set with KNN
if malicious activity detected then
Generate alert
Foreword it to web interface
end if
```

6.4.1.3 Isolation Forest

Instead of attempting to discover the most accurate description of regular points, the concept-based anomaly identification technique known as Isolation Forest isolates anomaly points. This approach is particularly powerful and useful for real-world data because anomalous points are much simpler to separate than regular points.

6.4.1.4 Hyperparameters Tunning of Isolation Forest

Hyperparameters regarding the Isolation Forest include contamination, max samples, and **n estimators**. Furthermore, we will use the exhaustive grid search technique for hyperparameters optimization (Liashchynskyi and Liashchynskyi 2019). Table 6.2 presents the abbreviations, acronyms, and symbols used in proposed model.

The decision-making process requires an anomaly score, much like with other outlier identification approaches. Any method for anomaly detection needs to have

Table 6.2 Abbreviations, acronyms, and symbols used in proposed model

S. No	Abbreviations/Acronyms/Symbols	Meaning
1	AI	Artificial intelligence
2	AIT	Austrian institute of technology
3	BST	Binary search trees
4	c(n)	Average path length of failed searches in a binary search tree
5	CBH	Current block hash
6	GBH	Genesis block hash
7	h(x)	Path length of the observation x
8	IDS	Intrusion detection system
9	IPFS	Inter planetary file system
10	PaaS	Platform as a service
11	PBH	Previous block hash
12	PoW	Proof-of-work
13	S(x, n)	Anomaly score

an anomaly score. The Isolation Tree's anomalous score is defined as it is difficult to calculate a score h (x) in order of log(n) when the greatest height rises in the order of n and the average height rises in the order of log(n).

Any of the aforementioned terms that equalize h (x) cannot be directly compared or are not bounded. The calculation of average h (x) for exterior node terminations is comparable to the ineffective search in the Binary Search Trees (BST), since iTrees and BST have the same structure. To calculate the isolation Tree's average path length, we use the BST approach. Therefore, the mathematical Eq. (6.1) defines the anomalous score of the isolation forest.

$$S(x,\ n) = 2(E(h(x))/c(n)) \tag{6.1}$$

where n is the number of external nodes, c(n) is the average path length of failed searches in a binary search tree, and h(x) represents the path length of the observation x. The algorithm for network log analysis is given in Algorithm 2.

Algorithm 2: Network log analysis methodology

Input: Network Devices Logs,packet capture file (pcap)
Output: Alert to user through web app
1: Collect pcap file and network logs
2: Download the KDDTrain + .arff file from the NSL-KDD dataset
3: Install the necessary packages and the KDDTrain + .arff file
4: Give the field's name a number
5: Make a column with the field name, then load the dataset using it
6: **if** Unprocessed Dataset **then**
7: Label values ranging from 0 to n classes should be encoded
8:using transmute standardize values
9:return fixed tags
10:reiterate process for all required labels
11: **end if**
12: Divide the dataset (data and labels) into the ratio of 70:30 for the train dataset and the test dataset
13: Choose Isolation Forest method
14: Train method using processed dataset
15: **if** malevolent activity spotted **then**
16:Generate an alert
17:Foreword it to web interface
20: **end if**

6.4.2 Blockchain

In order to store network logs on a blockchain, log entries must first be transformed into cryptographic hashes for data integrity, then they must be bundled and verified as transactions using the blockchain's consensus algorithm, and finally they must be added to blocks with cryptographic links for immutability. A tamper-resistant and transparent solution is provided for boosting network log security and dependability by decentralized storage across network nodes, safe and auditable access to historical data, and cryptographic keys.

Using well-known tools and technology, attackers clean the different types of access logs. The logs that were produced after abnormalities were discovered can be removed or changed. Most of the systems don't keep proper logs after spotting irregularities, the majority of these cyber-attacks go unnoticed.

We extant a method based on blockchain technology to store anomaly-based network and error logs in an immutable storage system while preserving the confidentiality and integrity of the logs. The algorithm for Blockchain methodology is given in Algorithm 3.

Algorithm 3: Blockchain methodology

Input: Log data in list format
Output: Immutable text file to user through web app
1: Extract log data in form of lists using pandas
2: Generate public and private key pair
3: Enter the public key in the web portal as proof-of-work
4: Generate Merkle Tree using the same log data
5: Verify the last block hash with merkle tree hash for integrity check
6: Enter the desired location for storing the immutable output file
7: The file will be linked to the distributed storage environment for easy access

6.4.2.1 Generating Keys

First, we will generate public and private key pairs. The public key will act as PoW, and it can be transferred in an encrypted format to entities who need to add the logs to the blockchain. The private key will be used to access the immutable log file extracted from the blockchain.

6.4.2.2 Generating Blocks

Each block in the blockchain will have five properties: index, timestamp, log data, PoW, and previous block hash. Here, each block's log data contains three distinct logs. Each individual block includes three distinct log entries. These log entries could represent various network events, activities, or data captured. The content of these logs would depend on the specific application and purpose of the blockchain network. The data collected from each log is converted into a string and Unicode. Finally, the Unicode value is hashed using the SHA-256 hashing algorithm.

6.4.2.3 Creating Merkle Tree

The Merkle Tree encodes the log data for easy integrity check and fault detection. Here the log data in the last block is hashed, then the log data in the second-last block is hashed; after this process, both hash values are combined and hashed using the SHA-256 hashing algorithm. This process goes on till we reach the genesis block.

For example, if we take a simple blockchain with three blocks, the first being the genesis block, and the second and third blocks will be actual log data. In the Merkle tree function, we perform hashing on each block individually, then we concatenate both hash values and again perform hashing. Here H represents hashing using the SHA-256 algorithm, GBH is the genesis block hash, PBH is the previous block hash, and CBH is the current block hash. Then, GBH can be computed using equation number 2.

$$\mathrm{GBH} = \mathrm{H}(\mathrm{H}(\mathrm{PBH}) + \mathrm{H}(\mathrm{CBH})) \tag{6.2}$$

Table 6.3 Description of Apache dataset used after prepossessing

Total samples	Total features	Training	Test
248,985	19	148,530	100,455

6.4.3 Flask Web App

Here in flask-based web app, we have integrated all the results from above all previous analyses and displayed them in the proper user interface that is effective and efficient in further analysis. In this implementation, all the essential details from logs and analysis of logs can be exported in report format that maintains the proper chain of custody for forensic analysis. The front-end part is well developed with bootstrap four and stored in the templates folder from where the jinja template (in built-in flask framework) renders it to the user and provides dynamic values to the user as per the code. Also, the graphical results are provided to the user with proper UI to explain the results from different machine learning algorithms.

6.5 Discussions and Results

Results from the analysis of the Apache logs and the network logs are shown in this section.

6.5.1 Apache Logs Analysis

Current section presents results for Apache Logs analysis under different headings such as Data Set, Metric, Evaluation, and Isolation Forest.

6.5.1.1 Data Set

The data set exploited in this study was created by the Austrian Institute of Technology (AIT) (2022). Six days were spent simulating four web servers. Labeled data sets are contained in the data file known as AIT-LDS-v1 0.zip.

Here for anomaly detection, Apache access logs have been used. For training, the mail.cup.com server Apache access logs are utilized, and mail.spiral.com Apache access logs have been used for testing. The Description of Apache dataset used after preprocessing is presented in Table 6.3.

6.5.1.2 Metric

Following metrics are taken into consideration while hyperparameters are tunning.

Accuracy: Measure of total correct classification of all classes in the model. Since the dataset we are utilizing has an unbalanced ratio of classes (i.e., just 0.06 contamination), accuracy is the most well-known performance indicator, but other metrics are also employed in this study.

Precision: It is a parameter that assesses the overall accuracy of anomaly predictions made based on the classification outcome. The higher the precision we obtain, the fewer resources will be needed to find and countermeasure the root cause of the anomaly.

Recall: It is a metric that assesses how accurately the resultant anomaly class was classified overall. The ideal classification model will perform admirably in each of the three performance measures given, but not always. In most situations, including this one, we must make compromises and pick a model depending on our requirements. More abnormalities will be found at the expense of more resources if recall is higher but precision is lower.

6.5.1.3 Evaluation

The accuracy of the model remained unchanged after employing the K-Nearest Neighbors (KNN) technique and tweaking the "k" parameter (number of neighbors) up to k = 10. At this point, k = 10 was selected, and the optimal set of hyperparameters for the KNN model was discovered using an exhaustive grid search. The grid search experimented with various values for "k" and distance measures such as Minkowski, Euclidean, and Manhattan.

The best result was obtained with a 93.7% accuracy using the hyperparameter values: "metric" = Minkowski, "n_neighbors" = 10, and "weights" = distance, after assessing the model with all feasible combinations of these hyperparameters (6 times in total). In order to enhance the model, several features with significant multicollinearity (correlation between features) were also eliminated, but when the accuracy was reevaluated using the same hyperparameter settings, it remained unchanged.

6.5.1.4 Isolation Forest

For hyperparameters in this particular algorithm, each parameter was tuned individually with other parameters as default and was then selected for the exhaustive grid search technique for the final optimization. Each value was selected with metric consideration. After the exhaustive grid search technique, the results were stored in a data frame. Data Insights showed that most models have high accuracy but low precision and recall. The data frame was sorted based on test accuracy, precision, and recall to ensure the model was chosen. There were some instances where the model was underfitting and some overfitting. We chose the top-performing model after ranking the models according to the recall score obtained from the test dataset (highlighted in Fig. 6.3).

6.5.1.5 Comparison Analysis

In the AIT Log Data Set V1.0, KNN and Isolation Forest Algorithm performance is compared. Figure 6.3 details the data set's description. Each algorithm is trained and tested in the same Google Colab environment.

From the Figure, it can be observed that KNN did outstanding in the Training phase with almost perfect in all three metrics with minimal time effort but failed to replicate its Test data set. The isolation Forest algorithm did super well in the

```
iforest_result_df.sort_values(by='testing_recall', ascending=False).head(20)
```

	contamination	n_estimators	max_samples	max_features	bootstrap	n_jobs	training_accuracy	training_precision	training_recall	testing_accuracy	testing_precision	testing_recall
413	0.06	50	200	10	True	40	0.983639	0.750806	0.948273	0.975947	0.762273	0.923332
318	0.06	10	200	10	False	30	0.981599	0.725735	0.944956	0.971995	0.726452	0.923030
405	0.06	50	200	3	True	40	0.981741	0.726349	0.948273	0.980875	0.812542	0.923030
334	0.06	10	700	10	False	30	0.987053	0.800179	0.946313	0.979760	0.802211	0.920012
365	0.06	30	200	10	True	40	0.990089	0.847970	0.947972	0.988611	0.908860	0.919559
381	0.06	30	700	10	True	40	0.896922	0.007379	0.009802	0.952243	0.588434	0.918352
428	0.06	50	700	10	True	30	0.898430	0.004919	0.006334	0.941451	0.532610	0.918201
383	0.06	30	700	10	False	40	0.898894	0.004959	0.006334	0.948629	0.568492	0.918201
382	0.06	30	700	10	False	30	0.900847	0.006341	0.007842	0.951635	0.585000	0.918201
399	0.06	50	100	10	False	40	0.897009	0.004913	0.006485	0.953557	0.596099	0.917899
296	0.06	10	100	5	True	30	0.908307	0.005939	0.006334	0.957688	0.621398	0.917748
328	0.06	10	700	5	True	30	0.898894	0.004842	0.006183	0.948599	0.568371	0.917748
430	0.06	50	700	10	False	30	0.901998	0.007947	0.009652	0.957449	0.619951	0.917296
380	0.06	30	700	10	True	30	0.904287	0.006763	0.007842	0.955558	0.608187	0.917145
319	0.06	10	200	10	False	40	0.905419	0.006914	0.007842	0.952581	0.590534	0.916994
366	0.06	30	200	10	False	30	0.902443	0.006406	0.007691	0.952531	0.590335	0.916239
297	0.06	10	100	5	True	40	0.898087	0.004889	0.006334	0.950162	0.577069	0.915334
317	0.06	10	200	10	True	40	0.926822	0.004443	0.002865	0.971228	0.723927	0.911410
95	0.02	30	700	10	False	40	0.944428	0.009668	0.002413	0.984031	0.855968	0.911259
309	0.06	10	200	3	True	40	0.922977	0.005958	0.004373	0.944288	0.546623	0.911259

Fig. 6.3 Result from exhaustive grid search technique of isolation forest

Table 6.4 Comparison of KNN and isolation forest algorithm performance in AIT Log Data Set V1.0

Algorithm	For training data			For test data			Training time	Testing time
	Accuracy	Precision	Recall	Accuracy	Precision	Recall		
KNN	0.99	0.99	0.99	0.94	0.99	0.95	46.7 ms	4 min 41 s
Isolation forest	0.99	0.85	0.95	0.98	0.9	0.91	2.9 s	1.32 s

Training dataset and even better on the Test data set. Though KNN took no time during Training the data, Isolation Forest was way faster in predicting the test data.

So we could conclude that Isolation Forest performs better on anomaly detection in an accurate data set with a higher result more efficiently. The results of Comparison of KNN and Isolation Forest Algorithm performance in AIT Log Data Set V1.0 are shown in Table 6.4.

6.5.2 Network Log Analysis

Hereafter examining our data to see that most of the traffic is normal, as expected, but a small amount is abnormal. The majority of the traffic is normal, as we predicted, and there is a small amount of aberrant traffic, according to our analysis of the data. We viewed all abnormal traffic as belonging to the same group, which we named "anomalous." In order to successfully detect anomalies, we used some default settings when training the model, such as the number of trees to build (100 trees) and other technical configurations.

Now from the proposed model, an anomaly score is assigned to each of the data points based on the depth of the tree required to arrive at that point. Here we used the decision function of the isolation forest to provide a score for the normal and abnormal (anomalous) training set and then examined the results. Figure 6.4 shows the normal observations and Fig. 6.5 shows the abnormal observations.

Knowing that the decision function measures how simple the point is to describe, we would like to separate simple points from complicated points by picking a numerical cut-off that gives clear separation.

Finally, we can use our model to make predictions and provide an assessment of its performance. As given in Table 6.5, the model could pick up 595 anomalies

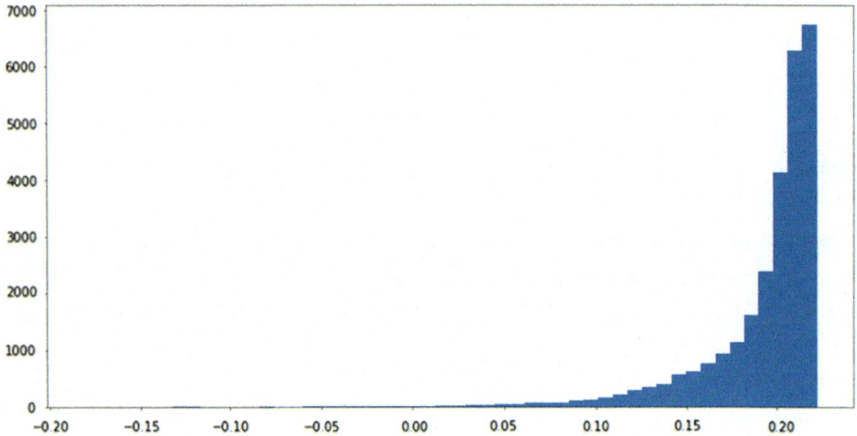

Fig. 6.4 Normal observations

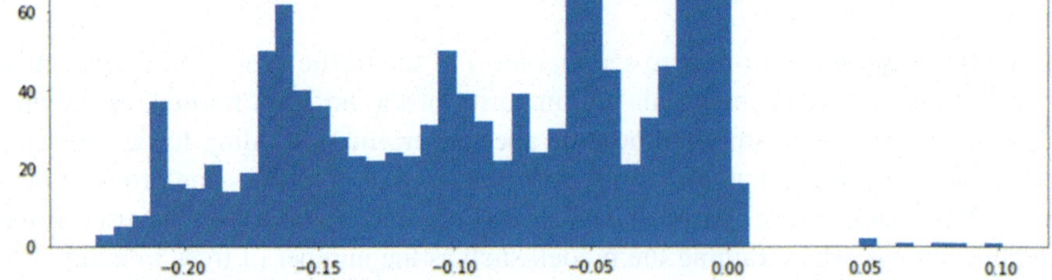

Fig. 6.5 Abnormal observations

Table 6.5 Distribution of anomalies

	Anomaly	Normal
Testing data	597	11,775
Results	595	23
Accuracy	99.67%	

out of 597 without triggering too many false positives that are instances of regular traffic, which are 23, so the absolute accuracy of the given model is 99.67% which is very high, considering the uniqueness of anomalies. Isolation Forest techniques have been shown to be quite good in detecting anomalies and are computationally considerably more efficient. Despite its benefits, some drawbacks include the fact that the final anomaly score depends on the contamination parameter supplied during model training. This suggests that in order to make more accurate predictions, we need to be aware of the proportion of anomalous data up front. Similar to how branching happens, the model has a bias.

6.5.3 Blockchain

Figure 6.8 shows an example of network log data stored in the blockchain. When we consider the network logs, each has the following data, IP address, Timestamp, Time zone, the Request method, Status code, Byte size, Referrer, and User-Agent.

The above Fig. 6.6 explains how we generated public and private key pairs using the OpenSSL command on Linux operating system locally. The public key is transferred to entities who want to add their data to the blockchain. The private key is used to access the output file from the IPFS file system. The Fig. 6.7 depicts a general scenario where the machine learning algorithm detects malicious network log data. The network log data is stored in the transaction part.

Lastly, we compare the hash value of the last block from the blockchain and the Merkle tree hash value for checking the integrity of the data. After an integrity check, the data is stored in the IPFS file system. The results of Blockchain after applying Merkle Hashing Function are shown in Fig. 6.8.

6.5.4 Web App

We also have integrated all the machine learning algorithms and blockchain results into a simple, user-friendly, easy-to-use Web Application using HTML, CSS, and Flask. This Web Application has four sections: Info, File Upload, Log Analysis, and View Report.

The Info tab gives the user an in-depth idea of the project and how to use the application effectively. The File upload section is where we can upload our log files. The Log Analysis tab contains three more subsections, Error Log analysis, Access Log Analysis, and Network Log Analysis; the user can select and use any

```
root@kali:~/Desktop/ITTCP# openssl req -new -x509 -newkey rsa:2048 -keyout Privkey.ou
t -pubkey -out Pubkey.out -days 365 -nodes -sha256
Generating a RSA private key
.......................................................+++++
..................................+++++
writing new private key to 'Privkey.out'
-----
You are about to be asked to enter information that will be incorporated
into your certificate request.
What you are about to enter is what is called a Distinguished Name or a DN.
There are quite a few fields but you can leave some blank
For some fields there will be a default value,
If you enter '.', the field will be left blank.
-----
Country Name (2 letter code) [AU]:
State or Province Name (full name) [Some-State]:
Locality Name (eg, city) []:
Organization Name (eg, company) [Internet Widgits Pty Ltd]:
Organizational Unit Name (eg, section) []:
Common Name (e.g. server FQDN or YOUR name) []:
Email Address []:
```

Fig. 6.6 Generating public and private key pair

```
Genesis block:

[{'index': 1, 'timestamp': 1645641270.1562915, 'transactions': [], 'proof': 100,
'previous_hash': 'Caught with wisdom'}, {'index': 2, 'timestamp':
1645641270.1562965, 'transactions': [{'IP_address': '192.168.10.190', 'Time_stamp':
'29/Feb/2020:00:00:02', 'Time_zone': '0000', 'Request_method': 'GET', 'Status_code':
'200', 'Byte_size': '2532', 'Referer': '-', 'User_agent': 'Mozilla/5.0'}, {'IP_address':
'192.168.10.4', 'Time_stamp': '29/Feb/2020:00:00:09', 'Time_zone': '0000',
'Request_method': 'POST', 'Status_code': '200', 'Byte_size': '402', 'Referer':
'http://mail.cup.com/kronolith/', 'User_agent': 'Mozilla/5.0'}, {'IP_address':
'192.168.10.190', 'Time_stamp': '29/Feb/2020:00:00:12', 'Time_zone': '0000',
'Request_method': 'POST', 'Status_code': '200', 'Byte_size': '402', 'Referer':
'http://mail.cup.com/kronolith/', 'User_agent': 'Mozilla/5.0'}], 'proof': 12345,
'previous_hash':
'95d238c9e86af3cd69b73de243df7d5ade08f4d2ed1c7ac1e139bee2b85088d9'}]

The above output is also stored in an immutable file 'output.txt'
```

Fig. 6.7 Blockchain in simple text format

section as per the originality of the log file. Figure 6.9 shows the descriptive error log overview and results.

Figure 6.9 is the screenshot of the Web App when we upload an Error Log file. The Web app provides a descriptive log overview, Anomaly information, and Blockchain data, representing malicious activities in pie chart format.

```
Inputs:
192.168.10.190 | 29/Feb/2020:00:00:02 | 0000 | GET | 200 | 2532 | - | Mozilla/5.0
192.168.10.4 | 29/Feb/2020:00:00:09 | 0000 | POST | 200 | 402 |
http://mail.cup.com/kronolith/ | Mozilla/5.0
192.168.10.190 | 29/Feb/2020:00:00:12 | 0000 | POST | 200 | 402 |
http://mail.cup.com/kronolith/ | Mozilla/5.0

Completely Hashed Blockchain:

[{'index': 1, 'timestamp': 1645640840.8294623, 'transactions': [], 'proof': 100,
'previous_hash': 'The Merkle Project'}, {'index': 2, 'timestamp': 1645640840.8294725,
'transactions': [{'transval':
'f28be82a5521478f010a5dfaa5b4b2ee5441af3d0cee57a6a6e12a47a1f45a08'}],
'proof': 12345, 'previous_hash':
'7b200a5d5f6725d3765b8a105d7936b0c993839358564cbc6a16446069272433'}]
```

Fig. 6.8 Blockchain after applying Merkle hashing function

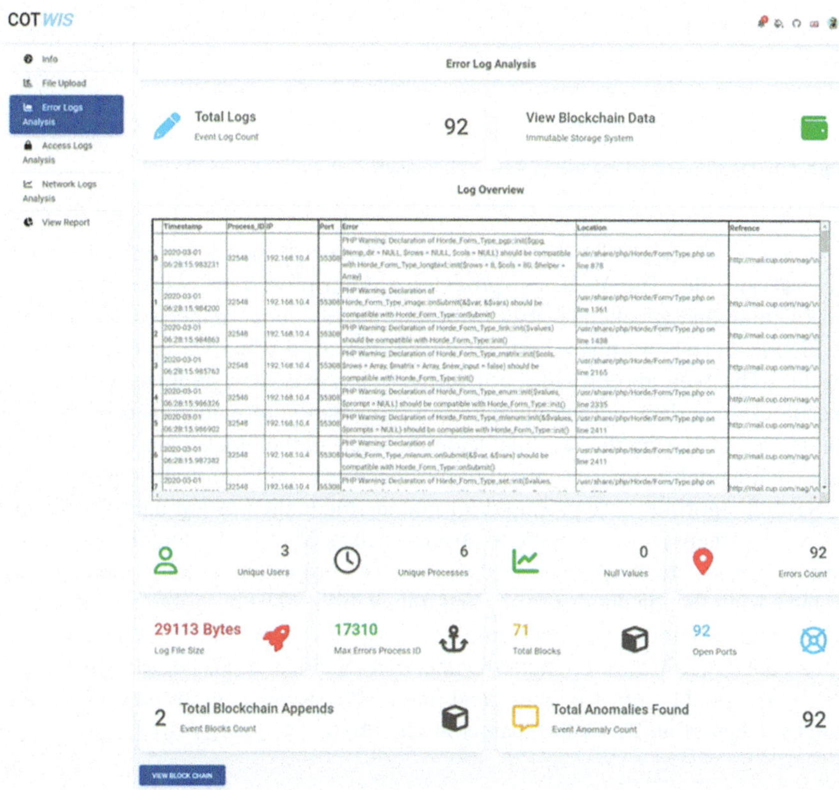

Fig. 6.9 Descriptive error log overview and results

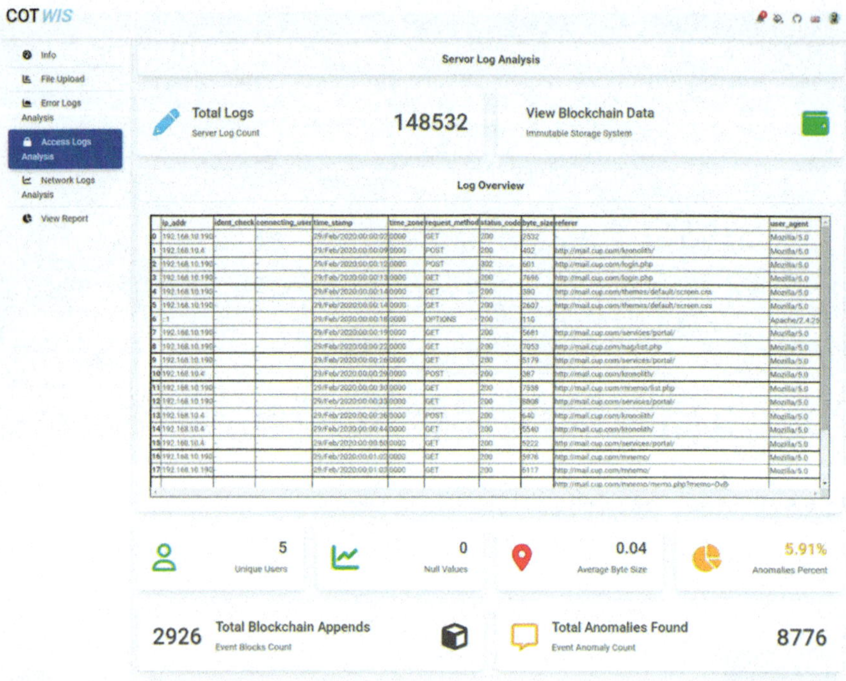

Fig. 6.10 Descriptive access log overview and results

Figure 6.10 shows the screenshot of the Web App when we upload a Server Access Log file. The Web app provides a descriptive Log overview, Anomaly information, and Blockchain data, with information about malicious activities.

Figure 6.11 shows the screenshot of the Web App when we upload a network log file. The Web app provides a descriptive log overview, anomaly information, and Blockchain data, with the graphical representation of malicious activities. The whole idea of storing the machine learning result in the blockchain is to make the data confidential, tamper-proof, and limit the accessibility of malicious logs.

Several algorithms are proposed for making the malicious log data confidential, but most of them are susceptible to attacks such as Man-in-the-middle, Log4j, Log forging, Cross-site scripting, and injection. The proposed model is more focused on increasing the log data's privacy and immutability. The proposed model is adept at preventing confidential data from attacks such as identity theft, system hacking, DNS Tunneling, Malware, and SQL injection (Rahman et al. 2020a; Mijwil et al. 2023). The double integrity check and immutable storage help to prevent spoofing attacks (Chen et al. 2007; Rahman et al. 2020b).

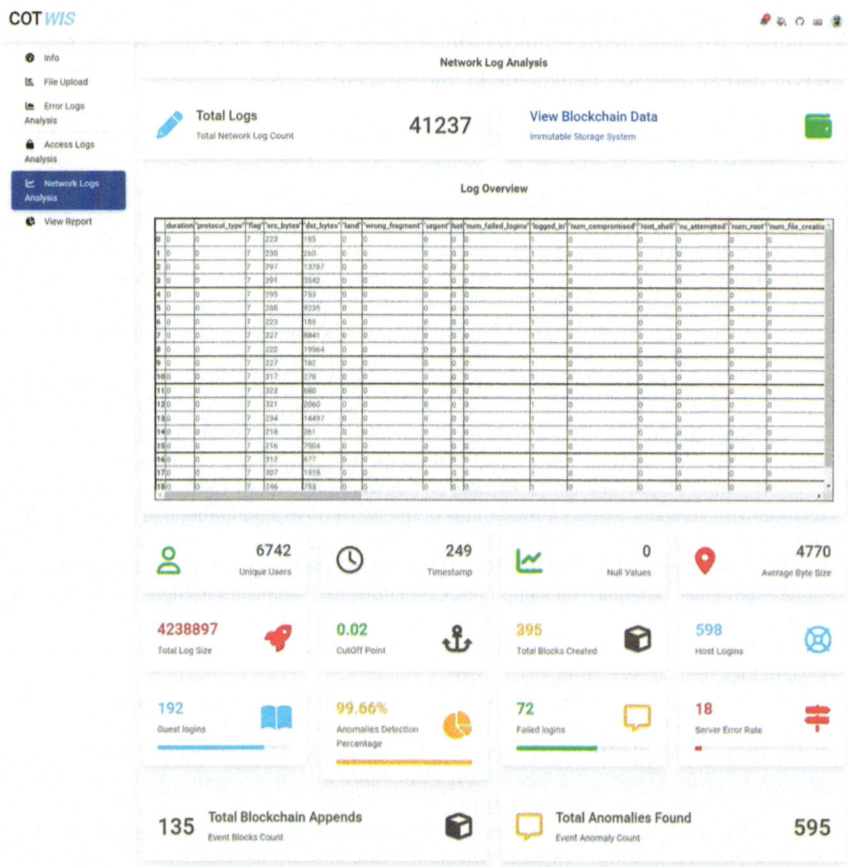

Fig. 6.11 Descriptive network log overview and results

6.6 Conclusion and Future Work

As malware is getting more advanced and persistent, their new obfuscation method is evading detection of malware. With all these advancements in persistent malware attacks, there is a need for a proper detection methodology that can reduce the triage time and hence help isolate the spread of malware and reduce the impact of the infection. To detect these threats effectively, we have analyzed the different types of logs, utilized multiple machine learning algorithms to detect the new threats accurately, and proposed a system to protect the logs from mutation or deletion from attackers and store them in distributed immutable storage.

Future work includes involving and comparing more Machine Learning algorithms, preferably LSTM Neural Network with a high-performance metric with efficient computational resources. The network log analysis architecture includes the Identity and Access Logs for insider threat detection. comparing certain log data with machine learning methods to improve root cause analysis and event correlation. We have built a public cloud service that is available everywhere for efficient and simple log management and analysis.

References

Abdullahi M, Baashar Y, Alhussian H, Alwadain A, Aziz N, Capretz LF, Abdulkadir SJ (2022) Detecting cybersecurity attacks in internet of things using artificial intelligence methods: a systematic literature review. Electronics 11(2):198

Albahar M, Alansari D, Jurcut A (2022) An empirical comparison of pen-testing tools for detecting web app vulnerabilities. Electronics 11(19):2991

Ambre A, Shekokar N (2015) Insider threat detection using log analysis and event correlation. Proc Comp Sci 45:436–445

Behera A, Panigrahi CR, Pati B (2022) Unstructured log analysis for system anomaly detection—a study. In Advances in Data Science and Management (pp 497–509). Springer

Bračevac O, Amin N, Salvaneschi G, Erdweg S, Eugster P, Mezini M (2018) Versatile event correlation with algebraic effects. Proceedings of the ACM on Programming Languages, 2(ICFP), pp 1–31

Candel JMO, Gimeno FJM, Mora Mora H (2023) Serverless security analysis for IoT applications. In International Conference on Ubiquitous Computing and Ambient Intelligence (pp 393–400). Springer

Catillo M, Pecchia A, Villano U (2022) AutoLog: anomaly detection by deep auto encoding of system logs. Expert Syst Appl 191:116263

Chen QX, Chang XH (2022) Resilient filter of nonlinear network systems with dynamic event-triggered mechanism and hybrid cyber attack. Appl Math Comput 434:127419

Chen Y, Trappe W, Martin RP (2007) Detecting and localizing wireless spoofing attacks. In *2007 4th Annual IEEE Communications Society Conference on sensor, mesh and ad hoc communications and networks* (pp 193–202). IEEE

Friedberg I, Skopik F, Settanni G, Fiedler R (2015) Combating advanced persistent threats: From network event correlation to incident detection. Comput Secur 48:35–57

Ghiasi M, Niknam T, Wang Z, Mehrandezh M, Dehghani M, Ghadimi N (2023) A comprehensive review of cyber-attacks and defense mechanisms for improving security in smart grid energy systems: past, present and future. Electric Power Syst Res 215:108975

Gill SS, Kumar A, Singh H, Singh M, Kaur K, Usman M, Buyya R (2022) Quantum computing: a taxonomy, systematic review and future directions. Soft Pract Exp 52(1):66–114

Huang W (2019) A blockchain-based framework for secure log storage. In 2019 IEEE 2nd International Conference on Computer and Communication Engineering Technology (CCET) (pp 96–100). IEEE

Jain S, Ahuja NJ, Srikanth P, Bhadane KV, Nagaiah B, Kumar A, Konstantinou C (2021) Blockchain and autonomous vehicles: Recent advances and future directions. IEEE Access 9:130264–130328

Jansen BJ (2022) *Understanding user-web interactions via web analytics*. Springer Nature

Kiayias A, Zindros D (2020) Proof-of-work sidechains. In *Financial Cryptography and Data Security: FC 2019 International Workshops, VOTING and WTSC, St. Kitts, St. Kitts and Nevis, February 18–22, 2019, Revised Selected Papers 23* (pp 21–34). Springer International Publishing

Kotenko IV, Levshun DS, Chechulin AA (2016) Event correlation in the integrated cyber-physical security system. In 2016 XIX IEEE International Conference on Soft Computing and Measurements (SCM) (pp 484–486). IEEE

Landauer M, Skopik F, Wurzenberger M, Hotwagner W, Rauber A (2020) Have it your way: generating customized log datasets with a model-driven simulation testbed. IEEE Transactions on Reliability

Lathkar M (2021) Building Web Apps with Python and Flask: Learn to Develop and Deploy Responsive RESTful Web Applications Using Flask Framework (English Edition). BPB Publications

Li L, Man Y, Chen M (2018) A method of large-scale log pattern mining. In *Human Centered Computing: Third International Conference, HCC 2017, Kazan, Russia, August 7–9, 2017, Revised Selected Papers 3* (pp 76–84). Springer International Publishing

Liashchynskyi P, Liashchynskyi P (2019) Grid search, random search, genetic algorithm: a big comparison for NAS. arXiv preprint arXiv:1912.06059

Landauer Max, Skopik Florian, Frank Maximilian, Hotwagner Wolfgang, Wurzenberger Markus, Rauber Andreas (2022) AIT Log Data Set V2.0 (v2_0) . Zenodo

Majd M, Najafi P, Alhosseini SA, Cheng F, Meinel C (2022) A comprehensive review of anomaly detection in web logs. In *2022 IEEE/ACM International Conference on Big Data Computing, Applications and Technologies (BDCAT)* (pp 158–165). IEEE

Mijwil M, Unogwu OJ, Filali Y, Bala I, Al-Shahwani H (2023) Exploring the top five evolving threats in cybersecurity: an in-depth overview. Mesopot J Cybersec 2023:57–63

Mohan AP, Gladston A (2020) Merkle tree and Blockchain-based cloud data auditing. Int J Cloud Appl Comput (IJCAC) 10(3):54–66

Nakamoto S (2008) Bitcoin: a peer-to-peer electronic cash system. *Decentralized business review*

Oliner A, Ganapathi A, Xu W (2012) Advances and challenges in log analysis. Commun ACM 55(2):55–61

Oprea SV, Bâra A (2021) Machine learning classification algorithms and anomaly detection in conventional meters and Tunisian electricity consumption large datasets. Comput Electr Eng 94:107329

Pourmajidi W, Miranskyy A (2018) Logchain: blockchain-assisted log storage. In 2018 IEEE 11th International Conference on Cloud Computing (CLOUD) (pp 978–982)

Pourmajidi W, Zhang L, Steinbacher J, Erwin T, Miranskyy A (2019) Immutable log storage as a service. In 2019 IEEE/ACM 41st International Conference on Software Engineering: Companion Proceedings (ICSE-Companion) (pp 280–281). IEEE

Boldyreva A, Kolesnikov V (Eds.) (2023) Public-Key Cryptography–PKC 2023: 26th IACR International Conference on Practice and Theory of Public-Key Cryptography, Atlanta, GA, USA, May 7–10, 2023, Proceedings, Part II (Vol 13941). Springer Nature

Rahman RU, Sahu DR, Tomar DS (2017) Challenges in securing ESB against web service attacks. In Exploring Enterprise Service Bus in the Service-Oriented Architecture Paradigm (pp 74–96). IGI Global

Rahman RU, Wadhwa D, Bali A, Tomar DS (2020) The emerging threats of web scrapping to web applications security and their defense mechanism. In *Encyclopedia of Criminal Activities and the Deep Web* (pp 788–809). IGI Global

Rahman RU, Verma R, Bansal H, Tomar DS (2020) Classification of spamming attacks to blogging websites and their security techniques. In *Encyclopedia of Criminal Activities and the Deep Web* (pp. 864–880). IGI Global

Reguieg H, Benatallah B, Nezhad HRM, Toumani F (2015) Event correlation analytics: scaling process mining using mapreduce-aware event correlation discovery techniques. IEEE Trans Serv Comput 8(6):847–860

Rout B, Natarajan B (2022) Impact of cyber attacks on distributed compressive sensing based state estimation in power distribution grids. Int J Electr Power Energy Syst 142:108295

Samuel AL (1967) Some studies in machine learning using the game of checkers. II—Recent progress. IBM J Res Devel 11(6):601–617

Schinckus C (2021) Proof-of-work based blockchain technology and Anthropocene: an undermined situation? Renew Sustain Energy Rev 152:111682

Simoes V, Maniar H, Abubakar A, Zhao T (2022) Deep learning for multiwell automatic log correction. In SPWLA 63rd Annual Logging Symposium. OnePetro

Wang B, Ying S, Yang Z (2020a) A log-based anomaly detection method with efficient neighbor searching and automatic k neighbor selection. Sci Program 2020:1–17

Wang J, Tang Y, He S, Zhao C, Sharma PK, Alfarraj O, Tolba A (2020b) LogEvent2vec: log event-to-vector based anomaly detection for large-scale logs in internet of things. Sensors 20(9):2451

Wang Z, Tian J, Fang H, Chen L, Qin J (2022) LightLog: a lightweight temporal convolutional network for log anomaly detection on the edge. Comput Netw 203:108616

Wang H, Yang D, Duan N, Guo Y, Zhang L (2018) Medusa: blockchain powered log storage system. In 2018 IEEE 9th International Conference on Software Engineering and Service Science (ICSESS) (pp 518–521). IEEE

Yasrab R (2018) Platform-as-a-service (paas): the next hype of cloud computing. *arXiv preprint* arXiv:1804.10811

Zhang X, Xu Y, Lin Q, Qiao B, Zhang H, Dang Y, Zhang D (2019) Robust log-based anomaly detection on unstable log data. In Proceedings of the 2019 27th ACM Joint Meeting on European Software Engineering Conference and Symposium on the Foundations of Software Engineering (pp. 807–817)

Zhu J, He S, Liu J, He P, Xie Q, Zheng Z, Lyu MR (2019) Tools and benchmarks for automated log parsing. In 2019 IEEE/ACM 41st International Conference on Software Engineering: Software Engineering in Practice (ICSE-SEIP) (pp 121–130). IEEE

Quantum Resilience and Distributed Trust: The Promise of Blockchain and Quantum Computing in Defense

7

Shalom Akhai and Vipul Kumar

Abstract

The combination of blockchain technology with quantum computing has opened up new doors for improving the safety and effectiveness of a diverse variety of business sectors. The most recent readily available technologies, such as blockchain and quantum computing, play a crucial role in improving both safety and efficiency, which in turn makes a wide range of previously unavailable possibilities available in a variety of industries. This chapter goes into one of the most fascinating industries, namely the defense industry and the armed forces. Both of these industries have only somewhat lately begun to make use of modern technology in order to address complex problems and enhance operational efficiency. These cutting-edge technologies find applications in a variety of fields, most notably the military and defense sectors, to improve safety standards in their respective industries. The technologies that are discussed in this chapter are now seeing widespread implementation across a variety of sectors, including the military, the defense industry, and even the food business. This article examines their uses in the context of using blockchain technology to guarantee safe communication and the storage of data. In addition to this, it highlights the benefits of applying the principles of quantum computing to matters pertaining to military and defense-related tasks. This chapter highlights a number of real-world situations that stand to profit from the combination of blockchain technology for secure data storage and transmission inside military networks and quantum computing for research and development activities. These scenarios are discussed in further detail throughout the chapter. These

S. Akhai (✉)
Chandigarh College of Engineering, CGC Jhanjeri, Mohali 140307, India
e-mail: shalom.akhai@gmail.com

V. Kumar
Quant Service, Dubai, United Arab Emirates

A. Kumar et al. (eds.), *Sustainable Security Practices Using Blockchain, Quantum and Post-quantum Technologies for Real Time Applications*, Contributions to Environmental Sciences & Innovative Business Technology,
https://doi.org/10.1007/978-981-97-0088-2_7

hypothetical situations are analyzed in great detail and presented. This chapter also provides some insight into where blockchain technology and quantum computing are headed in the future, taking into consideration their present level of use and the continuous resolution of technical obstacles. This article is full with useful information for researchers and professionals who are interested in improving the safety of military and defense operations by using blockchain technology and quantum computing. By integrating these technologies, the goal is to make a contribution to the development of sustainable security practices for real-time applications. This chapter comes to a close with a discussion of some of the most pressing issues facing society today, including questions about breaches of personal privacy and the possibility that these technologies may be used for military purposes. It does so by carrying out an ethical study of the consequences surrounding the usage of technologies in defense and military applications such as blockchain and quantum computing. The results of this analysis shed light on their influence on concerns relating to privacy and security. Thus, the purpose of this chapter is to provide a detailed analysis of the ways in which blockchain technology and quantum computing might work together to improve safety and productivity across a variety of industries, with a specific emphasis on defense and military applications. It covers ethical issues and provides a forward-looking viewpoint, making it an invaluable resource for academics and professionals who want to develop security practices in various fields.

Keywords

Blockchain technology • Data storage • Defense sector • Military applications • Quantum computation • Secure communication • Sustainable security practices

7.1 Introduction

Significant advancements are currently being achieved in the field of information technology due to the convergence of two state-of-the-art technologies: blockchain and quantum computing (Mosteanu and Faccia 2021). The aforementioned technologies possess a diverse range of potential applications, encompassing sectors such as the military, the financial industry, and the healthcare domain (Justinia 2019). The progression of technology has consistently been a pivotal determinant in shaping the trajectory of defense and military applications in forthcoming periods (Zhu et al. 2020). The significance of advancements in blockchain technology and quantum computing cannot be overstated, as they have enabled us to achieve unprecedented levels of efficiency and security (Kulkarni et al. 2022). The incorporation of these technologies has the potential to yield solutions to the complex challenges faced by defense and military entities, leading to enhanced operational efficiency (Srivastava et al. 2022). The phrase "blockchain technology" denotes a decentralized digital ledger that possesses qualities of trustworthiness and transparency, facilitating the recording of transactions occurring within a network of computers (Sarmah 2018). Due to its inherent resistance to modification or manipulation, the data stored within it is afforded a considerable level of security (Sheth and Dattani 2019). In contrast, quantum computing relies on the utilization of quantum states to perform operations on data and solve complex mathematical problems (Maslov et al. 2018). The aforementioned phenomenon possesses the capacity to fundamentally revolutionize our approach toward addressing significant mathematical challenges within the realm of defense and military applications (Krelina 2021). The concept of quantum gates and circuits, which are fundamental elements in quantum algorithms (Unogwu et al. 2022), holds significant importance in the field of quantum computing. The algorithms possess the capability to enhance the existing cryptography techniques employed in defense and military domains (Sultan et al. 2019). By employing quantum algorithms for cryptography, it is possible to develop encryption techniques that are resistant to attacks from quantum computers. This approach will enable us to preserve the integrity of our data in anticipation of future advancements in computer capacity (Rosch-Grace and Straub 2021). Furthermore, the application of quantum forensics is increasingly acknowledged as a valuable investigative tool within defense and military contexts (Mavroeidis et al. 2018). The application of quantum mechanics principles has enabled investigators to analyze data with an unprecedented level of precision and sensitivity (Ahmad et al. 2021). This aids investigators in the detection of security breaches and potential threats. The potential for significant technological advancements exists through the incorporation of quantum computing in various defense and military applications. For example, it can be utilized in the resolution of complex mathematical problems related to cryptography and encryption. The preservation of secure data is of utmost importance in defense and military operations, as they heavily rely on data security. The integration of quantum computing with existing defense systems has the potential to enhance encryption techniques, bolster data security, and improve the overall efficiency of defense and military operations. This measure will enable us to enhance the safeguarding of sensitive information.

7.1.1 Integration of Blockchain and Quantum Computing in Defense and Military Applications

Blockchain and quantum computing are two cutting-edge technologies that have the power to revolutionize how defense and military organizations function (Girasa and Scalabrini 2022a).

- Blockchain is a decentralized system that allows secure and transparent transactions without the need for intermediaries (Kulshrestha and Navy 2018). In defense and military applications, it can ensure secure communication, manage supply chains, and exchange data (Lilly and Lilly 2021).

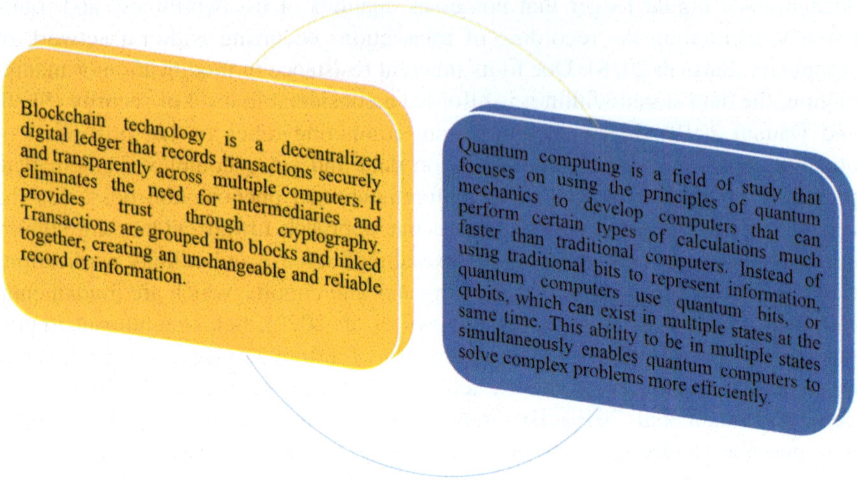

Fig. 7.1 Overview of blockchain technology and quantum computing

- Quantum computing is a type of computer that uses quantum–mechanical processes to perform computations. It can track military assets, verify information, and prevent cyber-attacks (Neumann et al. 2020). Quantum computing can solve complex problems that traditional computers cannot, like cracking encryption and modeling intricate systems. It has potential in cryptography, intelligence collection, and logistics optimization for defense and military applications (Neumann et al. 2021).

Figure 7.1 summarizes blockchain and quantum computing, two developing technologies that might transform defense and military applications (Dwivedi et al. 2023). These technologies may provide military institutions with extraordinary security and efficiency. Blockchain can encrypt quantum communication channels, whereas quantum computing can break and establish new cryptographic protocols. These technologies can encrypt communication, protect sensitive data, and solve defense and military problems. However, ethical questions concerning data confidentiality and abusive usage of these technologies must be addressed.

7.1.2 Importance of Sustainable Security Practices

Sustainable security is crucial in defense and military applications (D'Auria and Teller 2023). Military organizations require excellent security to combat rising threats and assaults. These solutions should persist and adapt to future difficulties (Ferenc 2020). Blockchain and quantum computing may be used to create security measures that can withstand current and future threats. Secure data storage and communication routes make it difficult for unauthorized individuals to access or alter sensitive data (Chen 2022). Defense and military activities need difficult cryptography and encryption mathematics issues, which quantum computers can solve (Fernandez-Carames and Fraga-Lamas 2020). Sustainable security measures improve military operations and safety (Derose 2023). Defense and military systems may improve efficiency and adaptability to changing threats and difficulties by using blockchain and quantum computing (Lindsay 2020).

There are three key goals that this effort aims to accomplish: First and foremost, the mission is to investigate and investigate the potential defense and military uses of blockchain technology and quantum computing. Understanding how these technologies may be combined to improve data security, communication, encryption, and intelligence gathering is required for this step. The second objective of this research is to determine the unique difficulties and constraints that can be presented during the implementation of blockchain technology and quantum computing in defense and military systems. The study aims to give insights into the practical feasibility and ethical concerns surrounding the deployment of these technologies by addressing these difficulties and attempting to find solutions to them. In conclusion, the purpose of this study is to offer sustainable security techniques that make use of blockchain and quantum computing to develop long-lasting and flexible defense and military systems that are capable of withstanding both present and future threats.

The contributions made by this study are important in a variety of different ways. In the first place, it provides light on the potential revolutionary effect that may be made by merging quantum computing with blockchain technology in defense and military applications. This research paves the way for innovative solutions to real-world challenges that are faced by defense and military organizations by demonstrating the benefits of using blockchain for secure data storage, supply chain management, and communication; demonstrating the capabilities of quantum computing in solving complex cryptographic problems; and showcasing the benefits of using quantum computing in solving complex cryptographic problems.

Second, the research contributes to the comprehension of sustainable security procedures in defense and military settings. The study presents a paradigm for defense systems that are able to grow with increasing threats. It does this by highlighting the significance of security measures that are both durable and flexible. A cutting-edge strategy for protecting sensitive information and guaranteeing the efficiency of military operations is to use a mix of blockchain technology and quantum computing as important components of sustainable security practices. This method has been recommended as a major aspect of sustainable security practices. In addition, the ethical issues that have been brought to light in this study help to the responsible adoption of these technologies. They encourage a methodical approach that strikes a balance between the need for security and the possible dangers and abuses of the technology.

7.2 Challenges Faced in Defense and Military Applications

In defense and military applications, there are several major challenges that need to be addressed. These applications involve the use of advanced technologies and systems to ensure the safety and security of nations and their citizens. However, they also face various obstacles that can significantly impact their effectiveness (Zhang et al. 2022; Li et al. 2021; Lindsay 2018; Bontekoe et al. 2022).

7.2.1 Secure Communication and Data Storage Challenges

Secure data storage and encrypted transmission are major difficulties in defense and military applications (Majot and Yampolskiy 2015). Military units and personnel must securely convey private information to conduct effective missions. Unauthorized access or interception may be disastrous.

Using traditional communication techniques like radio and satellite poses security risks, making safe communication a serious challenge. Hacking and interception by hostile actors may compromise the confidentiality and integrity of information during transmission over these methods (Tedeschi et al. 2022). Safe data storage is another issue. Military organizations provide abundant intelligence reports and operational plans daily. This information must be secured to avoid unauthorized access and data loss from system failures or natural catastrophes (Hasan et al. 2022; Dadhich and Tiwari 2022).

7.2.2 Overview of Existing Solutions and Their Limitations

Many techniques have been developed to safeguard data storage and transmission. Following are some existing solutions and their limitations:

- **Encryption**: Encryption is a method that protects data by transforming it into unreadable text via the application of complicated algorithms. This method is utilized extensively. It offers a high level of protection, but it is not entirely failsafe. Using techniques like brute force assaults or shared engineering, knowledgeable hackers may still be able to retrieve data that has been encrypted (Khalid and Askar 2021).
- **Virtual Private Networks (VPNs)**: Virtual Private Networks are often used to secure communication between military units and personnel. VPNs are an example of an encryption technology. They do this by establishing a safe and encrypted connection over the internet, which prevents any third parties from illegally obtaining the information that is being delivered. However, virtual private networks (VPNs) are not impenetrable to assaults, and the efficiency of VPNs may be hampered by security weaknesses in the design of the underlying network (Mashatan and Turetken 2020).

- **Blockchain**: It is a technology that has gained popularity as a dependable mechanism for storing and sending data. Blockchain technology is referred to as "blockchain." It is a decentralized ledger that is safe to use and secure, and it records transactions in a way that is both immutable and visible. Despite the fact that blockchain technology may have significant uses in defense and the military, its implementation on a broad scale is difficult owing to the complexity of the technology and the enormous amount of energy it requires (Kiktenko et al. 2018).
- **Quantum Key Distribution (QKD)**: QKD is an up-and-coming method that generates safe encryption keys by using concepts from quantum physics. In contrast to conventional encryption techniques, which are based on mathematical algorithms, QKD makes use of the physical characteristics of photons in order to generate keys. Because of this, it is now exceedingly difficult for hackers to steal the key without being discovered, and in certain cases, it may even be impossible. In spite of the fact that QKD is still in its infant phases of development, it has already shown encouraging outcomes in laboratory settings (Cui et al. 2020).

These solutions offer varying degrees of security, but it is important to acknowledge their limitations and ongoing efforts to enhance their effectiveness in defense and military applications. The convergence of quantum cryptography, post-quantum cryptography, quantum blockchains, and blind quantum computing holds the promise of introducing a new era characterized by unparalleled levels of security and efficiency in defense and military domains, within a rapidly evolving technological landscape. The potential for achieving unbreakable encryption, quantum-resistant cryptographic techniques, transparent transactional integrity, and safe data analysis becomes more evident as academics and innovators delve further into these intriguing domains. The pursuit of impregnable security and privacy is closely linked to the enigmatic principles of quantum physics, representing the domain in which the future of defense and military operations lies. The convergence of advanced technologies represents the locus of future developments in defense and military operations.

7.3 Blockchain Technology for the Securing of Data in Defense and Military Applications

Nowadays, protecting one's privacy and keeping one's data secure is of the utmost importance for people, businesses, and governments alike. Even government institutions that are responsible for defense and military activities confront the issue of securing sensitive and classified data from internet threats and assaults. This challenge is a challenge that even government agencies face. The blockchain technology provides the possibility of solutions by offering data storage that is both safe and decentralized. In this essay, we will investigate the inner workings of blockchain technology as well as its potential uses in the defense and military sectors.

7.3.1 Understanding Blockchain Technology

Because of its capacity to handle fundamental concerns including secure data storage, transparency, and immutability, blockchain technology has garnered substantial interest in a variety of industries, including defense and military applications. This is owing to the fact that it can be used in these areas. A blockchain, in its most basic form, is a decentralized and distributed digital ledger that may be used to record transactions or data across a network of computers known as nodes. A chain of blocks is created when each individual transaction or data entry, which is referred to as a block, is connected to the one before it using cryptographic hashes. Because of blockchain's decentralized structure, there is no one organization that can control the whole network. Because of this, it is very resistant to efforts at tampering and hacking. Additionally, cryptographic algorithms guarantee the provision of a safe and open-access system, allowing each participant to independently validate the data's validity (Fernandez-Carames and Fraga-Lamas 2020; Younan et al. 2021).

7.3.2 Examples of Blockchain's Potential Applications in Defense and Military Settings

Any compromise of the confidentiality of the information handled by the defense and military sectors may have serious repercussions. These businesses include the military and the defense sector. Blockchain technology may contribute to the solution of these problems by offering data storage choices that are both dependable and secure. The practical uses of blockchain technology in defense and military situations are highlighted in Fig. 7.2, with a particular emphasis on the use of the technology for secure data storage. In the defense and military sectors, it outlines a variety of potential applications for blockchain technology that aim to improve data security, integrity, and secrecy. The purpose of the picture is to provide a visual depiction of the possible benefits and advantages that may be realized as a result of deploying blockchain technology in these particular settings.

Following are a few interesting Blockchain Applications for Enhanced Defense and Military Operations:

- **Identity Management**: Blockchain, a distributed ledger technology, might provide a secure, decentralized identity management system. This would allow people and organizations to securely transmit identity information, reducing worries about illegal access. Blockchain-based identity management systems can verify the legitimacy of persons and organizations, which might help fight fraud and other crimes.
- **Data Integrity and Auditing**: Blockchain technology allows data validation auditing and verification. Blockchain records are immutable and unchangeable. Blockchain allows data audits by tracing data revisions and verifying its

Fig. 7.2 Use cases of blockchain in defense and military applications for secure data storage

integrity, guaranteeing no changes have happened. Thus, blockchain facilitates data audits.

- **Smart Contracts for Military Networks and Soldier's Networks**: Smart contracts are coming to the Military and Soldier's Networks. Smart contracts are blockchain-stored, automated contracts. These systems can automate financial processing, logistics, and supply chain administration.

7.3.2.1 Secure Communication and Collaboration

In defense and military operations, blockchain technology can provide encrypted communication and collaboration that ensures confidentiality. Traditional communication methods are vulnerable to eavesdropping and unauthorized access. Blockchain-based solutions enable secure communication where only authorized individuals can access and understand the transmitted data. It also allows for improved coordination between military units and automates established agreements or protocols, reducing the need for intermediaries and streamlining decision-making (Azzaoui et al. 2021; Koh et al. 2020).

7.3.2.2 Supply Chain Management and Logistics

Efficient supply chain management is crucial in defense and military activities. Blockchain technology enables transparency, traceability, and security by recording each stage of the supply chain on an immutable ledger. This helps track assets, equipment, and resources in real time, reducing the risks of theft, counterfeiting, and unauthorized tampering. Blockchain allows businesses in the military industry to verify the authenticity and safety of essential supplies and automate procurement, inventory management, and maintenance. This improves resource allocation and reduces inefficiencies (Peña et al. 2020; Cheung et al. 2021).

7.3.2.3 Identity Administration and Access Control

Verifying employee identities, managing access credentials, and ensuring secure authentication are important tasks in defense and military applications. Blockchain technology offers a decentralized and tamper-proof identity management system. Storing encrypted identification information on the blockchain allows military organizations to grant access only to authorized personnel for critical areas, equipment, or information. Blockchain-based identification systems simplify verification processes, reducing the need for multiple authentication phases and administrative work (Partida et al. 2021; Buchmann et al. 2017).

7.3.2.4 Cybersecurity for Information and Computer Networks

Cybersecurity is a critical concern for defense and military organizations dealing with sensitive information and facing sophisticated cyberattacks. Blockchain technology enhances cybersecurity by providing a decentralized and transparent system. It reduces risks associated with centralized data storage and single points of failure. Sensitive data can be encrypted and distributed among multiple nodes, making it extremely difficult for attackers to tamper with or modify the information. Blockchain-based systems can detect and report illegal attempts to access or alter data, improving overall cybersecurity (Abd El-Latif et al. 2021; Xu et al. 2023).

7.4 Potential Benefits of Quantum Computing in Defense and Military Applications

Computing based on quantum mechanics is an emerging field of technology that has the potential to be of significant use to defense and military endeavors. It has the potential to revolutionize a variety of areas, including cryptography, optimization, simulations, and the analysis of intelligence. The following are some real-world applications of its potential usefulness:

- **Cryptography and Secure Communication**: Quantum computing has the potential to significantly improve the level of security afforded to military and defense-related communication. Traditional encryption methods depend on huge numbers, which quantum computers are able to decipher with relative ease. Quantum computing does, however, provide a solution in the form of the development of encryption that is immune to quantum attacks. Even in the presence of very powerful quantum computers, this protects the privacy of critical data and conversations while still ensuring their legitimacy (Blahut 2014; Ko et al. 2021; Gupta et al. 2013; Mailloux et al. 2016).
- **Optimization and Resource Allocation**: The capacity of quantum computing to handle enormous volumes of data concurrently has the potential to dramatically enhance the optimization techniques used in defense and military operations. It is possible to be more efficient in planning routes, assigning resources, and scheduling tasks with its support. Quantum algorithms provide superior solutions to optimization issues compared to more conventional algorithms, which ultimately results in more effective resource allotment, logistics, and decision-making (Salau et al. 2022; Reding and Eaton 2020; Kumar et al. 2022a; Drager and Walsh 2005; Zhang et al. 2019; Symons et al. 2023; Wang and Alexander 2020).

- **Simulations and Modeling**: Modeling and simulations are very important in the field of defense and military applications since they allow for the testing of new technology, the training of personnel, and the development of plans. The capabilities of classical computers are surpassed by the capabilities of quantum computers, which permit more precise modeling of complex systems. It is capable of simulating materials, chemical interactions, and biological systems with a level of detail never before seen. This may result in the development of superior materials, enhanced weapon systems, and a deeper comprehension of intricate battlefield dynamics (Rosch-Grace and Straub 2022; Inglesant et al. 2018; Whitfield et al. 2011; Lucas et al. 2013; Lanzagorta and Uhlmann 2004).
- **Intelligence Analysis and Machine Learning**: The defense and military industries collect huge volumes of data from a variety of sources, such as sensors and satellites. The use of quantum computing improves machine learning methods, which in turn makes it possible to conduct more robust analyses of large datasets. This is very necessary in order to extract useful insights and patterns from the data in order to conduct intelligence analysis and make decisions. The use of quantum machine learning algorithms results in enhanced threat identification, anomaly detection, and predictive analysis (Mori 2018; Edwards and Rawat 2020; Chen and Chen31 2021; Raska 2022).

These benefits are a direct result of the approaches' increased accuracy and computing efficiency. As a result, quantum computing has a massive potential to revolutionize defense and military applications. This would result in advantages such as "improved simulations," "optimized resource allocation," "enhanced cryptography," and more accurate intelligence analysis.

7.5 Ongoing Developments

The military and the defense industry are actively researching the potential uses of quantum computing in the real world in order to enhance their capabilities and find solutions to difficult problems. Quantum cryptography and quantum machine learning are two particular areas of interest that are being examined for the following:

- In order to accomplish the goal of secure communication, quantum cryptography makes use of the ideas underlying quantum physics. This method, which was created in the 1990s, makes it possible to trade encryption keys while maintaining a level of security that can be independently verified using quantum key distribution (QKD) protocols. It has the potential to revolutionize secure communication in defense and military environments, assuring the confidentiality and dependability of essential information (Neumann et al. 2020; Padamvathi et al. 2016; Kumar et al. 2019; Zheng and Jiang 2015; Guoming et al. 2017).

- Quantum machine learning is another field of study that is becoming more important in the realms of both defense and military operations. Anomaly detection and predictive analysis are two examples of applications for the machine learning methods that are employed. Traditional machine learning algorithms may not be able to compete with the capabilities of quantum machine learning approaches, such as quantum support vector machines and quantum neural networks. They are able to do analyses on very large datasets more quickly and accurately (Huizing et al. 2019; Singh et al. 2022a; Bhatia et al. 2020; Acampora 2020; Jhanwar and Nene 2021).
- In addition, researchers are investigating the possibility of using quantum computers in order to mimic quantum chemistry. Quantum computers have the capability of accurately simulating the behavior of complicated molecules and the processes that they go through in the chemical industry. This has repercussions for the development of material science as well as the quest for novel pharmaceuticals. This study may be useful for the defense and military sectors in the investigation of chemical warfare agents, the development of innovative materials, and the advancement of research into therapeutic practices (Melagraki 2022; Tesvara 2021; Gupta 2015; Denison et al. 2005).
- Quantum computing is gaining popularity for contamination removal in environmental remediation studies. Quantum computers might change environmental issues by correctly simulating complicated chemical processes and molecular interactions. This aptitude might lead to catalysts and processes that transform harmful pollutants into non-toxic compounds. This breakthrough will greatly aid environmental rehabilitation and pollution prevention. Quantum computers may help scientists and researchers duplicate and enhance chemical processes that reduce hazardous waste, absorb carbon dioxide, and degrade contaminants. These technological developments might hasten efforts to minimize industrial pollution and address climate change. Quantum-enhanced environmental cleanup technologies are essential for ecosystem protection and long-term planet survival. Environmental science and quantum computers might revolutionize pollution control and provide a better, more sustainable world for future generations (Singh et al. 2022b; Chugh et al. 2020; Srikanth and Kumar 2022).

These solutions have the potential to be implemented immediately.

7.6 Technical Challenges Faced and Potential Solutions

The area of quantum computing has considerable potential for a variety of applications, including those in the defense and military sectors. However, in order to fully realize its potential, a number of significant technological obstacles must first be conquered. In order to gain a comprehensive comprehension of these obstacles, they are categorized as follows:

- **Preserving the coherence and stability of qubits**: Qubits, the fundamental building blocks of information in quantum computers, are very sensitive to environmental disturbances, which may result in mistakes and the destruction of quantum states. Methods for error correction are now being developed by researchers in an effort to mitigate the detrimental effects of noise and enhance the qubits' stability and dependability (Watson et al. 2018; Saffman 2016; Córcoles et al. 2019; Brandmeier et al. 2022; Biswas et al. 2017).
- **Scaling up**: Constructing a large-scale quantum computer with hundreds or thousands of qubits is not a simple undertaking. Scaling up involves increasing the number of qubits in the system. Researchers are looking at a variety of technologies, including superconducting circuits and trapped ions, in order to increase the processing capacity of qubits while also improving their quality and coherence. In addition to this, they are concentrating their efforts on the study and development of production techniques and materials (Varmantchaonala Moudina et al. 2000; Deventer et al. 2022; Uddin et al. 2022).
- **Developing quantum algorithms and software tools**: In order to fully harness the capabilities of quantum computers, specialized quantum algorithms will need to be built and optimized. This will need the development of software tools. Researchers are always looking into novel algorithms and developing software frameworks and programming languages with the goal of making quantum computing available to a larger range of users (Ur Rasool et al. 2023; Broughton et al. 2020; Javadi-Abhari 2017; Markidis 2023; Zhu et al. 2023).
- **Control and measurement systems**: In order for quantum computers to effectively conduct computations, the control and measurement of qubits in the computer must be very exact. To conquer these technical obstacles, advancements in control electronics, cryogenics, and measuring methods are very necessary (Leon et al. 2021; Acín et al. 2018; Rotta et al. 2017).

7.7 Future Directions

The realms of defense and military operations have always been pioneers in the adoption of new technologies. They are the driving force behind innovation and progress, both of which are ultimately beneficial to the civilian sector. There have been major breakthroughs made in a variety of fields of defense and military technology over the course of the previous several decades. The development of unmanned aerial vehicles (UAVs) and improved missile systems are examples of these breakthroughs. The last several years have seen a substantial movement toward the development of technologies that are increasingly self-sufficient, intelligent, and linked. The operations of militaries are undergoing a fundamental transformation as a result of this trend, which enables them to carry out their duties in a manner that is more efficient, rapid, and precise (Kumar et al. 2022a; Chamola et al. 2021; Mian 2022; Subramanian et al. 2022; Sharma et al. 2020; Nawaz et al. 2019; Ahammed et al. 2023).

- The expanding use of unmanned aerial vehicles (UAVs), sometimes known as drones, for surveillance and assault missions is one of the most significant recent developments in the defense and military industries. The method that contemporary armed forces carry out missions such as intelligence collection, reconnaissance, and targeted attacks has been fundamentally altered as a result of the proliferation of unmanned systems. Swarm technology has further improved their capabilities by enabling huge numbers of smaller drones to act together as a coordinated unit, communicating with each other in real time. This has allowed their capabilities to be expanded in more ways than one.
- The field of cyberwarfare has also seen considerable developments in recent years. Because of the significant risk that cyberattacks pose to national security, governments are making significant investments in both their offensive and defensive cybercapabilities. As societies become more dependent on digital infrastructure, the ability to engage in warfare in the digital realm, also known as cyber warfare, has become a crucial strategic skill. Artificial intelligence (AI) and machine learning (ML) are also playing a major role in the defense and military industries. These technologies are being used in the development of autonomous systems that are capable of functioning independently of direct human control. Artificial intelligence and machine learning increase decision-making processes, raise levels of situational awareness, and make it possible to operate unmanned vehicles and missile systems. The use of quantum computing in defense and military applications has the potential to usher in a new era of innovation, notably in the fields of cryptography, optimization, simulations, and intelligence analysis. It is anticipated that this will considerably improve both the efficiency and efficacy of the aforementioned domains. Despite the fact that quantum computing is yet in its infancy, significant strides have been achieved in the field, and a great number of military organizations are actively investigating its potential applications.
- At the moment, defense organizations are making investments in quantum computing programs and partnering with academic institutions, commercial partners, and government agencies in order to improve the technology. These programs aim to evaluate whether or not the use of quantum computing is feasible in a variety of applications and to build solutions that are quantum-ready.
- When it comes to the actual implementation, the government and its armed forces are only getting started with the process of incorporating cryptographic methods that are resistant to the effects of quantum computing into their respective information systems. They are investigating post-quantum cryptography techniques and assessing the resilience of these techniques against quantum assaults. They are able to secure the safety of sensitive data even while quantum computers are there if they take the preventative measure of adopting proactive measures.
- In addition, organizations in the defense sector are creating agreements with companies that specialize in quantum computing and investing in quantum research centers. The purpose of these alliances is to pool the expertise and resources that are already available in order to expedite the development and

deployment of quantum technologies that are suited for usage in the defense and military sectors.

7.8 Possible Future Directions and Developments

In the coming years, the defense and military sectors will witness several trends and innovations that will shape their operations. These include:

- **Data analytics and big data**: The increased use of data sources and the ability to store and interpret large volumes of information will provide insights into logistics, intelligence gathering, and war strategy (Song et al. 2015; Sun et al. 2021).
- **Hypersonic weaponry**: Advances in hypersonic weapons, which travel at speeds many times faster than the speed of sound, will pose challenges for detection and defense. Nations such as the United States, Russia, and China are actively developing these weapons (Panwar 2015; Nakamitsu 2019; Ellison et al. 2023).
- **Space-based capabilities**: Space is becoming an important battlefield, and investments are being made in space-based systems for communication, monitoring, and navigation. Disrupting or destroying these systems will become a strategic tool (Morgan 2015; Sweet 2021).
- **Nanotechnology:** The advancements in nanotechnology will enable the creation of materials with unique properties, including enhanced conductivity and tensile strength. These materials can be used in the development of new weapons and protective gear (Simić et al. 2018; Singh and Kaur 2023).
- **Integration of quantum computing**: As quantum computing technology progresses, military organizations may incorporate small-scale quantum computers into their infrastructure. These systems can be used to test quantum algorithms, develop applications suitable for quantum computing, and explore the potential of quantum computing in military operations (Michal 2021; Zaidi and Sushma 2022).
- **Quantum machine learning and AI**: Quantum machine learning algorithms can improve data analysis, pattern identification, and decision-making processes. Combining quantum computing with AI technologies can result in more sophisticated autonomous systems, enhanced situational awareness, and optimized resource allocation (Mathur et al. 2022; Kumar et al. 2023).
- **Quantum simulations and modeling**: Quantum simulators provide insights and enable judgments on defense technologies and plans by representing complex systems realistically. This can contribute to the development of novel materials, the discovery of new drugs, and the improvement of defensive systems (Szabadföldi 2021; Lele 2021).
- **Quantum communication networks**: Quantum communication methods, such as quantum key distribution, can establish secure and tamper-proof communication lines between defensive units, headquarters, and allied troops. Quantum

networks enhance communication infrastructure's resistance to cyberattacks and protect critical information (Broadbent and Schaffner 2016; Pirandola et al. 2020).

- **Enhanced cryptographic approaches**: Quantum computing can facilitate the development of improved cryptographic protocols, including quantum-resistant algorithms. These protocols offer increased security and privacy assurances, safeguarding vital information used in defense and military activities (Yaacoub et al. 2022; Hadi et al. 2023).

7.9 Recommendations for the Integration of Blockchain and Quantum Computing

In the past few years, blockchain technology has grown highly popular because it offers a safe and decentralized method of managing crucial activities in a variety of sectors, including banking, healthcare, and the administration of supply chains. On the other hand, quantum computing is a very new and interesting topic that has the potential to significantly boost computing capacity. This would make it possible for conventional computers to do computations that are both much quicker and more complicated. Combining blockchain technology with quantum computing presents a significant possibility to significantly improve the scalability, usability, and overall efficacy of blockchain-based systems. In this section, we review the advantages of integrating various technologies and analyze the obstacles and concerns that should be taken into account. In addition, we will evaluate the factors that should be taken into account.

7.9.1 Potential Benefits of Integrating These Technologies

This sub-section presents the potential benefits of integrating quantum computing and blockchain from a futuristic point of view. Details are presented as follows:

- **Improved Security**: One of the primary benefits that arises from integrating blockchain technology with quantum computing is an improvement in the system's overall level of security. Through decentralization and encryption, blockchain networks already offer an increased degree of safety for their users. Quantum computing, on the other hand, has the potential to substantially bolster the security of blockchain networks by providing encryption that is impervious to quantum assaults. This indicates that blockchain networks have the potential to grow more robust against possible threats posed by quantum computers, which have the power to break many of the cryptographic techniques employed in present blockchain systems. Quantum computers have been predicted to become commercially available in the next few years. Blockchain transactions can be kept private and safe even in the presence of powerful quantum computers if quantum-resistant cryptographic algorithms are included in the protocol. This protects both the integrity of the transactions and their privacy.

- **Increased Scalability**: One advantage of combining blockchain technology with quantum computing is an increase in the system's capacity to handle increasing use. Due to its limited processing speed and capacity, blockchain networks are already encountering difficulties when attempting to handle a high number of transactions per second. The validation process can be sped up by quantum computing, and the time needed to add new blocks to the network may be cut greatly by using this technique. As a result, blockchain networks will be able to process a much greater number of transactions. As a result of quantum computing's capacity to conduct complicated calculations at high rates, it may be possible to assist blockchain networks overcome the scalability challenges they now face, hence allowing for speedier consensus and transaction processing. The overall effectiveness of blockchain systems is improved as a result of this greater scalability.
- **Enhanced Productivity**: Interoperability is vital when implementing blockchain technology in military and defense applications. This is particularly true for exploiting technology's productivity gain. Blockchain networks generally function independently, making it difficult to develop seamless communication and data exchange across several blockchain networks. Quantum computing will enable more complex encryption and decoding, which might transform interoperability. This invention might make chain-wide communication and data transmission risk-free. This unique system allows safe data access and use across blockchain platforms, which might simplify and improve military operations. This might simplify and improve military operations. Layer-2 blockchain solutions provide standard blockchain network scalability enhancements. Because these technologies are based on current blockchains, quicker and more scalable transactions may be made without sacrificing data integrity. Quantum computing may improve layer-2 solutions. Quantum computing will enable more efficient execution of complicated calculations and transactions. Quantum computing and layer-2 technologies might boost military blockchain network performance. Thus, data processing and smart contract execution may be optimized and faster. Quantum computing, interoperability, and layer-2 solutions might revolutionize defense and military operations by increasing efficiency. Data handling capacity must be greatly improved to reach this aim.
- **New Use Cases**: The combination of blockchain technology with quantum computing opens the door to the possibility of new use cases and applications that were not previously possible with either technology on its own. In order to protect sensitive data in sectors such as healthcare and finance, blockchain networks that are immune to quantum computing may be used. In addition, the marriage of blockchain technology with quantum computing makes it possible to develop decentralized apps that make use of sophisticated levels of processing ability.

- **Improved Privacy**: The combination of blockchain technology with quantum computing has the potential to greatly improve users' ability to maintain their anonymity online. In order to offer capabilities for the protection of users' privacy, blockchain technology already makes use of zero-knowledge proofs and secure multiparty computation methods. The use of quantum computing in the creation of more sophisticated privacy solutions, which would protect the confidentiality of sensitive information held inside blockchain networks, is a potential use of this technology.

7.9.2 Challenges and Considerations to Be Addressed in Implementation

Combining blockchain technology with quantum computing has the potential to bring about significant improvements in various areas. However, there are several challenges and considerations that need to be addressed before successful integration can be achieved.

- Synchronizing the technologies is difficult. Blockchain networks use distributed consensus techniques, which may need to be adjusted for quantum computer computational power and latency. Quantum computations must match blockchain consensus mechanisms to preserve system integrity and security.
- Developing consensus algorithms that are resistant to the effects of quantum computing is another obstacle. Traditional consensus methods like proof-of-work and proof-of-stake could become vulnerable to attacks from quantum computers. Research and development efforts are required to design and implement quantum-resistant consensus protocols.
- Interoperability between blockchain platforms and quantum computing platforms is crucial. The designs, programming languages, and protocols of these technologies may be vastly different. Establishing compatibility and smooth interactions between them is essential for successful integration.
- Security standards and best practices are necessary to ensure the safe integration of blockchain technology with quantum computing. As these technologies advance, rules and standards need to be developed to establish frameworks and ensure the resilience of integrated systems. Collaboration among researchers, industry professionals, and regulatory agencies is vital in this process.
- Ethical and regulatory considerations must be addressed. The combination of quantum computing and blockchain technology brings extraordinary capabilities but also raises concerns about privacy, data ownership, and misuse. Legal frameworks and ethical criteria should be established to oversee the responsible and accountable use of integrated systems.
- Another significant challenge is the early stage of quantum computing. Quantum computers are still in development and not yet commercially accessible. Successfully merging blockchain technology with quantum computing may require substantial investment in research and development.

- Existing cryptographic algorithms could be vulnerable to attacks by quantum computers. While quantum computing can enhance blockchain network security, it can also compromise existing cryptographic protocols. Blockchain networks need to enhance the security of their cryptographic algorithms to withstand the impact of quantum computing.
- Regulatory and compliance issues may arise. The use of quantum-resistant encryption may not align with existing standards in some industries. Certain uses of quantum computing may require authorization from regulatory organizations.
- Integrating blockchain and quantum computing requires significant modifications to existing infrastructure. Network design, hardware, and software changes are necessary for successful integration. Organizations need to invest in infrastructure modernization to effectively integrate these technologies.

7.9.3 Ethical Implications

The increasing capabilities of advanced AI systems to collect, store, and analyze vast amounts of data about individuals raise concerns about privacy. People may not be aware of what information is being collected, how it will be used, or who can access it. This lack of transparency creates worries about potential misuse, such as identity theft or targeted advertising. While advancements in AI have led to improvements in healthcare, education, transportation, and other areas, they also raise ethical concerns that need careful consideration. Two significant ethical issues currently are the potential for weaponization and invasion of privacy. These advancements have the potential to be used in harmful ways or intrude upon individuals' private lives, highlighting the importance of addressing these ethical concerns (Kop 2021; Johnson 2019; Sajwan and Jayapandian 2019; García et al. 2022; Prewett et al. 2020; Girasa and Scalabrini 2022b).

7.9.4 Discussion of Privacy Concerns

Integrating blockchain technology with quantum computing raises significant ethical concerns, particularly when it comes to protecting users' personal information. While blockchain technology offers immutability and transparency, there are uncertainties about the privacy of personal data recorded on the distributed ledger. Facial recognition technology is a prime example of the privacy issues surrounding AI. This technology uses algorithms to analyze and compare facial traits with a database of photos. While it may have potential benefits, such as aiding law enforcement in suspect identification, it also raises serious privacy concerns. Misuse of this technology can lead to wrongful arrests or convictions, which is unacceptable.

To address these concerns, collaboration between legislators and technologists is crucial to develop protections that preserve individuals' privacy rights while still allowing for innovation. This may involve enacting legislation that requires consent before collecting personal data and developing technologies that minimize the collection of personal information. Additionally, businesses may need to obtain individuals' consent before collecting their data.

The risk of private information being exposed is a significant aspect of privacy protection. Once data is recorded on a blockchain network, it cannot be erased or edited due to the permanent and public nature of the ledger. While transparency can be beneficial, especially in supply chain management, it can also pose risks to privacy, particularly when sensitive or personal data is involved. Measures must be implemented to ensure effective protection of sensitive information and to limit access to relevant data only to authorized individuals.

Furthermore, the integration of quantum computing introduces new challenges to privacy. Quantum computers have the potential to crack traditional encryption methods, jeopardizing the confidentiality of personal information stored on blockchains. To secure user privacy, it is crucial to develop and deploy encryption methods that are resistant to quantum computing. Privacy-enhancing techniques like zero-knowledge proofs and secure multiparty computing can also be employed to safeguard users' privacy within blockchain-based systems (Quach et al. 2022; Le and Hsu 2021; Ukpabi et al. 2022; Banerjee et al. 2020; Akhai 2023b).

7.9.5 Consideration of Potential Weaponization of These Technologies

The integration of artificial intelligence into the military gives rise to ethical considerations pertaining to accountability and the delegation of decisions about matters of life and death to computer systems. Ethical considerations have been brought out regarding the moral implications associated with the use of autonomous weapon systems, which possess the capability to make judgments without human intervention. Moreover, the use of artificial intelligence has the potential to facilitate the creation of altered videos, often referred to as deepfakes. These deepfakes may be employed with malicious intent, such as manipulating an audience or disseminating deceptive content. In order to tackle these issues, it is imperative for legislators and engineers to collaborate in formulating laws and developing technological solutions capable of detecting deepfakes and preventing the development and use of autonomous weaponry.

In the context of defense and military applications, striking a balance between innovation and safety is essential when harnessing advanced technologies like blockchain and quantum computing to mitigate possible issues. Although the blockchain technology has the capability to securely store data in a decentralized and immutable way, it also harbors the potential for exploitation in illegal activities. On the other hand, excessive use of quantum computing poses distinct risks to both national security and individuals' privacy. The use of blockchain networks' openness and transparency may prove advantageous; nevertheless, it also renders them susceptible to potential misuse. Furthermore, the computational capabilities of quantum computing have the potential to compromise the security of critical computer systems. The establishment of regulatory frameworks, implementation of quantum-resistant encryption, and engagement in ongoing research are necessary in order to safeguard the integrity of systems amidst escalating technological challenges. By doing so, it will ensure responsible use and mitigate any hazards. To optimize the potential benefits arising from emerging technologies while mitigating the potential for unexpected repercussions, it is essential for several stakeholders to collaborate together (Gill et al. 2022, 2020; Kumar et al. 2022b, 2022c; Kaushik and Kumar 2023; Akhai 2023a; Jain et al. 2021).

7.9.6 Conclusion

Defense and military applications of blockchain and quantum computing are of great interest. Blockchain protects data storage, communication, and supply chain management. This improves efficiency, information integrity, and military alliances. Quantum computing might transform military operations. Encryption and secure communication protocols may safeguard data, while resource allocation tactics can improve logistics and decision-making. Quantum simulations and models may improve intelligence analysis, strategic planning, and machine learning decision-making. Quantum cryptography, machine learning, and quantum chemical modeling studies in the defense and military might lead to major advances. These technologies improve data analysis, communication security, material science, and medical research. Quantum computing research must address scalability, algorithm design, qubit stability, and control system issues to meet defense and military needs. These sectors may give cutting-edge technology and methods to solve challenging military problems. To utilize AI technologies like autonomy and quantum computing ethically, cooperation must comply with ethics and regulations. User privacy, cryptographic security, and abuse avoidance need strict precautions. Thus following are the conclusions drawn:

- Blockchain technology offers secure and reliable solutions for data storage and transmission in the defense and military sectors. It addresses challenges related to secure communication, supply chain management, identity administration, and cybersecurity. Its applications can enhance operational efficiency, safeguard vital information integrity, and improve communication between military units and allied troops.
- Quantum computing has the potential to revolutionize defense and military operations. It can improve cryptography, secure communication, optimize resource allocation, enhance simulations and modeling, and enable more effective intelligence analysis and machine learning. Further research and development are needed to fully realize the capabilities of quantum computing in these sectors.
- Practical applications of quantum computing in defense and military sectors are being actively researched and developed. These include quantum cryptography

for secure communication, quantum machine learning for improved analysis, and quantum chemistry modeling for advancements in material science and medicinal research.
- The continuous progress in quantum computing research brings us closer to unlocking its full potential for defense and military applications. Overcoming challenges related to scalability, algorithm design, qubit stability, and control systems is essential for harnessing the transformative power of quantum computing in these sectors.
- Responsible and ethical utilization of technological advancements, including autonomy, AI, cyber warfare, and quantum computing, is crucial for maintaining global peace and security in defense and military industries. Quantum technologies, such as quantum machine learning, simulations, communication networks, and cryptographic methods, will shape the future of these sectors, leading to safer, more effective, and technologically advanced defense systems and tactics.
- Integrating blockchain technology with quantum computing offers significant potential but also poses challenges that need to be addressed. Organizations should carefully evaluate benefits and risks, invest in research and development, and upgrade infrastructure to ensure successful integration. Collaborative efforts, including the development of quantum-resistant encryption and consensus algorithms, interoperability solutions, standardized security frameworks, and adherence to ethical and regulatory norms, are crucial for overcoming these challenges.
- Ethical considerations are paramount when integrating blockchain technology and quantum computing. Protecting user privacy, addressing cryptographic security threats, and establishing regulatory frameworks and security measures to prevent misuse and defend against harm are essential. Responsible and beneficial use of AI technologies requires collaboration and the establishment of safeguards that protect individuals' rights.

Thus, the combination of blockchain technology and quantum computing holds great potential for enhancing the defense and military sectors. By addressing security concerns, optimizing operations, and embracing ethical practices, these technologies can contribute to safer, more efficient, and technologically advanced defense systems and tactics. Continued research, collaboration, and responsible implementation are key to harnessing the benefits of these powerful technologies.

References

Abd El-Latif AA, Abd-El-Atty B, Mehmood I, Muhammad K, Venegas-Andraca SE, Peng J (2021) Quantum-inspired blockchain-based cybersecurity: securing smart edge utilities in IoT-based smart cities. Inf Process Manage 58(4):102549

Acampora G (2020) Quantum computing meets Artificial Intelligence. In: AISEM annual conference on sensors and microsystems, pp 151–155. Springer International Publishing, Cham

Acín A, Bloch I, Buhrman H, Calarco T, Eichler C, Eisert J, Esteve D, Gisin N, Glaser SJ, Jelezko F, Kuhr S, Wilhelm FK (2018) The quantum technologies roadmap: a European community view. New J Phys 20(8):080201

Ahammed TB, Patgiri R, Nayak S (2023) A vision on the artificial intelligence for 6G communication. ICT Express 9(2):197–210

Ahmad RW, Hasan H, Yaqoob I, Salah K, Jayaraman R, Omar M (2021) Blockchain for aerospace and defense: opportunities and open research challenges. Comput Ind Eng 151:106982

Akhai S (2023a) From black boxes to transparent machines: the quest for explainable AI. SSRN 4390887

Akhai S (2023b) Healthcare record management for healthcare 4.0 via blockchain: a review of current applications, opportunities, challenges, and future potential. In: Blockchain for healthcare 4.0, pp 211–223

Azzaoui AE, Kim TW, Pan Y, Park JH (2021) A quantum approximate optimization algorithm based on blockchain heuristic approach for scalable and secure smart logistics systems. Hum-Centric Comput Inf Sci 11(1)

Banerjee S, Das D, Biswas M, Biswas U (2020) Study and survey on blockchain privacy and security issues. In: Cross-industry use of blockchain technology and opportunities for the future, pp 80–102. IGI Global

Bhatia A, Bibhu V, Lohani BP, Kushwaha PK (2020) An application framework for quantum computing using Artificial intelligence techniques. In: 2020 Research, innovation, knowledge management and technology application for business sustainability (INBUSH), pp 264–269. IEEE

Biswas R, Jiang Z, Kechezhi K, Knysh S, Mandra S, O'Gorman B, Perdomo-Ortiz A, Petukhov A, Realpe-Gómez J, Rieffel E, Venturelli D, Wang Z (2017) A NASA perspective on quantum computing: opportunities and challenges. Parallel Comput 64:81–98

Blahut RE (2014) Cryptography and secure communication. Cambridge University Press

Bontekoe TH, Neumann NM, Phillipson F, Wezeman S (2022) Quantum computing for radar and sonar information processing, vol 12093. In: Quantum information science, sensing, and computation XIV, pp 90–101. SPIE

Brandmeier RA, Heye JA, Woywod C (2022) Future development of quantum computing and its relevance to NATO. Connect: Q J 20:89–110

Broadbent A, Schaffner C (2016) Quantum cryptography beyond quantum key distribution. Des Codes Crypt 78:351–382

Broughton M, Verdon G, McCourt T, Martinez AJ, Yoo JH, Isakov SV, Massey P, Halavati R, Niu MY, Zlokapa A, Peters E, Mohseni M (2020) Tensorflow quantum: a software framework for quantum machine learning. arXiv:2003.02989

Buchmann N, Rathgeb C, Baier H, Busch C, Margraf M (2017) Enhancing breeder document long-term security using blockchain technology. In: 2017 IEEE 41st annual computer software and applications conference (COMPSAC), vol 2, pp 744–748. IEEE

Chamola V, Kotesh P, Agarwal A, Gupta N, Guizani M (2021) A comprehensive review of unmanned aerial vehicle attacks and neutralization techniques. Ad Hoc Netw 111:102324

Chen G (2022) Discussion on blockchain system attack and defense technology. Acad J Comput & Inf Sci 5(2):87–89

Chen L, Chen31 J (2021) Global military machine learning technology development tracking and evaluation. In: ECCWS 2021 20th European conference on cyber warfare and security, p 61. Academic Conferences Inter Ltd.

Cheung KF, Bell MG, Bhattacharjya J (2021) Cybersecurity in logistics and supply chain management: an overview and future research directions. Transp Res Part E Logist Transp Rev 146:102217

Chugh N, Sharma DK, Singhal R, Jain S, Srikanth P, Kumar A, Aggarwal A (2020) Blockchain-based decentralized application (DApp) design, implementation, and analysis with healthcare 4.0 trends, vol 126. In: Basic & clinical pharmacology & toxicology, pp 139–140. Wiley, 111 River St, Hoboken 07030-5774, NJ, USA

Córcoles AD, Kandala A, Javadi-Abhari A, McClure DT, Cross AW, Temme K, Nation PD, Steffen M, Gambetta JM (2019) Challenges and opportunities of near-term quantum computing systems. arXiv:1910.02894

Cui W, Dou T, Yan S (2020) Threats and opportunities: blockchain meets quantum computation. In: 2020 39th Chinese control conference (CCC), pp 5822–5824. IEEE

D'Auria V, Teller M (2023) What are the priorities and the points to be addressed by a legal framework for quantum technologies? Res Dir Quantum Technol 1:e9

Dadhich M, Tiwari H (2022) Quantum blockchain for smart society: applications, challenges, and opportunities. Adv Quantum Blockchain R-Time Appl 178–198

De Leon NP, Itoh KM, Kim D, Mehta KK, Northup TE, Paik H, Palmer BS, Samarth N, Sangtawesin S, Steuerman DW (2021) Materials challenges and opportunities for quantum computing hardware. Science 372(6539):eabb2823

Denison MK, Montgomery CJ, Sarofim AF, Bockelie MJ, Webster AG (2005) Computational modeling of a chemical demilitarization deactivation furnace system. Environ Eng Sci 22(2):232–240

Derose K (2023) Establishing the legal framework to regulate quantum computing technology. Cathol Univ J Law Technol 31(2):145–172

Drager SL, Walsh LD (2005) Quantum computing at the air force research laboratory information directorate, vol 5815. In: Quantum information and computation III, pp 1–12. SPIE

Dwivedi A, Saini GK, Musa UI (2023) Cybersecurity and prevention in the quantum era. In: 2023 2nd International conference for innovation in technology (INOCON), pp 1–6. IEEE

Edwards D, Rawat DB (2020) Quantum adversarial machine learning: Status, challenges and perspectives. In: 2020 Second IEEE international conference on trust, privacy and security in intelligent systems and applications (TPS-ISA), pp 128–133. IEEE

Ellison B, Evans D, Lytwyn M, Schwalbe J (2023) Without precedent: global emerging trends in nuclear and hypersonic weapons. J Policy & Strat 3(1):21–50

Ferenc K (2020) Security of encryption procedures and practical implications of building a quantum computer. AARMS–Academic Appl Res Mil Public Manag Sci 19(3):5–22

Fernandez-Carames TM, Fraga-Lamas P (2020) Towards post-quantum blockchain: a review on blockchain cryptography resistant to quantum computing attacks. IEEE Access 8:21091–21116

García DP, Cruz-Benito J, García-Peñalvo FJ (2022) Systematic literature review: quantum machine learning and its applications. arXiv:2201.04093

Gill SS, Kumar A, Singh H, Singh M, Kaur K, Usman M, Buyya R (2020) Quantum computing: a taxonomy systematic review and future directions. ACM Comput Surv

Gill SS, Kumar A, Singh H, Singh M, Kaur K, Usman M, Buyya R (2022) Quantum computing: a taxonomy, systematic review and future directions. Softw Pract Exp 52(1):66–114

Girasa R, Scalabrini GJ (2022) Regulation of innovative technologies: blockchain, Artificial Intelligence and quantum computing. Springer Nature

Girasa R, Scalabrini GJ (2022b) States' regulation of Artificial Intelligence. Regulation of innovative technologies: blockchain, Artificial Intelligence and quantum computing. Springer International Publishing, Cham, pp 115–143

Guoming L, Sheng L, Guowei Y (2017) The application of quantum cryptography in military GIS security. J Simul 5(5):61

Gupta RC (ed) (2015) Handbook of toxicology of chemical warfare agents. Academic Press

Gupta R, Sultania K, Singh P, Gupta A (2013) Security for wireless sensor networks in military operations. In: 2013 Fourth international conference on computing, communications and networking technologies (ICCCNT), pp 1–6. IEEE

Hadi HJ, Cao Y, Nisa KU, Jamil AM, Ni Q (2023) A comprehensive survey on security, privacy issues and emerging defence technologies for UAVs. J Netw Comput Appl 213:103607

Hasan MK, Alkhalifah A, Islam S, Babiker NB, Habib AA, Aman AHM, Hossain MA (2022) Blockchain technology on smart grid, energy trading, and big data: security issues, challenges, and recommendations. Wirel Commun Mob Comput 2022:1–26

Huizing A, Heiligers M, Dekker B, de Wit J, Cifola L, Harmanny R (2019) Deep learning for classification of mini-UAVs using micro-doppler spectrograms in cognitive radar. IEEE Aerosp Electron Syst Mag 34(11):46–56

Inglesant P, Jirotka M, Hartswood M (2018) Responsible innovation in quantum technologies applied to defence and national security. NQIT (Networked Quantum Information Technologies)

Jain S, Ahuja NJ, Srikanth P, Bhadane KV, Nagaiah B, Kumar A, Konstantinou C (2021) Blockchain and autonomous vehicles: recent advances and future directions. IEEE Access 9:130264–130328

Javadi-Abhari A (2017) Towards a scalable software stack for resource estimation and optimization in general-purpose quantum computers. Doctoral dissertation, Princeton University

Jhanwar A, Nene MJ (2021) Enhanced machine learning using quantum computing. In: 2021 Second international conference on electronics and sustainable communication systems (ICESC), pp 1407–1413. IEEE

Johnson WG (2019) Governance tools for the second quantum revolution. Jurimetrics 59(4):487–522

Justinia T (2019) Blockchain technologies: opportunities for solving real-world problems in healthcare and biomedical sciences. Acta Informatica Medica 27(4):284

Kaushik K, Kumar A (2023) Demystifying quantum blockchain for healthcare. Secur Priv 6(3):e284

Khalid ZM, Askar S (2021) Resistant Blockchain cryptography to quantum computing attacks. Int J Sci Bus 5(3):116–125

Kiktenko EO, Pozhar NO, Anufriev MN, Trushechkin AS, Yunusov RR, Kurochkin YV, Lvovsky AI, Fedorov AK (2018) Quantum-secured blockchain. Quantum Sci Technol 3(3):035004

Ko Y, Kim J, Duguma DG, Astillo PV, You I, Pau G (2021) Drone secure communication protocol for future sensitive applications in military zone. Sensors 21(6):2057

Koh L, Dolgui A, Sarkis J (2020) Blockchain in transport and logistics–paradigms and transitions. Int J Prod Res 58(7):2054–2062

Kop M (2021) Establishing a legal-ethical framework for quantum technology. Yale Law School, Yale Journal of Law & Technology (YJoLT), The Record

Krelina M (2021) Quantum technology for military applications. EPJ Quantum Technol 8(1):24

Kulkarni AU, Jain S, Kumar A (2022) Quantum computing and quantum blockchain: recent advancements, analysis and future directions. Quantum and blockchain for modern computing systems: vision and advancements: quantum and blockchain technologies: current trends and challenges. Springer International Publishing, Cham, pp 311–339

Kulshrestha RDS, Navy I N D I A N (2018) Military applications of blockchain technology. Age Blockchain: Collect Artic 21

Kumar A, Dadheech P, Singh V, Poonia RC, Raja L (2019) An improved quantum key distribution protocol for verification. J Discret Math Sci Cryptogr 22(4):491–498

Kumar A, Ottaviani C, Gill SS, Buyya R (2022c) Securing the future internet of things with post-quantum cryptography. Secur Priv 5(2):e200

Kumar A, de Jesus Pacheco DA, Kaushik K, Rodrigues JJ (2022) Futuristic view of the internet of quantum drones: review, challenges and research agenda, p 100487. Vehicular Communications

Kumar A, Gill SS, Abraham A (2022) Quantum and blockchain for modern computing systems: vision and advancements. Springer, Cham

Kumar A, Krishnamurthi R, Sharma G, Jain S, Srikanth P, Sharma K, Aneja N (2023) Revolutionizing modern networks: advances in AI, machine learning, and blockchain for quantum satellites and UAV-based communication. arXiv:2303.11753

Lanzagorta M, Uhlmann JK (2004) Quantum computational geometry, vol 5436. In: Quantum information and computation II, pp 332–339. SPIE

Le TV, Hsu CL (2021) A systematic literature review of blockchain technology: security properties, applications and challenges. J Internet Technol 22(4):789–802

Lele A (2021) Quantum technologies and military strategy. Springer International Publishing

Li W, He M, Haiquan S (2021) An overview of blockchain technology: applications, challenges and future trends. In: 2021 IEEE 11th international conference on electronics information and emergency communication (ICEIEC), pp 31–39. IEEE
Lilly B, Lilly S (2021) Weaponising blockchain: military applications of blockchain technology in the US, China and Russia. RUSI J 166(3):46–56
Lindsay J (2018) Why quantum computing will not destabilize international security: the political logic of cryptology. SSRN 3205507
Lindsay JR (2020) Demystifying the quantum threat: infrastructure, institutions, and intelligence advantage. Secur Stud 29(2):335–361
Lucas RF, Wagenbreth G, Tran JJ, Pratt DR, Davis DM (2013) Practical adiabatic quantum computing: implications for the simulation community. In: The proceedings of the interservice/industry simulation, training and education conference, Orlando, Florida
Mailloux LO, Lewis CD II, Riggs C, Grimaila MR (2016) Post-quantum cryptography: what advancements in quantum computing mean for it professionals. IT Prof 18(5):42–47
Majot A, Yampolskiy R (2015) Global catastrophic risk and security implications of quantum computers. Futures 72:17–26
Markidis S (2023) Programming quantum neural networks on NISQ systems: an overview of technologies and methodologies. Entropy 25(4):694
Mashatan A, Turetken O (2020) Preparing for the information security threat from quantum computers. MIS Q Exec 19(2)
Maslov D, Nam Y, Kim J (2018) An outlook for quantum computing [point of view]. Proc IEEE 107(1):5–10
Mathur GCA, Srivastava MSK, Prabu MI (2022) Leveraging technological advances in C4ISR to enhance situational awareness and decision making. Synergy 59
Mavroeidis V, Vishi K, Zych MD, Jøsang A (2018) The impact of quantum computing on present cryptography. arXiv:1804.00200
Melagraki G (2022) Reducing health & environmental impacts of chemical warfare agents: computational chemistry contributions. Chemosphere 288:132564
Mian S (2022) Foundations of artificial intelligence and applications. J Artif Intell Technol 2(1):1–2
Michal K (2021) Quantum technology for military applications. EPJ Quantum Technol 8(1)
Morgan J (2015) A web of drones: a 2040 strategy to reduce the United States dependance on space based capabilities. Doctoral dissertation, AIR University
Mori S (2018) US defense innovation and artificial intelligence. Asia-Pac Rev 25(2):16–44
Mosteanu NR, Faccia A (2021) Fintech frontiers in quantum computing, fractals, and blockchain distributed ledger: paradigm shifts and open innovation. J Open Innov Technol, Mark, Complex 7(1):19
Nakamitsu I (2019) The role of emerging technologies in military conflicts. Inter Press Service
Nawaz SJ, Sharma SK, Wyne S, Patwary MN, Asaduzzaman M (2019) Quantum machine learning for 6G communication networks: state-of-the-art and vision for the future. IEEE Access 7:46317–46350
Neumann NM, van Heesch MP, de Graaf P (2020) Quantum communication for military applications. arXiv:2011.04989
Neumann NM, van Heesch MP, Phillipson F, Smallegange AA (2021) Quantum computing for military applications. In: 2021 International conference on military communication and information systems (ICMCIS), pp 1–8. IEEE
Padamvathi V, Vardhan BV, Krishna AVN (2016) Quantum cryptography and quantum key distribution protocols: a survey. In: 2016 IEEE 6th international conference on advanced computing (IACC), pp 556–562. IEEE
Panwar LGDR (2015) Disruptive military technologies: an overview. Studies 4
Partida A, Criado R, Romance M (2021) Identity and access management resilience against intentional risk for blockchain-based IOT platforms. Electronics 10(4):378

Peña M, Llivisaca J, Siguenza-Guzman L (2020) Blockchain and its potential applications in food supply chain management in Ecuador. In: Advances in emerging trends and technologies: volume 1, pp 101–112. Springer International Publishing

Pirandola S, Andersen UL, Banchi L, Berta M, Bunandar D, Colbeck R, Englund D, Gehring T, Lupo C, Ottaviani C, Pereira JL, Wallden P (2020) Advances in quantum cryptography. Adv Opt Photonics 12(4):1012–1236

Prewett KW, Prescott GL, Phillips K (2020) Blockchain adoption is inevitable—barriers and risks remain. J Corp Account & Finance 31(2):21–28

Quach S, Thaichon P, Martin KD, Weaven S, Palmatier RW (2022) Digital technologies: tensions in privacy and data. J Acad Mark Sci 50(6):1299–1323

Raska M (2022) The AI wave in military affairs: enablers and constraints. Technological innovation and security: the impact on the strategic environment in East Asia, p 89

Reding DF, Eaton J (2020) Science and technology trends 2020–2040: exploring the S and T edge. NATO S and T Organization

Rosch-Grace D, Straub J (2021) Analysis of the necessity of quantum computing capacity development for national defense and homeland security. In: 2021 IEEE international symposium on technologies for homeland security (HST), pp 1–8. IEEE

Rosch-Grace D, Straub J (2022) Analysis of the likelihood of quantum computing proliferation. Technol Soc 68:101880

Rotta D, Sebastiano F, Charbon E, Prati E (2017) Quantum information density scaling and qubit operation time constraints of CMOS silicon-based quantum computer architectures. NPJ Quantum Inf 3(1):26

Saffman M (2016) Quantum computing with atomic qubits and Rydberg interactions: progress and challenges. J Phys B At, Mol Opt Phys 49(20):202001

Sajwan P, Jayapandian N (2019) Challenges and opportunities: quantum computing in machine learning. In: 2019 Third international conference on I-SMAC (IoT in social, mobile, analytics and cloud) (I-SMAC), pp 598–602. IEEE

Salau BA, Rawal A, Rawat DB (2022) Recent advances in artificial intelligence for wireless internet of things and cyber–physical systems: a comprehensive survey. IEEE Internet Things J 9(15):12916–12930

Sarmah SS (2018) Understanding blockchain technology. Comput Sci Eng 8(2):23–29

Sharma A, Vanjani P, Paliwal N, Basnayaka CMW, Jayakody DNK, Wang HC, Muthuchidambaranathan P (2020) Communication and networking technologies for UAVs: a survey. J Netw Comput Appl 168:102739

Sheth H, Dattani J (2019) Overview of blockchain technology. Asian J Converg Technol (AJCT). ISSN-2350-1146

Simić D, Marjanović M, Vitorović-Todorović M, Bauk S, Lazić D, Samolov A, Ristović N (2018) Nanotechnology for military applications: a survey of recent research in military technical institute. Sci Tech Rev 68(1):59–72

Singh H, Kaur K (2023) Role of nanotechnology in research fields: medical sciences, military & tribology-A review on recent advancements, grand challenges and perspectives. Mater Today Proc

Singh M, Dhara C, Kumar A, Gill SS, Uhlig S (2022) 13 Quantum Artificial Intelligence for the science of Artificial Intelligence, machine learning and blockchain in quantum satellite, drone and network, p 199

Singh M, Dhara C, Kumar A, Gill SS, Uhlig S (2022) Quantum artificial intelligence for the science of climate change. In: Artificial Intelligence, machine learning and blockchain in quantum satellite, drone and network, pp 199–207. CRC Press.

Song X, Wu Y, Ma Y, Cui Y, Gong G (2015) Military simulation big data: background, state of the art, and challenges. Math Probl Eng 2015

Srikanth P, Kumar A (2022) Secure quantum computing for healthcare sector: a short analysis. arXiv:2211.10027

Srivastava T, Bhushan B, Bhatt S, Haque AB (2022) Integration of quantum computing and blockchain technology: a cryptographic perspective. In: Multimedia technologies in the internet of things environment, vol 3, pp 197–228. Springer, Singapore

Subramanian T, Dhyani A, Kumar A, Gill SS (eds) (2022) Artificial Intelligence, machine learning and blockchain in quantum satellite, drone and network. CRC Press

Sultan A, Mushtaq MA, Abubakar M (2019) IOT security issues via blockchain: a review paper. In: Proceedings of the 2019 international conference on blockchain technology, pp 60–65

Sun L, Zhang H, Fang C (2021) Data security governance in the era of big data: status, challenges, and prospects. Data Sci Manag 2:41–44

Sweet KM (2021) Space-based offensive weapons: have policymakers discussed this enough? Online J Space Commun 3(6):5

Symons BC, Galvin D, Sahin E, Alexandrov V, Mensa S (2023) A practitioner's guide to quantum algorithms for optimisation problems. arXiv:2305.07323

Szabadföldi I (2021) Artificial Intelligence in military application-opportunities and challenges. Land Forces Acad Rev 26(2):157–165

Tedeschi P, Sciancalepore S, Di Pietro R (2022) Satellite-based communications security: a survey of threats, solutions, and research challenges. Comput Netw 109246

Tesvara C (2021) Modeling the reactivity of chemical warfare agents on metal oxides using computational chemistry methods. University of California, Los Angeles

Uddin W, Khan B, Dewan S, Das S (2022) Silicon-based qubit technology: progress and future prospects. Bull Mater Sci 45(1):46

Ukpabi D, Karjaluoto H, Bötticher A, Nikiforova A, Petrescu D, Schindler P, Valtenbergs V, Lehmann L, Yakaryilmaz A (2022) Framework for understanding quantum computing use cases from a multidisciplinary perspective and future research directions. arXiv:2212.13909

Unogwu OJ, Doshi R, Hiran KK, Mijwil MM (2022) Introduction to quantum-resistant blockchain. In: Advancements in quantum blockchain with real-time applications, pp 36–55. IGI Global

Ur Rasool R, Ahmad HF, Rafique W, Qayyum A, Qadir J, Anwar Z (2023) Quantum computing for healthcare: a review. Future Internet 15(3):94

van Deventer O, Spethmann N, Loeffler M, Amoretti M, van den Brink R, Bruno N, Comi P, Farrugia N, Gramegna M, Jenet A, Kassenberg B, Wilhelm-Mauch FK (2022) Towards European standards for quantum technologies. EPJ Quantum Technol 9(1):33

Varmantchaonala Moudina C, Fendji Kedieng Ebongue JL, Atemkeng M (2000) Quantum algorithms for combinatorial optimization problems: a comprehensive survey from 2000 to 2022. Jean Louis and Atemkeng, Marcel, quantum algorithms for combinatorial optimization problems: a comprehensive survey from

Wang L, Alexander CA (2020) Quantum science and quantum technology: progress and challenges. Am J Electr Electron Eng 8(2):43–50

Watson TF, Philips SG, Kawakami E, Ward DR, Scarlino P, Veldhorst M, Savage DE, Lagally MG, Friesen M, Coppersmith SN, Eriksson MA, Vandersypen LMK (2018. A programmable two-qubit quantum processor in silicon. Nature 555(7698):633–637

Whitfield JD, Biamonte J, Aspuru-Guzik A (2011) Simulation of electronic structure Hamiltonians using quantum computers. Mol Phys 109(5):735–750

Xu M, Ren X, Niyato D, Kang J, Qiu C, Xiong Z, Wang X, Leung VC (2023) When quantum information technologies meet blockchain in web 3.0. IEEE Network

Yaacoub JPA, Noura HN, Salman O, Chehab A (2022) Robotics cyber security: vulnerabilities, attacks, countermeasures, and recommendations. Int J Inf Secur 1–44

Younan M, Elhoseny M, Ali AA, Houssein EH (2021) Quantum chain of things (QCoT): a new paradigm for integrating quantum computing, blockchain, and Internet of Things. In: 2021 17th International computer engineering conference (ICENCO), pp 101–106. IEEE

Zaidi T, Sushma BS (2022) An overview of future applications of quantum computing. Artif Intell, Mach Learn Blockchain Quantum Satell, Drone Netw 127–138

Zhang H, Xie J, Shi J, Zhang Z, Fu X (2019) Sensor scheduling and resource allocation in distributed MIMO radar for joint target tracking and detection. IEEE Access 7:62387–62400

Zhang S, Li Y, Ge W, Shen X (2022) Military application of blockchain technology for future battlefield operations. In: International conference on cloud computing, Internet of Things, and computer applications (CICA 2022), vol 12303, pp 352–362. SPIE

Zheng X, Jiang D (2015) The quantum cryptography communication and military application. In: Proceedings of the 15th international conference on man–machine–environment system engineering, pp 267–273. Springer, Berlin

Zhu Y, Zhang X, Ju ZY, Wang CC (2020) A study of blockchain technology development and military application prospects. J Phys Conf Ser 1507(5):052018. (IOP Publishing)

Zhu S, Yu T, Xu T, Chen H, Dustdar S, Gigan S, Gunduz D, Hossain E, Jin Y, Lin F, Liu B, Pan Y (2023) Intelligent computing: the latest advances, challenges, and future. Intell Comput 2:0006

Integration of Blockchain and the Internet of Things in Healthcare Sector

8

Rajesh Singh, Anita Gehlot, Shaik Vaseem Akram, Ravindra Sharma, and Praveen Kumar Malik

Abstract

The benchmarking of blockchain-based Internet of Things (IoT) healthcare Industry 4.0 systems falls under the multi-criteria decision-making (MCDM) predicament owing to the analysis, relevance, and attribute nature of several security and privacy factors. IoT technology aims to simplify the collection of distributed data inside healthcare practices, speed information exchange, and increase knowledge processing across various collaborative organizations. IoT blockchain architecture offers an excellent platform for the secure dependable and efficient collection, delivery, and retention of healthcare-related data from networked devices. Blockchain's particular attributes, including Decentralized governance, security, privacy, transparency, and inviolability, are some of the primary grounds for using it in healthcare systems. Incorporating this technology into a healthcare facility involves the establishment of an extra blockchain components framework that is tailored to the healthcare institution's architecture. Integrating unique security solutions for edge networks inside the Internet of Medical Things (IoMT) has been created as a cutting-edge bio-analytical tool for optimizing human health by incorporating network-connected biomedical equipment with software applications. As an outcome of the combined

R. Singh · A. Gehlot
Uttaranchal Institute of Technology, Uttaranchal University, Dehradun, Uttarakhand 248007, India

S. V. Akram
Department of ECE, School of Engineering, SR University, Warangal, Telangana 506371, India

R. Sharma
Swami Rama Himalayan University, Dehradun, Uttarakhand, India

P. K. Malik (✉)
School of Electronics and Electrical Engineering, Lovely Professional University, Phagwara, Punjab, India
e-mail: pkmalikmeerut@gmail.com

A. Kumar et al. (eds.), *Sustainable Security Practices Using Blockchain, Quantum and Post-quantum Technologies for Real Time Applications*, Contributions to Environmental Sciences & Innovative Business Technology,
https://doi.org/10.1007/978-981-97-0088-2_8

efficacy of IoMT and blockchain, healthcare personnel's duties in diagnosing and securely sustaining patient data are simplified. It has the potential to revolutionize clinical research in the healthcare industry. Wearable sensors and an IoMT network allow patients, medical professionals, and caregivers to collaborate. IoT-driven fog computing is being developed in the healthcare industry to accelerate the delivery of services and facilities to the public at large, possibly safeguarding countless lives and radically enhancing the quality of healthcare. Healthcare, being one of the key businesses, has experienced several cybersecurity difficulties, including malware assaults and Distributed Denial-of-Service (DDoS) attacks. The increased demand for medical equipment and drugs has unfortunately given rise to illegal activities such as black marketing and counterfeiting. Blockchain technology has the potential to make it easier to develop a standardized database for collecting data across clinical trials while preserving the greatest degree of patient data privacy. Blockchain technology is used to securely store digital health records (DHR) while maintaining the integrity of the source data in order to protect patient identities and privacy. Data management, security, data sharing, and patient privacy are just a few of the Electronic Health Records (EHR)-related challenges that blockchain can address. Next-generation Healthcare IoT (HIoT) applications might be one of the sectors transformed by the blockchain network as a technical improvement.

Keywords

Personal health records (PHRs) • Internet of medical things (IoMT) • Blockchain-as-a-utility (BaaU) • Digital health records (DHR) • Electronic health records (EHR) • MedRec • Physically unclonable functions (PUF) • Cybersecurity • Distributed denial-of-service (DDoS)

8.1 Introduction

The healthcare sector is leveraging the capabilities of AI, IoT, blockchain, and big data technologies to successfully meet the difficulties posed by the COVID-19 pandemic (Bazel et al. 2021). To address this issue, the creation of a one-of-a-kind system characterized by the combination of blockchain and AI technologies, dubbed Blockchain- and AI-enabled Secure Medical Data Transfer (BAISMDT), offers a promising novel approach to effectively address the security concerns raised by Internet of Things (IoT) networks in the healthcare area (Alzubi et al. 2021). IoT technology seeks to ease the collecting of dispersed data inside healthcare practices, streamline information interchange, and improve knowledge processing across varied collaboration entities. This is accomplished by using appropriate corporate information systems to support these functions (Pal 2022). Numerous technologies, including IoT and blockchain, play critical roles in this dynamic ecosystem. These technologies are used in a variety of daily healthcare operations, contributing to their advancement and innovation (Azbeg et al. 2022).

In this regard, our focus is on the architecture that combines IoT and blockchain for healthcare monitoring applications. This IoT blockchain architecture provides a strong foundation for the secure, dependable, and efficient gathering, transfer, and storage of healthcare-related data from networked devices (Attia et al. 2019). In recent years, a large amount of healthcare data has been obtained from a multitude of sources, including varied equipment, smart chips, small sensors, and wearable devices. These interconnected tools and items, known as elements of healthcare technology, are linked via the internet and are vital components of the current healthcare setting (Abounassar et al. 2022). IoT has the ability to expedite disruptive changes in the healthcare landscape and boost patient participation and outcomes. It has the capacity to switch healthcare from a reactive to a proactive strategy, altering access and care delivery (Salih et al. 2019). IoT is moving the healthcare business forward by facilitating device connectivity regardless of location, resulting in a highly integrated ecosystem. Nonetheless, in an IoT-enabled healthcare infrastructure, ensuring healthcare data security and sustaining user privacy remain top priorities (Das and Namasudra 2022). IoT promises numerous advantages in healthcare, including accurate patient monitoring and data-driven analytics. In the context of medical device integration, the attention is turning to the consumer sector, encompassing devices such as blood pressure cuffs, glucose meters, and multiple devices designed for collecting patient vitals and health data (Chacko and Hayajneh 2018). IoT will be critical in satisfying the significant data transfer demands of many cognitive healthcare applications. Predicting security performance is critical for mobile IoT-enabled healthcare networks to tackle complicated healthcare security situations in real time (Xu et al. 2021).

8.1.1 Internet of Things (IoT) and Blockchain-Integrated Healthcare

The benchmarking of blockchain-based IoT healthcare Industry 4.0 systems falls under the multi-criteria decision-making (MCDM) predicament owing to the analysis, relevance, and attribute nature of several security and privacy factors (Qahtan et al. 2022). A peer-to-peer, secure, and adaptive transaction system can be developed with blockchain technology. It has an enormous opportunity to affect healthcare systems as well since it is a transversal technology that has transformed numerous sectors (Soltanisehat et al. 2020). Medical data is presently stored in the cloud, but there are concerns with security, privacy, and trust that have not been addressed. Therefore, blockchain-based cloud technology may tackle multiple issues with the retention of data relevant to the healthcare sector (Fatima et al. 2022). The majority of conventional machine learning algorithms and techniques for IoT in healthcare lack concrete applications for secure data sharing. Blockchain is therefore introduced for secure and reliable interactions in IoT for healthcare (Shukla et al. 2021). The Blockchain's distinctive characteristics, notably Decentralization, Security, Privacy, Integrity, and Inviolability are a few of the principal justifications to employ it in healthcare systems (Ratta et al. 2021). Blockchain and

IoT technologies have the potential to improve essential management, operational efficiency, data security, accountability, and the expansion of possible business avenues in the healthcare sector (Farouk et al. 2020). Blockchain technology has emerged as a critical tool for strengthening user privacy and confidentiality, demonstrating its significance in current healthcare systems. Its use in optimizing medical record administration has the ability to reduce the healthcare system's complexity and expenses (Awotunde et al. 2022a). Incorporating this technology into a hospital demands the development of an additional blockchain components framework that is adapted to the infrastructure of the medical institution. The goal is to ensure all data's cryptographic secrecy, prohibiting unauthorized access (Abiodun et al. 2022). IoT and blockchain technologies have enormous potential in areas such as remote patient monitoring, full patient medical records, prescription fulfilment monitoring, and secure access and storage of crucial medical data (Saxena et al. 2022). With a rising number of applications in the healthcare industry, the implementation of blockchain technology must be done with extreme caution to avoid any negative effects on human health concerns (Schinckus 2022). Incorporating medical professionals into a blockchain system may offer them access to patient data while also ensuring a secure and traceable patient journey. Individual product tracking via blockchain could benefit pharmaceutical companies (Haleem et al. 2021). This combination capitalizes on blockchain's particular characteristics, such as decentralized management, consistency, accountability, and integrity, which contribute to a strong framework for healthcare data management (Arbabi et al. 2022). Blockchain technology creates a self-governing, decentralized, and authority-independent ecosystem. Given its potential to manage the multifarious concerns effectively, blockchain is widely lauded as the superior solution to complex security and interoperability challenges in the healthcare sector (Huang and Al Foysal 2021). Among other application areas, healthcare stands out as a crucial field where blockchain is ready to make a significant influence. This technology is bringing in a plethora of prospects and potential in today's healthcare systems (Kumar et al. 2018). Complicated intermediary networks and operational accountability gaps frequently raise issues in healthcare systems. Blockchain technology has the potential to alleviate these challenges by creating incentive structures and transparency mechanisms that can improve system efficiency and accountability (Akbar et al. 2022). Many emerging technologies, such as Physically Unclonable Functions (PUF), Blockchain, Artificial Intelligence (AI), and Software-Defined Networking (SDN), are expected to be essential in addressing numerous challenges in the field of e-healthcare. These challenges include issues like privacy, security, efficiency, and precision (Razdan and Sharma 2022) (Fig. 8.1).

8.1.2 Blockchain-Integrated Internet of Medical Things (IoMT)

Leveraging AI enabled by IoT-based technologies in the healthcare sector offers the potential to speed up disease detection and tracking. This technique can greatly

Fig. 8.1 Advantages of blockchain technology

reduce the burden on medical procedures, notably in the areas of disease diagnosis, forecasting, and management (Awotunde et al. 2022b). Blockchain technology has been identified as a transformational innovation capable of integrating into unique security solutions for edge networks inside the Internet of Medical Things (IoMT). It has the potential to be critical in improving the security of IoMT devices and preventing unwanted data transmission (Pelekoudas-Oikonomou et al. 2022). IoMT was created as a cutting-edge bio-analytical tool, combining network-connected biomedical devices with software applications to improve human health (Manickam et al. 2022). IoMT is assisting the healthcare business by boosting the accuracy, dependability, and efficiency of electronic devices. While IoT encompasses many disciplines, the currently specific focus is on IoT research contributions in the healthcare sector (Joyia et al. 2017). These applications are mobile, allowing them to move between sites while using multiple wireless networks. Nonetheless, when compared to the more straightforward coarse-grained and fine-grained workloads generally associated with IoMT, the operational operations of healthcare applications are far more complex (Lakhan et al. 2021). The aim of IoMT and digital healthcare systems is to provide individuals with the ease of getting high-quality medical treatment from the comfort of their homes. Improving these systems' intelligence and efficiency to enable precise forecasts of critical illnesses has the potential to save countless lives while relieving pressure on traditional healthcare frameworks (Ashfaq et al. 2022). As a result, the combined effectiveness of IoMT and blockchain simplifies the duty of healthcare personnel in diagnosing and securely managing patient data. IoMT incorporates AI to enable data-driven decision-making (Almalki et al. 2022). IoMT provides numerous benefits, including the ability to provide remote healthcare by remotely monitoring patients' vital health signs, providing healthcare assistance to senior citizens, and conducting large-scale health tracking within a region or country to effectively identify and manage epidemics (Naresh et al. 2020). IoMT application integration has the potential to transform healthcare by enabling remote patient monitoring, individualized treatment plans, and early illness identification. According to research, IoMT can improve patient outcomes, decrease hospital readmissions, and streamline healthcare delivery (Dwivedi et al. 2022). It has the potential to change the course of clinical research in the healthcare business. Patients, medical professionals, and carers can cooperate via wearable sensors and an IoMT network that incorporates data, software, and communication technologies (Al-Sit et al. 2023). Accurate prediction of human disorders, particularly life-threatening conditions like cancer, is critical for human welfare in the age of Smart Healthcare Industry 5.0. Because of the substantial expansion and implementation of IoMT networks, security, and privacy are critical aspects of the IoMT environment (Abbas et al. 2023). IoMT has realized and continues to realize P4 Medicine (Personalized, Participatory, Predictive, and Preventive) delivery, reaching even remote locations with the use of interconnected sensors and equipment that enable real-time patient care (Venkatesh 2019). IoMT is critical in providing rapid emergency responses and improving the quality of medical services while reducing costs. However, as modern technology advances, the rising popularity of medical devices brings serious security and data privacy problems. These reservations originate from the device's resource constraints and open connectivity (Nayak et al. 2022). IoMT and machine intelligence domains have been critical in enabling the implementation of intelligent tele-healthcare and smart healthcare approaches, resulting in advancements such as expedited diagnosis, improved disease management, and enhanced treatment options for various health conditions, even during difficult and uncertain times (Qureshi et al. 2022). Furthermore, wearable healthcare equipment can be programmed to generate emergency notifications for carers. When these IoMT-powered devices are connected and matched with the patient's healthcare monitoring or hospital system, they hold the possibility of giving essential assistance in the therapeutic decision-making process (Mitra et al. 2022).

8.1.3 Fog/Edge Computing e-Healthcare Management

Fog computing emerges as a response to the issues brought by cloud computing, providing computing and storage services at the network's edge (Awaisi et al. 2020). Edge computing has made significant contributions in a variety of domains, including lowering latency, establishing extensive device connectivity, and achieving increased data transmission rates within networks, all of which are critical components for ensuring highly reliable communication in the context of a smart healthcare system (Awotunde et al. 2022c). Healthcare is a latency-sensitive application domain. As a result, the strategic integration of fog computing within this sector is critical. Thorough analytics and research have the potential to result in better care, improved treatment techniques, and increased patient satisfaction (Kaur et al. 2021). The decreasing of latency is an important feature for computing systems that enable flawless healthcare procedures. In this sense, fog computing outperforms cloud computing due to its use of low-end computers, mobile phones, and personal devices, all of which contribute to lower latency (Elhadad et al. 2022). Making essential decisions involving vital resources such as medical conditions, ambulances, hospitals, and doctors, which directly affect patients' lives, might be hampered in specific emergency scenarios. To address these challenges, optimal computing solutions such as cloud computing, edge computing, and fog computing have been proposed (Quy et al. 2022). The evolving paradigms of fog and edge computing present novel solutions by bringing resources closer to users. This technique distinguishes itself from traditional cloud-based domains by providing low-latency and energy-efficient data processing solutions (Tuli et al. 2020). Since its beginnings, fog computing has a distinctive computing paradigm that garnered substantial attention. Within the healthcare industry, IoT-driven fog computing is being developed to accelerate the delivery of services and facilities to the general population, potentially saving countless lives and radically enhancing healthcare results (Mani et al. 2020). A fog device strategically located at the network's peripheral, facilitating data collection and processing close to its intended end-users. Many systems, particularly in the field of e-health, rely on real-time data to monitor patients, surroundings, and medical workers, with the overriding goal of improving procedures and safety measures (Shukla et al. 2021). It is now possible to reduce mortality rates and increase average lifespans by implementing a Smart Healthcare System. This holistic approach includes eHealth, mHealth services, smart home services, electronic databases, and intelligent medical equipment, all of which work together to improve healthcare outcomes (Tiwari and Sharma 2022). The use of mobile-edge computing (MEC) and blockchain technology in healthcare has resulted in significant changes in e-health services, with the goal of streamlining healthcare processes. MEC has an authentication method based on smart contracts, allowing for decentralized user access verification at the network edge without the requirement for a central governing authority (Nguyen et al. 2021). Patient data is extremely valuable, making safe transmission an essential component of smart healthcare systems. Recent advancements in digital and wireless communication have signaled a watershed moment in the e-healthcare business, allowing for the efficient and secure sharing of patient data (Pirbhulal et al. 2019) (Fig. 8.2).

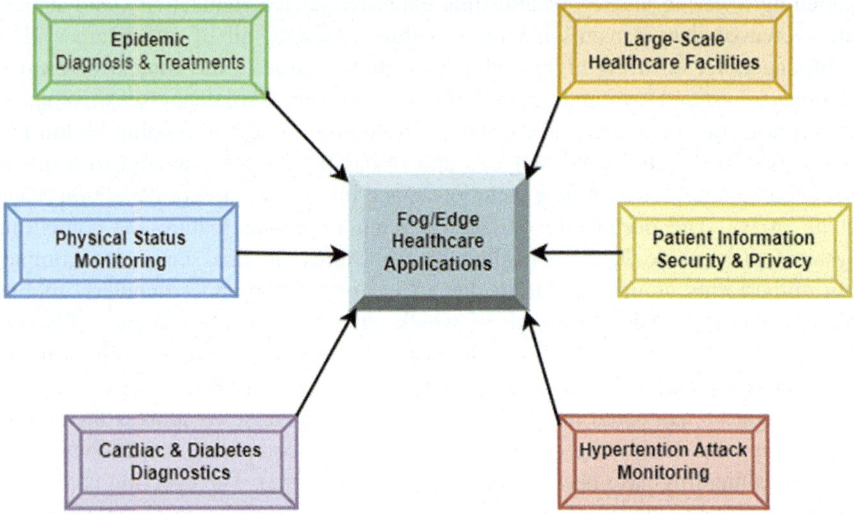

Fig. 8.2 Application of fog/edge technology in healthcare sector

8.1.4 Blockchain-Integrated Healthcare

As one of the major industries, healthcare has faced numerous cybersecurity challenges, including malware attacks and Distributed Denial-of-Service (DDoS) attacks. To overcome this problem, the healthcare industry must upgrade its existing security methods and embrace blockchain-based security solutions that offer more protection against such threats (Nwosu et al. 2021). The substantial quantity of sensitive data incorporated in these current records undoubtedly draws the attention of hostile hackers. As a result, securing the dependability and security of servers and processes is critical. Blockchain technology is emerging as a key step forward in the evolution of personal health information, promising greater security and data integrity (Girardi et al. 2020). eHealth services are not immune to the security threats that traditional services face. However, the goal of blockchain technology is to provide a secure and orderly system for protecting an organization's data. Medical records can be conveniently structured and securely exchanged using blockchain, contributing to improved dissemination and accessibility (Srivastava et al. 2022). The rising demand for medical equipment and medications has unfortunately given rise to criminal activities such as black marketing and counterfeiting. However, the use of technology may provide a viable alternative by eliminating direct human interaction and thus the risk of contamination (Kumar and Pundir 2020). In an alarming pattern, counterfeiters are investing considerable time, resources, and effort into infiltrating and abusing healthcare data. Blockchain technology can make it easier to create a uniform database for aggregating data throughout clinical investigations while maintaining the highest level

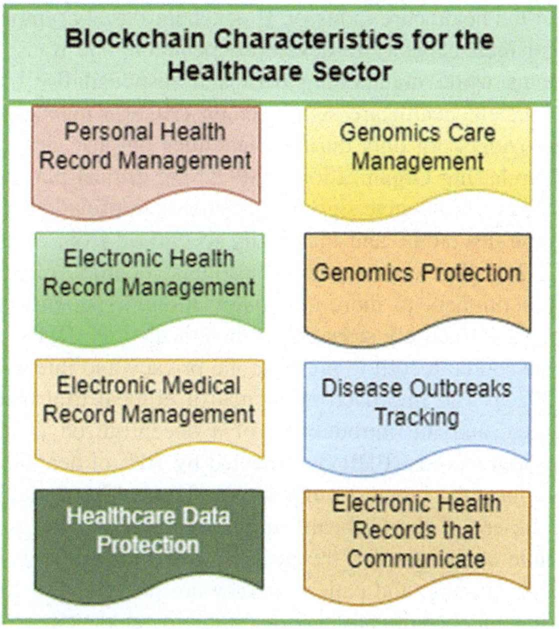

Fig. 8.3 Blockchain capabilities for healthcare sector

of patient data protection (Quasim et al. 2020). Blockchain has tremendous potential as a transformational technology that can be integrated into healthcare systems to improve capabilities across all aspects of the sector. (Teli and Masoodi 2021) (Fig. 8.3).

The pharmaceutical supply chain (PSC) has witnessed rapid alterations and increased vulnerability to disruptions. While the integration of IoT and blockchain technology has been intensively researched in the areas of supply chain and logistics management, there is growing hope for future breakthroughs in the pharmaceutical supply chain (Chen et al. 2023). Digital supply chain integration has turned into a critical factor for business competitive differentiation. Enterprises at the vanguard of this digital supply chain approach are quickly maximizing their assets and bringing in fresh, data-driven contributions that fuel their success (Pundir et al. 2019). Since the healthcare supply chain is fragmented, it can be tricky to put into practice and carry out these activities in a secure, reputable, effective, economical, and easily attainable manner. A revolutionary method called blockchain, which functions in coordination with IoT, might be able to deliver a viable solution to these challenges (Jayaraman et al. 2019). Considering IoT devices have restricted memory, performance, and computational power, relying only on traditional cryptographic algorithms for security is not a viable option. Smart contracts help with authorization and access control, data management, and device authentication. In doing so, they present a more practical and forward-thinking solution to IoT security challenges (Satamraju 2020). Inadequate data management interrupts supply chain continuity, causing long-term consequences in the healthcare industry. Blockchain-based cryptocurrencies are well recognized for their capacity to develop dependable and transparent solutions to these problems while maintaining trust and responsibility (Jadhav and Deshmukh 2022). Efficient healthcare supply chain (HCSC) management is critical for healthcare providers not only during pandemics but also during everyday operations. Group Purchasing Organizations (GPOs) are critical players in the HCSC system. These organizations give significant benefits to providers, including cost reductions, volume discounts, and the ability to choose from a variety of vendors (Omar et al. 2021). Blockchain allows hospitals, doctors, pharmacists, and other healthcare practitioners to more effectively monitor patient data. Blockchain technology is used to securely store digital health records (DHR) while keeping the integrity of the source record to preserve the privacy and integrity of patient identities (Samad 2022). The improved efficiency of clinical trial management, regulatory compliance, and the introduction of a decentralized framework for sharing electronic health records (EHR) is expected by 70% of healthcare leaders, according to IBM, to have the greatest influence on the healthcare industry (Hasselgren et al. 2020). The era of digital health transformation and the rise of EHR records pose multiple challenges for the healthcare systems of today. Data management, security, data sharing, and patient privacy are just a few of the issues associated with EHR that blockchain has the potential to tackle (Bennacer et al. 2023). The implementation of substantial redesigns for EHRs has typically been impeded by a long-standing commitment to compliance. MedRec manages identification, confidentiality, transparency, and data sharing—critical components when handling critical information—by employing unique blockchain features (Ekblaw et al. 2016). With low-spent hierarchical key distribution, a blockchain-based hierarchical data sharing framework (BHDSF) provides immediate retrieval of encrypted Personal Health Records (PHRs) and highly refined access control (Zhang et al. 2022). Next-generation Healthcare IoT (HIoT) applications may constitute one of the industries that the blockchain network is likely to transform as a technological advancement. Blockchain-as-a-utility (BaaU) for HIoT systems, however, necessitates a structured framework (Almagrabi et al. 2021). Integrating cloud servers for data storage and cloud computing has culminated in more effective and economical medical resource management as well as enhanced results for patients (Mehbodniya et al. 2022).

8.2 Recommendations

Elderly Care: IoT devices can be used to monitor the senior population's health and well-being, allowing them to live independently while yet receiving critical medical care. Some uses in this setting include fall detection, prescription reminders, and vital sign monitoring.

Personalized Treatment: Data obtained from IoT devices can be utilized to formulate patient-specific medicines. Machine learning algorithms can examine data determine trends and anticipate potential health issues, allowing for preventive measures.

Data-Driven Insights: IoT devices produce huge amounts of data, which can be used to acquire insights into population health patterns, medication efficacy, and prescription adherence. This information can help progress medical studies and healthcare legislation.

Remote Patient Monitoring: The Internet of Medical Things (IoMT) enables healthcare providers to observe the vital signs and symptoms of their patients in real-time. Individuals with chronic conditions, post-operative patients, and the elderly could reap the rewards from constant surveillance without the need for recurrent hospital visits.

Epidemic Detection and Mitigation: IoMT can help identify possible epidemics upfront by collecting and analyzing health data from a large number of people in a certain region or country. Patterns and abnormalities in health data can provide critical insights for public health officials to take preventive steps.

Efficient Healthcare Delivery: IoMT aids in the simplification of healthcare delivery processes. Medical personnel can remotely view patient data, decreasing the requirement for physical presence and allowing for faster decision-making.

Reduced Healthcare Costs: IoMT has the potential to safeguard both patients and healthcare systems financially by reducing hospital readmissions, enabling early intervention, and improving overall health management.

Telemedicine Advancement: IoMT promotes the expansion of telemedicine and virtual consultations, allowing patients to speak with healthcare practitioners remotely, share vital signs, and receive medical advice without having to travel.

Chronic Disease Management: Patients with chronic diseases like as diabetes, hypertension, and heart disease can employ IoMT devices to better manage their symptoms and reduce the risk of consequences.

Improved Patient Engagement: IoMT allows patients to be more involved in their treatment. Access to their health data in real time helps individuals to make more informed decisions and live better lifestyles.

8.3 Conclusion

Remote health monitoring technology, telemedicine platforms, intelligent devices, robotics, and drones have all enhanced the healthcare system tremendously. These developments are critical in disease prevention because they enable improved quality of life, early diagnosis, screening, and effective therapy. This paper presents an overview of recent developments and discoveries, mainly in AI-supported Internet of Medical Things (IoMT) devices developed for extremely efficient bio-sensing, a critical prerequisite for illness treatment success. A healthcare industry 5.0 IoMT-enabled intelligent system is now being developed with the goal of rapidly and precisely predicting malignant illnesses while maintaining the highest level of patient privacy. AI-enabled IoMT devices serve a key role in essential medical areas such as surgical operations, diabetes control, cardiac monitoring, and cancer surveillance. Cloud computing has shown tremendous promise in monitoring cardiac electrophysiology and medical imaging. IoMT provides the ability to alleviate these constraints by reducing the amount of time healthcare staff engage in repetitive tasks. This frees them up to focus on other activities, such as seeing more patients. Ambient Assisted Living (AAL) is a new industry that employs AI to offer unique goods, procedures, and services. These advancements are meant to provide safe, high-quality healthcare support while increasing patient autonomy in their everyday activities. The integration of blockchain technology in healthcare research has established itself as a recognized academic domain, with both the quantity and quality of publications expanding significantly. Blockchain has immense potential to transform the present healthcare environment in areas like medical record storage and retrieval, as well as treatments and pharmaceutical supply chain management. It enhances the security of electronic medical records, encourages the economic use of health data, improves collaboration and interoperability among healthcare facilities, and supports the effort against spurious pharmaceuticals. Blockchain's key traits of decentralization, anonymity, and communal vigilance have the potential to alter the business. As a result, it is anticipated that it will cause a shift in the healthcare business, perhaps strengthening process quality while minimizing costs. Further efforts are being made to expand the blockchain-MEC paradigm to many other healthcare systems, including responsibilities such as interpreting voice and video data and safeguarding the security of real-time health monitoring systems.

References

Abbas T, Fatima A, Shahzad T, Alissa K, Ghazal TM, Al-Sakhnini MM, et al. (2023) Secure IoMT for disease prediction empowered with transfer learning in healthcare 5.0, the concept and case study. IEEE Access

Abiodun KM, Adeniyi EA, Awotunde JB, Chakraborty C, Aremu DR, Adebiyi AA, Adebiyi MO (2022) Blockchain and internet of things in healthcare systems: prospects, issues, and challenges. In: Digital health transformation with blockchain and artificial intelligence. CRC Press, pp 1–22

Abounassar EM, El-Kafrawy P, Abd El-Latif AA (2022) Security and interoperability issues with internet of things (IoT) in healthcare industry: a survey. Security and Privacy Preserving for IoT and 5G Networks: Techniques, Challenges, and New Directions, pp 159–189

Akbar MA, Leiva V, Rafi S, Qadri SF, Mahmood S, Alsanad A (2022) Towards roadmap to implement blockchain in healthcare systems based on a maturity model. J Softw: Evol Process 34(12):e2500

Akram SV, Malik PK, Singh R, Gehlot A, Juyal A, Ghafoor KZ, Shrestha S (2022) Implementation of digitalized technologies for fashion industry 4.0: opportunities and challenges, scientific programming, Hindawi, 2022/08/17, https://doi.org/10.1155/2022/7523246

Almagrabi AO, Ali R, Alghazzawi D, AlBarakati A, Khurshaid T (2021) Blockchain-as-a-utility for next-generation healthcare internet of things. Comput Mater Continua 68(1)

Almalki J, Al Shehri W, Mehmood R, Alsaif K, Alshahrani SM, Jannah N, Khan NA (2022) Enabling blockchain with IoMT devices for healthcare. Information 13(10):448

Al-Sit WT, Al-Dmour NA, Ghazal TM, Issa GF (2023) Iomt-based healthcare framework for ambient assisted living using a convolutional neural network. CMC Comput Mater Contin 74(3):6867–6878

Alzubi OA, Alzubi JA, Shankar K, Gupta D (2021) Blockchain and artificial intelligence enabled privacy-preserving medical data transmission in Internet of Things. Trans Emerg Telecommun Technol 32(12):e4360

Arbabi MS, Lal C, Veeraragavan NR, Marijan D, Nygård JF, Vitenberg R (2022) A survey on blockchain for healthcare: Challenges, benefits, and future directions. IEEE Commun Surv Tutor

Ashfaq Z, Rafay A, Mumtaz R, Zaidi SMH, Saleem H, Zaidi SAR et al (2022) A review of enabling technologies for Internet of Medical Things (IoMT) ecosystem. Ain Shams Eng J 13(4):101660

Attia O, Khoufi I, Laouiti A, Adjih C (2019) An IoT-blockchain architecture based on hyperledger framework for health care monitoring application. In: NTMS 2019–10th IFIP international conference on new technologies, mobility and security. IEEE Computer Society, pp 1–5

Awaisi KS, Hussain S, Ahmed M, Khan AA, Ahmed G (2020) Leveraging IoT and fog computing in healthcare systems. IEEE Internet Things Mag 3(2):52–56

Awotunde JB, Misra S, Ayoade OB, Ogundokun RO, Abiodun MK (2022a) Blockchain-based framework for secure medical information in internet of things system. Blockchain Applications in the Smart Era. Springer International Publishing, Cham, pp 147–169

Awotunde JB, Folorunso SO, Ajagbe SA, Garg J, Ajamu GJ (2022) AiIoMT: IoMT-based system-enabled artificial intelligence for enhanced smart healthcare systems. In: Machine learning for critical internet of medical things: applications and use cases, pp 229–254

Awotunde JB, Ijaz MF, Bhoi AK, AbdulRaheem M, Oladipo ID, Barsocchi P (2022) Edge-IoMT-based enabled architecture for smart healthcare system. In: 5G IoT and edge computing for smart healthcare. Academic Press, pp 1–27

Azbeg K, Ouchetto O, Andaloussi SJ, Fetjah L (2022) A taxonomic review of the use of IoT and blockchain in healthcare applications. Irbm 43(5):511–519

Bazel MA, Mohammed F, Alsabaiy M, Abualrejal HM (2021) The role of internet of things, blockchain, artificial intelligence, and big data technologies in healthcare to prevent the spread of the Covid-19. In: 2021 international conference on innovation and intelligence for informatics, computing, and technologies (3ICT). IEEE. pp 455–462

Bennacer SA, Sabiri K, Aaroud A, Akodadi K, Cherradi B (2023) A comprehensive survey on blockchain-based healthcare industry: applications and challenges. Indones J Electr Eng Comput Sci 30:1558–1571

Chacko A, Hayajneh T (2018) Security and privacy issues with IoT in healthcare. EAI Endorsed Trans Pervasive Health Technol 4(14)

Chen X, He C, Chen Y, Xie Z (2023) Internet of Things (IoT)—blockchain-enabled pharmaceutical supply chain resilience in the post-pandemic era. Front Eng Manag 10(1):82–95

Das S, Namasudra S (2022) A novel hybrid encryption method to secure healthcare data in IoT-enabled healthcare infrastructure. Comput Electr Eng 101:107991

Dwivedi R, Mehrotra D, Chandra S (2022) Potential of Internet of Medical Things (IoMT) applications in building a smart healthcare system: a systematic review. J Oral Biol Craniofacial Res 12(2):302–318

Ekblaw A, Azaria A, Halamka JD, Lippman A (2016) A case study for blockchain in healthcare: "MedRec" prototype for electronic health records and medical research data. In: Proc IEEE Open Big Data Conference, Vol 13, p 13

Elhadad A, Alanazi F, Taloba AI, Abozeid A (2022) Fog computing service in the healthcare monitoring system for managing the real-time notification. J Healthcare Eng

Farouk A, Alahmadi A, Ghose S, Mashatan A (2020) Blockchain platform for industrial healthcare: Vision and future opportunities. Comput Commun 154:223–235

Fatima N, Agarwal P, Sohail SS (2022) Security and privacy issues of blockchain technology in health care—A review. ICT Anal Appl 193–201

Girardi F, De Gennaro G, Colizzi L, Convertini N (2020) Improving the healthcare effectiveness: The possible role of EHR IoMT and Blockchain. Electronics 9(6):884

Haleem A, Javaid M, Singh RP, Suman R, Rab S (2021) Blockchain technology applications in healthcare: an overview. Int J Intell Netw 2:130–139

Hasselgren A, Kralevska K, Gligoroski D, Pedersen SA, Faxvaag A (2020) Blockchain in healthcare and health sciences—a scoping review. Int J Med Informatics 134:104040

Huang G, Al Foysal A (2021) Blockchain in Healthcare

Jadhav JS, Deshmukh J (2022) A review study of the blockchain-based healthcare supply chain. Soc Sci HumIties Open 6(1):100328

Jayaraman R, Salah K, King N (2019) Improving opportunities in healthcare supply chain processes via the internet of things and blockchain technology. Int J Healthc Inf Syst Inform (IJHISI) 14(2):49–65

Joyia GJ, Liaqat RM, Farooq A, Rehman S (2017) Internet of medical things (IoMT): applications, benefits and future challenges in healthcare domain. J Commun 12(4):240–247

Kaur J, Verma R, Alharbe NR, Agrawal A, Khan RA (2021) Importance of fog computing in healthcare 4.0. Fog Computing for Healthcare 4.0 Environments: technical, societal, and future implications, pp 79–101

Kimothi S, Thapliyal A, Gehlot A, Aledaily AN, Gupta A, Bilandi N, Singh R, Malik PK, Akram SV (2023) Spatio-temporal fluctuations analysis of land surface temperature (LST) using Remote Sensing data (LANDSAT TM5/8) and multifractal technique to characterize the urban heat Islands (UHIs). Sustain Energy Technol Assess 55:102956. ISSN 2213-1388, https://doi.org/10.1016/j.seta.2022.102956

Kumar S, Pundir AK (2020) Blockchain—Internet of Things (IoT) enabled pharmaceutical supply chain for COVID-19. In: Proceedings of the NA international conference on industrial engineering and operations management Detroit, Detroit, MI, USA, pp 10–14

Kumar T, Ramani V, Ahmad I, Braeken A, Harjula E, Ylianttila M (2018) Blockchain utilization in healthcare: Key requirements and challenges. In: 2018 IEEE 20th International conference on e-health networking, applications and services (Healthcom). IEEE, pp 1–7

Lakhan A, Mohammed MA, Kozlov S, Rodrigues JJ (2021) Mobile-fog-cloud assisted deep reinforcement learning and blockchain-enable IoMT system for healthcare workflows. Trans Emerg Telecommun Technol e4363

Mani N, Singh A, Nimmagadda SL (2020) An IoT guided healthcare monitoring system for managing real-time notifications by fog computing services. Proc Comput Sci 167:850–859

Manickam P, Mariappan SA, Murugesan SM, Hansda S, Kaushik A, Shinde R, Thipperudraswamy SP (2022) Artificial intelligence (AI) and internet of medical things (IoMT) assisted biomedical systems for intelligent healthcare. Biosensors 12(8):562

Marwah GPK, Jain A, Malik PK, Singh M, Tanwar S, Safirescu CO, Mihaltan TC, Sharma R, Alkhayyat A (2022) An improved machine learning model with hybrid technique in VANET for robust communication. Mathematics 10(21):4030. https://doi.org/10.3390/math10214030

Mehbodniya A, Webber JL, Neware R, Arslan F, Pamba RV, Shabaz M (2022) Modified Lamport Merkle Digital Signature blockchain framework for authentication of internet of things healthcare data. Expert Syst 39(10):e12978

Mitra A, Roy U, Tripathy BK (2022) IoMT in healthcare industry—concepts and applications. Next Gener Healthc Inform 121–146

Naresh VS, Pericherla SS, Murty PSR, Reddi S (2020) Internet of Things in healthcare: architecture, applications, challenges, and solutions. Comput Syst Sci Eng 35(6)

Nayak J, Meher SK, Souri A, Naik B, Vimal S (2022) Extreme learning machine and bayesian optimization-driven intelligent framework for IoMT cyber-attack detection. J Supercomput 78(13):14866–14891

Nguyen DC, Pathirana PN, Ding M, Seneviratne A (2021) BEdgeHealth: a decentralized architecture for edge-based IoMT networks using blockchain. IEEE Internet Things J 8(14):11743–11757

Nwosu AU, Goyal SB, Bedi P (2021) Blockchain transforming cyber-attacks: healthcare industry. In: Innovations in bio-inspired computing and applications: proceedings of the 11th international conference on innovations in bio-inspired computing and applications (IBICA 2020) held during December 16–18, 2020 11. Springer International Publishing, pp 258–266

Omar IA, Jayaraman R, Debe MS, Salah K, Yaqoob I, Omar M (2021) Automating procurement contracts in the healthcare supply chain using blockchain smart contracts. IEEE Access 9:37397–37409

Pal K (2022) Blockchain-integrated internet-of-Things architecture in privacy preserving for large-scale healthcare supply chain data. In: blockchain technology and computational excellence for society 5.0. IGI Global, pp 80–124

Pelekoudas-Oikonomou F, Zachos G, Papaioannou M, de Ree M, Ribeiro JC, Mantas G, Rodriguez J (2022) Blockchain-based security mechanisms for IoMT Edge networks in IoMT-based healthcare monitoring systems. Sensors 22(7):2449

Pirbhulal S, Pombo N, Felizardo V, Garcia N, Sodhro AH, Mukhopadhyay SC (2019) Towards machine learning enabled security framework for IoT-based healthcare. In: 2019 13th international conference on sensing technology (ICST). IEEE, pp 1–6

Pundir AK, Jagannath JD, Chakraborty M, Ganpathy L (2019) Technology integration for improved performance: a case study in digitization of supply chain with integration of internet of things and blockchain technology. In: 2019 IEEE 9th annual computing and communication workshop and conference (CCWC). IEEE, pp 0170–0176

Qahtan S, Sharif KY, Zaidan AA, Alsattar HA, Albahri OS, Zaidan BB, et al. (2022) Novel multi security and privacy benchmarking framework for blockchain-based IoT healthcare industry 4.0 systems. IEEE Trans Ind Inform 18(9):6415–6423

Quasim MT, Algarni F, Radwan AAE, Alshmrani GMM (2020) A blockchain based secured healthcare framework. In: 2020 international conference on computational performance evaluation (ComPE). IEEE, pp 386–391

Qureshi A, Batra S, Vats P, Singh S, Phogat M, Sharma AK (2022) A review of machine learning (ML) in the internet of medical things (IOMT) in the construction of a smart healthcare structure. J Algebr Stat 13(2):225–231

Quy VK, Hau NV, Anh DV, Ngoc LA (2022) Smart healthcare IoT applications based on fog computing: architecture, applications and challenges. Complex Intell Syst 8(5):3805–3815

Ratta P, Kaur A, Sharma S, Shabaz M, Dhiman G (2021) Application of blockchain and internet of things in healthcare and medical sector: applications, challenges, and future perspectives. J Food Qual 2021:1–20

Razdan S, Sharma S (2022) Internet of medical things (IoMT): overview, emerging technologies, and case studies. IETE Tech Rev 39(4):775–788

Rokade A, Singh M, Malik PK, Singh R, Alsuwian T (2022) Intelligent data analytics framework for precision farming using IoT and regressor machine learning algorithms. Appl Sci 12(19):9992. https://doi.org/10.3390/app12199992

Salih FI, Bakar NAA, Hassan NH, Yahya F, Kama N, Shah J (2019) IOT security risk management model for healthcare industry. Malays J Comput Sci 131–144

Samad A (2022) Internet of things integrated with blockchain and artificial intelligence in healthcare system. Res J Comput Syst Eng 3(1):01–06

Satamraju KP (2020) Proof of concept of scalable integration of internet of things and blockchain in healthcare. Sensors 20(5):1389

Saxena N, Chakravarthi DS, Venkatesh AN, Soni N, Kant S (2022) The future of blockchain technology and the internet of things in healthcare. In: 2022 International conference on innovative computing, intelligent communication and smart electrical systems (ICSES). IEEE, pp 1–9

Schinckus C (2022) A Nuanced perspective on blockchain technology and healthcare. Technol Soc 71:102082

Shukla S, Thakur S, Hussain S, Breslin JG, Jameel SM (2021) Identification and authentication in healthcare internet-of-things using integrated fog computing based blockchain model. Internet of Things 15:100422

Soltanisehat L, Alizadeh R, Hao H, Choo KKR (2020) Technical, temporal, and spatial research challenges and opportunities in blockchain-based healthcare: a systematic literature review. IEEE Trans Eng Manage 70(1):353–368

Srivastava S, Pant M, Jauhar SK, Nagar AK (2022) Analyzing the prospects of blockchain in healthcare industry. Comput Math Methods Med

Teli TA, Masoodi F (2021) Blockchain in healthcare: challenges and opportunities. SSRN Electron J 1–6

Tiwari S, Sharma N (2022) Idea, architecture, and applications of 5G enabled IoMT systems for smart health care system. ECS Trans 107(1):5499

Tuli S, Basumatary N, Gill SS, Kahani M, Arya RC, Wander GS, Buyya R (2020) HealthFog: an ensemble deep learning based smart healthcare system for automatic diagnosis of heart diseases in integrated IoT and fog computing environments. Futur Gener Comput Syst 104:187–200

Venkatesh DAN (2019) Reimagining the future of healthcare industry through internet of medical things (IoMT), artificial intelligence (AI), machine learning (ML), big data, mobile apps and advanced sensors. Artificial Intelligence (AI), Machine Learning (ML), Big Data, Mobile Apps and Advanced Sensors

Xu L, Zhou X, Tao Y, Liu L, Yu X, Kumar N (2021) Intelligent security performance prediction for IoT-enabled healthcare networks using an improved CNN. IEEE Trans Industr Inf 18(3):2063–2074

Zhang J, Yang Y, Liu X, Ma J (2022) An efficient blockchain-based hierarchical data sharing for Healthcare Internet of Things. IEEE Trans Industr Inf 18(10):7139–7150

Quantum-Resistant Cryptography to Prevent from Phishing Attack Exploiting Blockchain Wallet

9

Divya Rishi Sahu, Harsh Tiwari, Deepak Singh Tomar, and R. K. Pateriya

Abstract

Blockchain technology is a way to maintain a decentralized and distributed digital ledger that allows multiple parties to maintain a shared database without the need for a central authority. Common application areas of blockchain where security is a significant concern are cryptocurrency, supply chain management, financial services, healthcare, voting systems, and intellectual copyright protection. But, attackers are exploiting phishing attacks to impersonate the blockchain wallet by tricking novice users. Phishing is one of the top ten attack scenarios to mimic the user by taking him in confidence as a legitimate one to reveal their private keys, seed phrases, and login credentials. Users can protect themselves from phishing attacks and their consequences by adopting or following best practices such as verifying the authenticity before taking any action, being cautious of hyperlinks, using secure website connections, enabling two-way authentication, raising awareness. But, due to the human nature of forgiveness and a lack of knowledge, users are still exploited by phishing attacks. Service providers are also responsible for managing secure transactions and protecting novice users from any fraudulent attack, including phishing. Quantum computing can contribute to strong security and prevent users from phishing attacks that target blockchain wallets. Potential methods of quantum computing to secure against phishing attacks include quantum cryptography, quantum-resistant cryptography, quantum random number generation, enhanced digital signatures, and public key infrastructure (PKI). Additionally, quantum-resistant

D. R. Sahu (✉) · H. Tiwari
Samrat Ashok Technological Institute (Degree), Vidisha (M.P.), India
e-mail: drsahu.cse@satiengg.in

D. S. Tomar · R. K. Pateriya
Maulana Azad National Institute of Technology, Bhopal (M.P.), India

A. Kumar et al. (eds.), *Sustainable Security Practices Using Blockchain, Quantum and Post-quantum Technologies for Real Time Applications*, Contributions to Environmental Sciences & Innovative Business Technology,
https://doi.org/10.1007/978-981-97-0088-2_9

cryptography can contribute specifically to preventing phishing attacks by providing secure key generation, robust authentication, resilient encryption, etc. Blockchain is a decentralized, immutable, and transparent technology, while the properties of quantum computing are quantum entanglement, quantum superposition, and quantum parallelism. Blockchain and quantum computing, both technologies, have the potential to bring significant advancements in securing against phishing attacks, irrespective of challenges and opportunities. Implementing quantum-resistant cryptography in blockchain networks would help mitigate the potential risks of quantum computing. Hence, the quantum-resistant cryptography method is proposed in this chapter to prevent phishing attacks that may exploit blockchain wallets.

Keywords

Quantum computing • Phishing attack • Blockchain • Cryptography • Quantum-resistant cryptography • Quantum random number generation • Enhanced digital signatures • Security

9.1 Introduction

Researchers are exploring new and fast computing technologies that can fit with fast-moving digital world. Standard computers are working on binary digits zeros and ones that can represent one of the states at a time. While the qbits can represent multiple possible states simultaneously with the help of a linear combination of bit 0 and bit 1. These qbits are called quantum bits which is the fundamental unit and concept behind quantum computing and quantum computers. This property to work in multiple states simultaneously is called the superposition, which enables a more efficient parallel computation and excels the processing power.

Quantum computing is the branch of computer science that uses mathematics and physics with quantum mechanics to solve problems that are more complex to solve by standard computers. Quantum computers are working on the concepts of quantum mechanics. It is the new era of computing where complex and time-consuming mathematical and cryptographic problems can be solved faster, enabling the doors for researchers to explore the deep-down knowledge in the field of science.

Quantum computing also excels in the power of other fields including cryptography, Security, Blockchain, Machine Learning, quantum systems, material science, simulation of molecular interaction, etc. It excels the power of cryptographic algorithms as well as supports to develop new secure cryptographic methods. Six commonly used quantum Cryptography methods are Quantum-Resistant, Lattice-based Cryptography, Code-based Cryptography, Hash-based Cryptography, and Multivariate Polynomial Cryptography. Quantum resistance cryptography is also known as the post-quantum cryptography (PQC). It can be used to maintain data security for a long time, to secure big data, to maintain integrity and authorization in communication between IoT devices, to secure sensitive data from zero-day vulnerabilities, quantum attacks, phishing attacks, etc.

Quantum-resistant cryptography does not deal directly with phishing attacks. Because phishing attacks are tricks to lure novice users by providing duplicate malicious GUI. In phishing attacks, hackers lure the end users and reveal sensitive information like passwords, account PNI number, and personal details to further harm the user.

Phishing attacks are possible due to human vulnerabilities as well as minimum security on the service provider side. Human vulnerabilities include the nature of forgiveness, easy trust, silly mistakes in curiosity or urgency, obliviousness, lack of knowledge, and ignorance. That can be overcome from both sides that is from the user side as well as from the service provider side. On the user side, awareness and education of novice users about the phishing attacks or tricks of hackers, to know about list of fake websites, and to know best practices for a secure way of communication. Also, service providers can overcome phishing attacks by implementing the Multi-Factor Authentication (MFA), enhancing the security of websites, enabling end-to-end encryption, maintaining the confidentiality and integrity by adopting cryptographic techniques, etc.

To warn end users about phishing attacks, service providers spread information from time to time through mail, messages, graphics, website pop-ups, etc. As well as Indian Government is also supporting various consortiums, departments, public sectors, and private sectors from time to time to spread awareness to novice users by different means like campaigns, workshops, courses, guidelines, etc. Standardization Testing and Quality Certification (STQC) Directorate, an attached office of the Ministry of Electronics and Information Technology, Government of India jointly with the National Informatics Centre (NIC) drafted and implemented the Guidelines for Indian Government Websites (GIGW) (NIC & SQTC 2023) and tools to evaluate website vulnerabilities (India Portal Team 2023). Hence, this book chapter focuses on the prevention of phishing attacks by quantum resistance cryptography algorithm.

The quantum resistance cryptography algorithm can also overcome phishing attacks that can exploit blockchain wallets. Quantum resistance cryptography to prevent phishing attacks is more suitable due to its various properties including fast and parallel processing, compatibility with existing environments, security against quantum attacks, and its processing is based on complex mathematics that is hard to break even by the quantum computers itself.

9.2 Cryptocurrency and Blockchain Wallet

Cryptocurrency is a digital currency that has been secured by blockchain technologies and can be operated by blockchain wallets. Most of us are remarkably familiar with blockchain technology, which is the backbone behind cryptocurrency also known as digital currency. These cryptocurrencies are managed by the GUI

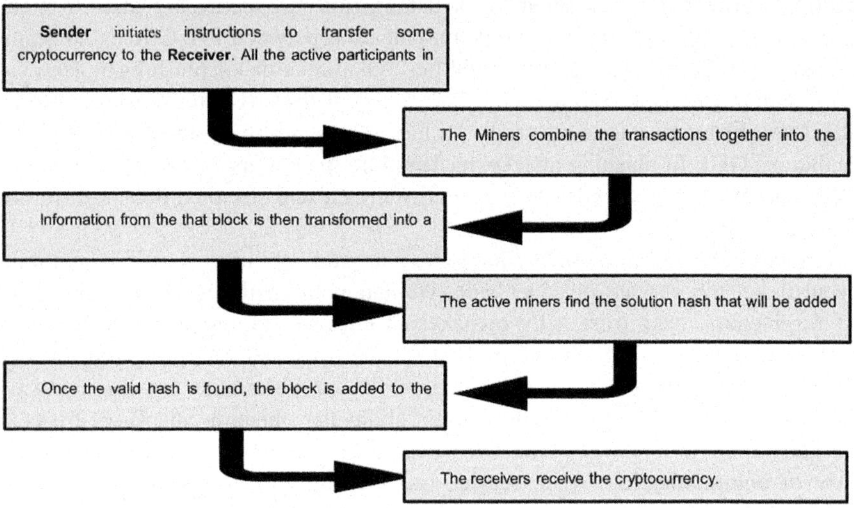

Fig. 9.1 Process of cryptocurrency transaction

called wallet. For example, cryptocurrency coins are stored in a database, and it can change by transaction. It is like a transfer from one bank account to another but digitally. For the transactions, processing, and visualization of cryptocurrency coins, we need an application that is a wallet.

In the recent digital world, two main cryptocurrency markets are bitcoins and Ethereum. Bitcoin technology uses blockchain to maintain transactions, manage the authorization, and maintain integrity by cryptography. Cryptocurrency can be used as a means of payment, to store value. The process of cryptocurrency transactions is described in the next section.

9.2.1 Process of Cryptocurrency Transaction

Education wing of the Reserve Bank of Australia (Reserve Bank of Australia, Education, 2023) provided a common process used in cryptocurrency exchange or transaction as shown in Fig. 9.1.

9.2.2 Possible Attacks on Blockchain Network

Hackers are targeting blockchain technology to exploit vulnerabilities and launch attacks with various intentions including to make a source of money, to disturb humans, and to down the business of competitors. Mihail and Anna (2023) classified the five types of attack vectors for blockchain technology as depicted in Fig. 9.2.

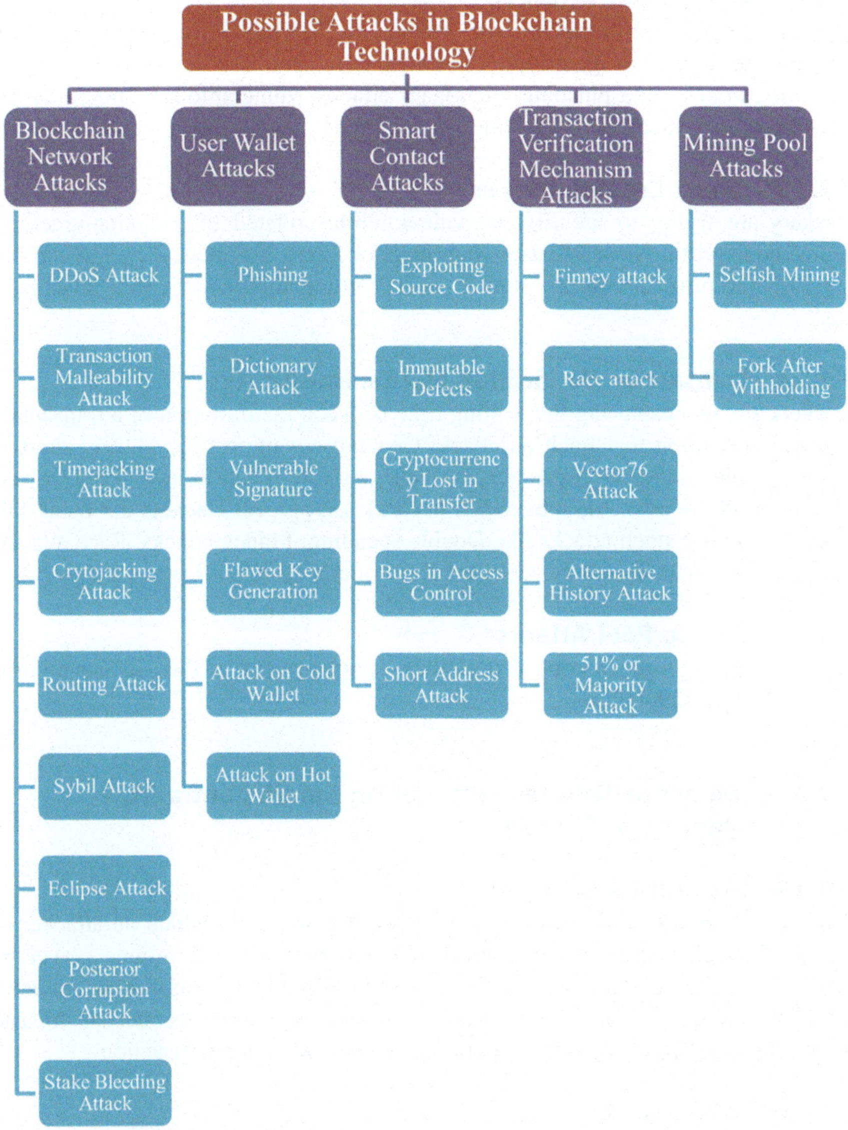

Fig. 9.2 Hierarchal representation of attacks possible on the blockchain network

9.2.2.1 Blockchain Network Attacks

Hackers are identifying the vulnerabilities in blockchain networks and exploiting these by attacks like Distributed Denial of Service, Transaction malleability attacks, Timejacking, Routing attacks, Sybil attacks, Eclipse attacks, Posterior Corruption, Stake Bleeding, etc.

9.2.2.2 User Wallet Attacks

Humans are ignoring the security or weakness of blockchain or unaware about it. Hackers take benefit of ignorance or unawareness to gain access to user's wallet by launching attacks like phishing, dictionary attacks, vulnerable signatures, flawed key generation, attack on cold wallet, attack on hot wallet, etc.

9.2.2.3 Smart Contact Attacks

Hackers are trying to identify the vulnerabilities related to smart contracts in source code, virtual machine used, Interface for Smart contracts, etc., and exploiting vulnerabilities in contract source code or vulnerabilities in virtual machines, etc.

9.2.2.4 Transaction Verification Mechanism Attacks

Hackers are also capturing the waiting time between the initiation of a transaction and its verification to launch an attack. Generally, waiting time is truly short in a fast communication network but in blockchain, the transaction will not be verified until all nodes of the blockchain network are agreed. Hackers are using this waiting time to launch attacks like Double spending, Finney attacks, Race attacks, Vector76, Alternative history attacks, 51% or majority attacks, etc.

9.2.2.5 Mining Pool Attacks

Hackers are targeting to control the mining pool by attacks like Selfish mining, and Fork after withholding.

9.2.3 Impact of Different Attacks on Smart Contracts, Identity, or Finance

9.2.3.1 51% Attack

This type of attack is also known as the majority attack, in which an attacker or group of attackers gathers the control of blockchain network mining power by more than 50 percent, i.e., getting the majority of the blockchain transaction entry under their control. This allows them to control the consensus mechanism and tempering the data of the transactions. They can reverse the transaction.

9.2.3.2 Sybil Attack

It is a type of attack where the hacker installs malicious software in the system and tries to alter the function of the nodes. It is a community-oriented attack to infect a bunch of active network nodes. In this type of attack, the attacker creates multiple compromised nodes and tries to influence the network, manipulates the consensus process, and distorts the network.

9.2.3.3 Eclipse Attack

In this type of attack, the attacker creates fake surroundings by isolating the user with his own modified malicious nodes in the network. This manipulates the user to think of a genuine aesthetic environment and steals its confidential information.

9.2.3.4 Smart Contract Exploits

Smart Contracts are computer instructions or programs for automating the transactions that are stored on the blockchain network and performed according to the condition. Exploiting those can make the system work differently than intended and produce non-useful, or resource wastage. This could interrupt the entire system or create a deadlock situation.

9.2.3.5 Distributed Denial of Service (DDoS)

In this type of attack, the attacker tries to overlook transaction nodes by flooding them with a vast number of requests and transactions making the network operation slow and time-consuming and even making them unavailable.

9.2.3.6 Double Spending

In this type of attack the end user sends two continuous conflicting transactions in a truly short period of time with the same bitcoin to different.

Attacks always lead to destruction and loss of property. Damages are always subject to the type of environment and application, which are targeted. Based on the research (Vokerla et al. 2019), losses and their effect on the above-mentioned attack scenario are given in Table 9.1.

This book chapter focuses on the possibilities of phishing attacks on blockchain networks. Given examples are near to around the possibilities to exploit the blockchain wallet of cryptocurrency coins.

Table 9.1 Impact of different attacks on smart contracts, identity, and finance (Vokerla et al. 2019)

Attack type	Smart contracts	Identity	Financial
51% attack	Effect	Does not affect	Effect
Sybil attack	Effect	Does not affect	Effect
Eclipse attack	Effect	Does not affect	Effect
Smart contract exploit	Effect	Does not affect	Does not affect
DDoS	Effect	Does not affect	Does not affect
Double spending	Effect	Does not affect	Effect
Phishing attack	Effects all three		

9.3 Phishing Attacks to Exploit Blockchain Wallet

Blockchain wallet does not hold cryptocurrency coins but provides GUI to the user to manage it. These cryptocurrency coins are stored in blockchain or distributed ledger. The wallet uses cryptographic keys to authenticate users related to cryptocurrency coins and perform transactions on it. If these cryptographic keys are leaked or stolen by the hacker, then cryptocurrency coins can be transferred without the user's concern.

Hence, the backbone of cryptocurrency is blockchain and the cryptography methods used in it. Blockchain is a non-mutable, decentralized form of processing and managing transactions. It is adopted globally and at any level of scale with a variety of real-life implementations. Blockchain can be represented as a distributed and secured digital ledger, which is secured using cryptographic algorithms.

Day by day, the use of blockchain and adoption of cryptocurrencies is increasing which is attracting hackers to penetrate it or luring users to transfer cryptocurrency into other accounts illegally. For this, hackers can exploit different attacks like phishing, sybil, Distributed Denial of Service (DDoS), Smart Contract Exploits, Eclipse Attack, zero-day attack, etc.

Phishing attacks are possible by exploiting human vulnerabilities as discussed in the introduction section. Five consequences of these attacks are reduced end-user trust in the service provider, negative impact on the blockchain network, loss of financial assets, service outrage, and innovation setback. Hence, the security of it is the concern of all the stakeholders associated with it.

Phishing attacks are not only tricking end users to steal the private cryptographic keys, it also can trick end users to authorize the transaction deliberately created by the hacker to transfer cryptocurrency coins to other accounts. Researchers called such phishing technique Ice Phishing (Microsoft Threat Intelligence, 2022).

9.3.1 Exploitation of Cialdini's Six Principles to Influence Humans for Phishing Attacks in Blockchain Technology

In phishing attacks, hackers play with the end users' minds and senses to influence them or to lure them to follow instructions according to the hacker's intentions. Cialdini's (Gupta et al. 2018) gives six principles to influence humans. These six principles are exploited by hackers to lure the end user via social engineering attacks or phishing attacks. Cialdini's Six Principles of Influence are depicted in Fig. 9.3.

9.3.1.1 Principle of Reciprocation

It states that human nature is to pay back to others for what they have received from them. This principle can be exploited by the hackers in phishing attacks to lure end users or gain the trust of end users by providing gifts in advance as payback.

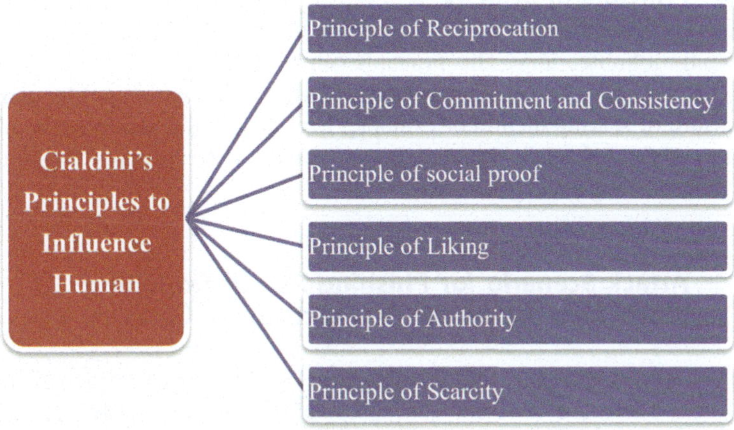

Fig. 9.3 Cialdini's six principles of influencing humans

9.3.1.2 Principle of Commitment and Consistency

It states that humans behave according to their commitments and consistent habits. This principle can be exploited by hackers in phishing attacks to gain trust of the end user by fulfilling the small initial commitments. Thereafter hacker servers the real request according to the malicious intention of the attackers and realizes the commitments.

9.3.1.3 Principle of Social Proof

It states that humans are behaving and believing similarly to the beliefs or behavior of surrounding humans. This principle can be exploited by hackers in phishing attacks to gain the trust of the end users by providing fake information that your colleagues trust us and already accepted the attacker's request.

9.3.1.4 Principle of Liking

It states that humans prefer to accept the request of the people they like or know them. It is to strengthen social bonding. Physical attraction is another common factor in linking someone else. This principle can be exploited by hackers in phishing attacks to maintain trust for a long time by playing with the desires of the end users which returns the like back.

9.3.1.5 Principle of Authority

It states that most humans are following and relying on their authorities. Authority is created in two ways. First via expertise and second via seniority. This principle can be exploited by hackers in phishing attacks to lure users and bind end users to behave according to the hacker's intention. For example, an attacker asks the end user to share cryptographic keys or credentials to resolve the specific issues as per the instructions of his/her authority. Authority simply be forged by showing or representing the same uniforms, images, badges, ID cards, titles, images, etc.

9.3.1.6 Principle of Scarcity

It states that humans realize more value for the items that are available less or available for a limited time. This principle can be exploited by hackers in phishing attacks to create pressure on the end user to follow their instructions by generating the pressure of scaring them. For example, hackers say that this offer is available for a limited time or available for a few users only. It creates pressure on the end user's mind to follow the instructions.

9.3.2 Attack Scenario for Phishing Attack Exploiting Blockchain Wallet for Cryptocurrency Coins

The attack scenario represents the steps used by the attacker to exploit the vulnerabilities. There are different attack scenarios that are possible and may be changed in the future to exploit the same vulnerability. One of the attack scenarios for phishing attacks to exploit the vulnerabilities in blockchain wallets or cryptocurrency systems is given in Fig. 9.4.

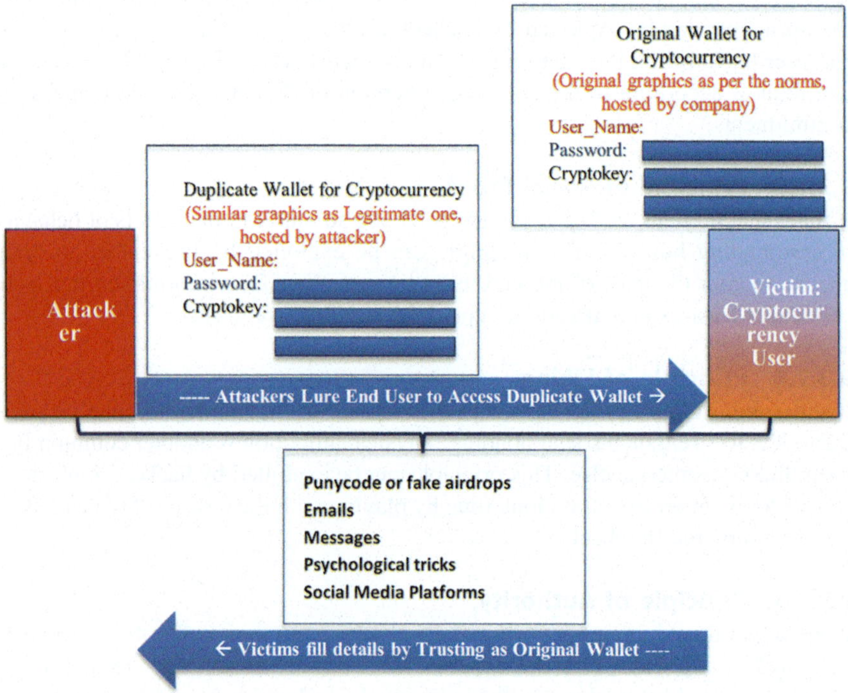

Fig. 9.4 Attack scenario for phishing attack exploiting blockchain wallet or cryptocurrency

The attacker exploits the human vulnerabilities as discussed in the introduction section to launch a phishing attack and to harm the victim. Steps for phishing attacks to exploit blockchain wallet or cryptocurrency are as follows-

- **Step 1**: Attacker creates a duplicate website with a similar GUI as the original website or wallet.
- **Step 2**: Hosted with another domain that looks like the original one or hide the link into hypertext.
- **Step 3**: Lure the end user by trick and tact to access the duplicate website by choice or mistakenly.
- **Step 4**: Share or spread the tempting messages with links to duplicate wallets to end users. Attackers may use one or more channels like Punycode or fake airdrops, Emails, Messages, Psychological tricks, Social Media Platforms, Advertisements, etc.
- **Step 4**: End user is trapped in the pseudo trap and follows the instructions given on the duplicate wallet by choice or mistakenly and becomes the victim of the attack.
- **Step 5**: Confidential, secret keys, usernames, passwords, sensitive data, and cryptographic keys are transferred to the attacker's database. This information can be utilized to harm the end users.

To take the end user into confidence, attackers use tricks like using a fake SSL certificates (Holub et al. 2018), using the anonymity feature of blockchain platforms to hide identity (Banwari et al. 2021), using similar GUI and instructions in duplicate wallets, etc. Attackers also use psychological tricks (Weber et al. 2020) and social engineering (Ivanov et al. 2021) to gain the trust of the users and lure the end user to follow their instructions.

9.4 Literature Work: Detection Methods for Phishing Attacks Exploiting Cryptocurrency

Kabla et al. (2022) use machine learning techniques to detect the Ethereum Phishing Scam. Resolved the issues phased by the other researchers like minimum detection accuracy and imbalanced dataset.

Chen et al. (2020) proposed an artificial intelligence and graph-based method to detect phishing accounts. The author applied the cascade feature extraction method to the record of blockchain transactions and applied the dual-sampling ensemble framework to identify the suspected account.

Zhou et al. (2023) proposed a two-layer Edge Aggregated Graph Attention Network (EGAT) classification method to classify the Ethereum phishing accounts. It is based on the static subgraph representation of the transaction network. Finally compared with the Graph2Vec and Deep Walk methods.

Hu et al. (2022) proposed a deep learning method to detect the smart contract as fraudulent or not. One of the features of the algorithm is the bytecode of smart contracts. The presented SCSGuard method extracted the feature automatically, providing a unified solution.

Wu et al. (2022) proposed a one-class support vector machine (SVM) to identify and classify the account into normal and phishing ones. In this work transaction amount and timestamp are used as features extracted from labeled phishing transaction records.

Bartoletti et al. (2021) identified that the dataset for cryptocurrency attacks is not sufficient and available datasets are providing misleading data. Also, there is not available a standard taxonomy for cryptocurrency attacks. It makes it difficult to classify and detect cryptocurrency attacks into normal and malicious ones by any classifier algorithm based on dataset training. Hence, thousands of cryptocurrency attacks and literature have been analyzed and a systematic dataset is created. Finally, a taxonomy has been proposed for the policymakers to overcome the cryptocurrency attacks by following it.

Badawi and Jourdan (2020) reviewed 1,221 publications related to cryptocurrency threats and detection methods. After review, authors listed the type of cryptocurrency attacks, detection methods, and publicly available dataset. Also listed websites, tools, and libraries used for data analysis and collection.

Phillips and Wilder (2020) listed the blockchain dataset available online in public. Applied the Density-Based Spatial Aggregation for Applications with Noise (DBSCAN) clustering algorithm to identify the typology of phishing scams including exchanges, gambling sites, and mixers. The author identified that a single identity may run more than one instance of attacks.

Alyami et al. (2023) identified that most investors are unaware about phishing attacks on cryptocurrency and how to protect themselves from such attacks. The author also identified the security risks related to the investment into cryptocurrency. Hackers are floating cryptocurrency investment transactions without the knowledge of authorized identity or physical wallet because it is a form of peer-to-peer process. Listed the types of cryptocurrency phishing attacks.

Weber et al. (2020) identified that attackers are using social engineering methods to trick the end users without compromising the blockchain security mechanism. The target of the social engineering methods is the end user of an information system to gain unauthorized access of sensitive information or credentials. Social engineering methods may exploit cryptocurrency by taking illegitimate access to the credentials. The author analyzes the five cryptocurrency attacks using ontological models for social engineering attacks. Analyses the psychological tricks used by the hackers to exploit cryptocurrency.

Quiel (2013) author defined the social engineering for their work as the attack. Here, the end user is illegally influenced end user to work according to the hacker's intention using various methods like persuasion, impersonation, fingerprinting, misinterpretation, etc.

Gupta et al. (2018) analyzed the various possible taxonomies of phishing attacks and its solutions. And identified the impact of phishing on the Internet of Things.

Trozze et al. (2022) et al. review the forty-nine types of attacks exploiting cryptocurrency. Ivanov et al. (2021) identified that Phishing, Cryptojacking, and ransomware are the recent attacks exploiting cryptocurrency. Further analyzes the phishing mails, malicious PDFs, trojans, and ransomware.

Wen et al. (2023) proposed a phishing detection method by using a hybrid deep neural network to identify the phishing attacks in cryptocurrency coin wallets. Autor uses the Long Short-Term Memory Fully Convolutional Network (LSTM-FCN) method for the extraction of transactional features of transaction records. Therefore, applied Back Propagation (BP) neural network method was to establish a chain among identified features.

Farrugia et al. (2020) et al. identified more than 40,000 phishing fraud wallets of Ethereum cryptocurrency coin in two years from 2016 to 2018. These all suffered from approximately 2,000 phishing attack vectors.

Wang et al. (2023) et al. identified that commonly researchers applied the standard machine learning methods on temporal features of the transaction account. Thereafter creates the graph of extracted features or creates the temporal random wandering to analyze the graph. The author proposed the PDTGA method to identify phishing attacks on cryptocurrency accounts or networks. The backbone of PDTGA method is the self-attentive mechanism, time encoding function, and multilayer perceptron.

With the increased use of blockchain and global extensive network, fast processing, anonymity, etc., his book chapter focuses on utilizing the power of quantum computing to detect phishing attacks possible on blockchain networks. There are various vulnerabilities as discussed in the above sections to exploit the wallet of cryptocurrency coins through phishing attacks. In the next section, Quantum-Resistant Cryptography methods are discussed to prevent phishing attacks exploiting blockchain wallets of cryptocurrency coins.

9.5 Use of Quantum Computing to Protect from Phishing Attacks

Quantum computing is the field that combines the power of computer science, physics, and mathematics. Google developed the quantum processor that is 54-qubit Sycamore processor. It can complete a task in 3.33 min that a supercomputer can process in approximately 10,000 years (Martinis and Boixo 2019).

A combination of quantum computing with blockchain is transforming digital world into a new edge to boost the potential of its networking, better random number generation, harder and more complex algorithms, better consensus mechanisms, and optimized searching algorithms.

In addition, quantum computing with blockchain leads to the road of new discoveries and other opportunities. Although challenges exist after crossing the barrier, the other side the paths such as-

- Enhanced Security: The most essential element of any process execution and transaction is security, quantum computing with Blockchain, adds the additional layer

- Quantum-safe blockchain
- Optimized Consensus Blockchain
- Improved Data Analysis
- Industry-Specific Applications

But with great power comes great responsibility. Having great processing power and advancements could lead to challenges that pose a potential threat to the system. Both are the most advanced and useful technologies that will achieve and unlock the ease and better implementation of technology, but they also come with challenges which are very essential to point out-

1. Quantum Security
2. Scalability
   (a) Setting-up Standards
   (b) Adoption and Integration
   (c) Awareness and Education
   (d) Regulation and Management
3. Quantum Compatible Hardware

9.5.1 Application Areas of Quantum Computing

Quantum computing has a large set of properties including able to solve complex problems, fast processing, work in multiple states simultaneously, and support parallel computation, entanglement: qbits may become correlated such that the state of one qbit cannot be defined without another qbit, end so on.

With large properties and principles fit the newly fast-moving digital worlds and solutions to its problems. The potential application areas of quantum computing are not limited. Nine application areas are Cryptography or Quantum Key Distribution, Security, Error Correction, Machine Learning, Optimization, Material Science, Quantum Simulation, Quantum Data Analysis as shown in Fig. 9.5.

This chapter focuses on quantum cryptography methods to protect from phishing attacks. So, the next section will provide guidelines about quantum cryptography, one of the principal application areas of quantum computing.

9.5.2 Quantum Cryptography

Quantum cryptography is based on the following two principles of quantum mechanics. First is Heisenberg Uncertainty principle and second is the principle of photon polarization. The Heisenberg Uncertainty principle states that the accurate calculation of any quantum state for a system is not possible. Similarly, the principle of photon polarization states that the detection of polarized photons is only possible by correct polarization. In case of incorrect polarization, photon will

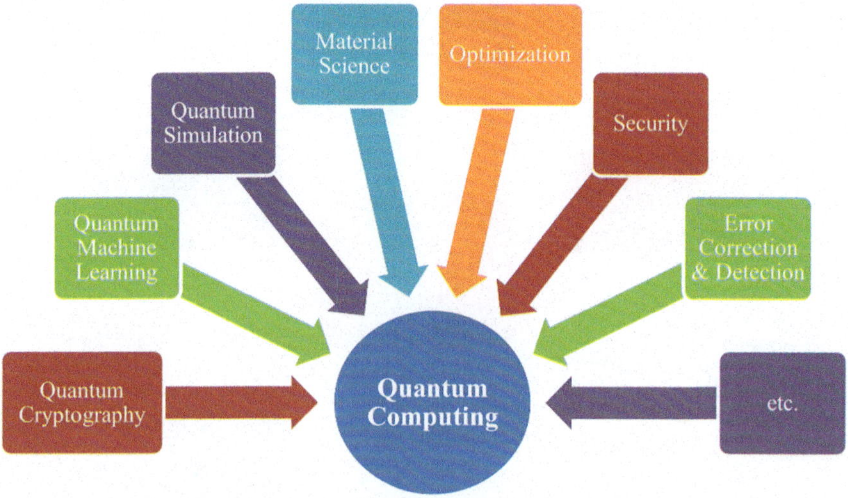

Fig. 9.5 Potential application area of quantum computing

break. These principles are the backbone to minimize eavesdropping and maintain the privacy of data via quantum cryptography technologies.

Quantum Cryptography is a branch of computer science that involves cryptography with the principles of quantum mechanics to provide an enhanced and advanced form of communication and transaction, unlike the traditional method where we use complex mathematical equations. It used the fundamentals of quantum physics to get correct and genuine data.

In comparison with standard cryptography methods, quantum cryptography supports the secure distribution of quantum keys among sender and receiver in terms of qbits via a quantum channel. The sender and receiver use these quantum keys to encrypt and decrypt the messages and share messages securely and maintain authenticity, integrity, and authorization. In addition to the standard cryptography techniques quantum cryptography supports detecting eavesdropping. Quantum computers can disintegrate the traditional cryptography methods like RSA, elliptic curve cryptography, etc. by exploiting its vulnerabilities.

Quantum cryptography provides various methods for providing secure communication that is secure from attacks by quantum computers. Four methods or components are Quantum Key Distribution (QKD), Quantum Coin Flipping, Quantum Secret Sharing, and Quantum Resistance Cryptography as shown in Fig. 9.6.

Different quantum cryptography methods have their own goals, properties, and facilities. The goal of quantum resistance computing is to establish secure communication and prone toward securing from attacks possible from quantum computers and quantum attacks as well. Quantum resistance computing can generate new methods in comparison to other methods. This work proposed quantum resistance cryptography to protect from phishing attacks in blockchain networks to protect cryptocurrency wallets.

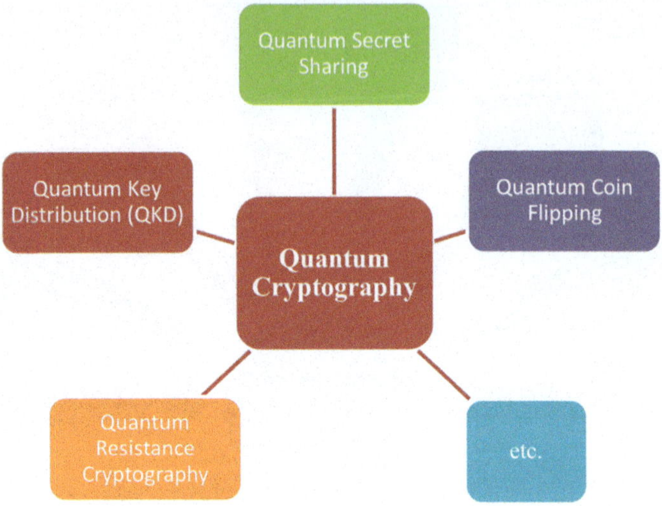

Fig. 9.6 Quantum cryptography methods and components

9.5.3 Quantum-Resistant Cryptography

Quantum-Resistant Cryptography is a type of cryptography in which cryptographic protocols and operations can withstand securely even if a large-scale quantum computer cannot break its algorithm making it quantum-proof security. These are based on mathematical problems that are believed to be hard to solve even for the powerful quantum computers and their systems.

Post-Quantum Cryptography (PQC) also known as Quantum-Resistance public-key cryptographic algorithm that provides security for both standard computer and quantum computers. The National Institute of Standard and Technology (NIST), US Department of Commerce (NIST 2022), approved the CRYSTALS-KYBER as a Public-key Encryption and Key-establishment Algorithms Standard. Also approved following three Digital Signature Algorithms CRYSTALS-DILITHIUM, FALCON, and SPHINCS+. Fernández-Carames and Fraga-Lamas (2020) classify the post-quantum cryptosystems into five categories for encryption as given in Table 9.2. These technologies are suitable for blockchain systems.

Based on the quantum standards and technologies, we can apply Quantum-Resistant Cryptography to prevent phishing attacks by implementing the Multi-factor Authentication, Quantum Resistant Digital Signature, Quantum Key Distribution (QKD), or Post Quantum Cryptography (PQC) to encrypt the coins and communication shared in the channel to protect cryptocurrency coins in the wallet from phishing attacks.

Table 9.2 Five classes of post-quantum cryptosystems for encryption methods

Blockchain public key cryptosystem				
Encryption methods				
Post quantum cryptosystem				
Code based cryptosystems	Lattice-based cryptosystems	Multivariate-based cryptosystems	Super-singular elliptic curve isogeny cryptosystems	Hybrid cryptosystem
• HQC • Classic McEiece • BIKE • LEDACrypt • NTS • ROLLO- II • RQC	• Kyber • FrodoKEM • LAC • New Hope • NTRUEncrypt • NTRUPrime • Round5 • SABER • Three Bears	• Based on Multivariate quadratic equations • Matsumoto-Imai's • Hidden Field Equations (HFE)	• SIKE	• CECPQ1 • CECPQ2

9.5.3.1 Multifactor Authentication

One of the most common authentication methods is based on a multi-factor authentication system. Quantum-resistant cryptography can be used in conjunction with multifactor authentication techniques to provide an additional layer of security. By combining different authentication factors, such as something you know (e.g., password), something you have (e.g., hardware token), and something you are (e.g., biometrics), the risk of successful phishing attacks can be significantly reduced.

Multi-Factor Authentication requires the generation of OTP which are commonly random, which should not be predictable and should be completely unbiased. Quantum computing can produce the **Quantum Random Number Generation (QRNG)**, which generates the numbers using the principle of quantum mechanics.

Based on the device's dependence for generation of random numbers, it can be classified into two types.

Device Dependent Quantum Random Number Generation (DD-QRNG)

It generates the number based on the physical properties and characteristics of the device used for generation. The generated randomness is obtained from its features and behaviors like fluctuations and quantum noise. It can be achieved by photon polarization, quantum phase, and quantum entanglement.

Device Independent Quantum Random Number Generation (DI-QRNG)

It aims to generate random numbers without specifically depending on the features and characters of the device. It uses the core quantum mechanics phenomena like Bell's inequality and non-locality. The security of this is based on the violation of Bell's inequality (Liu et al. 2018), which gives the 100% surety of real randomness that can be modified or manipulated by external factors, making it more secure and reliable. It provides a higher level of security as it is not vulnerable to any sort of flaws of the devices.

9.5.3.2 Quantum Resistant Digital Signature

Another most common way of secure authentication and integration is digital signature, with Enhanced digital signatures which are advanced cryptographic scheme that provides improved security, and more functionality compared the traditional digital signature. Four enhanced digital signature schemes are Quantum-Resistant Digital Signature, Threshold Digital Signatures, Blind Signatures and Ring Signatures.

Traditional Signatures algorithms like RSA and ECC may become vulnerable to quantum attack (Roy and Kalita 2019), by using quantum Resistant Digital Signatures, which provide more security by using the mathematical problems that are hard to solve even for the quantum computers such as lattice-based or code-based cryptographic algorithms.

Quantum-resistant signature schemes can be employed to ensure the integrity and authenticity of transactions within a blockchain network. These schemes can help prevent unauthorized parties from tampering with transactions or initiating fraudulent transactions by forging digital signatures.

9.5.3.3 Quantum Resistant Key Exchange

Quantum-resistant cryptographic algorithms can be used for secure key exchange protocols. By utilizing quantum-resistant key exchange mechanisms, the risk of an attacker intercepting and gaining unauthorized access to sensitive cryptographic keys is reduced.

Quantum-resistant key exchange algorithms are designed to withstand attacks from both classical and quantum computers. They typically utilize mathematical problems that are believed to be hard to solve even for quantum computers, ensuring the security of the key exchange process. Examples of quantum-resistant key exchange algorithms include:

New Hope: New Hope is a key exchange algorithm based on lattice cryptography. It provides a secure key exchange mechanism that is resistant to attacks from both classical and quantum computers.

Super-singular Isogeny Diffie-Hellman (SIDH)

It is another key exchange algorithm based on isogeny-based cryptography. It relies on the difficulty of computing isogenies between supersingular elliptic curves to establish a shared secret key.

FrodoKEM

It is a key exchange algorithm based on the learning with errors (LWE) problem. It provides a secure key exchange mechanism that is believed to be resistant to attacks from quantum computers.

9.6 Conclusion and Future Work

In the current digital world, the use of cryptocurrency coins is growing with tremendous speed. It is a digital currency that uses cryptography and security technologies to secure transactions or wallet, integration, authentication, and authorization. On the other side, the continuous new inventions in quantum computing and quantum computers with their fast-processing speed make cracking of standard security technologies easy for hackers. Also new attack vectors like quantum attacks on modern technologies may be launched by hackers to exploit the vulnerabilities in modern technology.

Hackers are exploiting vulnerabilities in technologies by launching different attack vectors and attack scenarios. Few of the attack scenarios do not need to exploit the vulnerabilities in technology like phishing. In phishing attacks, hacker tricks end users and lure them to grape credentials and gain access to the information system without the knowledge of authentic users. There are different attack scenarios that use email, messages, phone calls, social media, etc., to share fake messages and influence the novice end user to work according to the intention of the hacker. Hackers use human vulnerabilities, psychological tricks, and various principles to influence humans. Cialdini's summarized the six influencing principles that hackers use for their malicious intensions to launch phishing attacks.

It needs new security technologies to provide protection to this new digital world like to protect cryptocurrency coin and novice end user from hackers launching phishing attacks and taking benefits of the human vulnerabilities and psychological tricks. Modern technology like quantum computing and quantum-resistant cryptography can protect end users from these attacks. Quantum-resistant cryptography methods can protect novice users and cryptocurrency coins in wallets from human vulnerabilities, psychological tricks, quantum attacks, phishing attacks.

Quantum resistance computing in conjunction with Multifactor Authentication, Quantum Resistant Digital Signature, Quantum Key Distribution (QKD) or Post Quantum Cryptography (PQC) is capable to protect from phishing attack exploiting blockchain wallet and cryptocurrency coin. Recently a new standard for quantum public-key encryption and key-establishment algorithm was approved by NIST, which is CRYSTALS-KYBER. Also, approve the standard for quantum digital signature algorithms these are CRYSTALS-DILITHIUM, FALCON and SPHINCS+. Implementation of these quantum resistance cryptography and quantum digital signature algorithms can protect cryptocurrency wallets from phishing attacks.

References

Alyami M, Alhotaylah R, Alshehri S, Alghamdi A (2023) Phishing attacks on cryptocurrency investors in the Arab States of the Gulf. J Risk Financ Manag 16:271. https://doi.org/10.3390/jrfm16050271

Badawi E, Jourdan G-V (2020) Cryptocurrencies emerging threats and defensive mechanisms: a systematic literature review. IEEE Access 8:200021–200037. https://doi.org/10.1109/ACCESS.2020.3034816

Banwari L, Agarwal R, Shukla S (2021) Understanding money trails of suspicious activities in a crypto currency based Blockchain. https://doi.org/10.48550/arXiv.2108.11818

Bartoletti M, Lande S, Loddo A, Pompianu L, Serusi S (2021) Cryptocurrency scams: analysis and perspectives. IEEE Access 9:148353–148373. https://doi.org/10.1109/ACCESS.2021.3123894

Chen W, Guo X, Chen Z, Zheng Z, Lu Y (2020) Phishing scam detection on ethereum: towards financial security for blockchain ecosystem. In: ACM, proceedings of the twenty-ninth international joint conference on artificial intelligence (IJCAI-20) special track on AI in FinTech, pp 4506–4512 (2020)

Farrugia S, Ellul J, Azzopardi G (2020) Detection of illicit accounts over the Ethereum blockchain. Elsevier, Expert Syst Appl 150 (2020)

Fernández-Caramès TM, Fraga-Lamas P (2020) Towards post-quantum blockchain: a review on blockchain cryptography resistant to quantum computing attacks. IEEE Access 8:21091–21116. https://doi.org/10.1109/ACCESS.2020.2968985

Gupta BB, Arachchilage NAG, Psannis KE (2018) Defending against phishing attacks: taxonomy of methods, current issues and future directions. Springer, Telecommun Syst 67:247–267. https://doi.org/10.1007/s11235-017-0334-z

Holub, O'Connor J (2018) COINHOARDER: tracking a Ukrainian bitcoin phishing ring DNS style. In: 2018 APWG symposium on electronic crime research (eCrime), San Diego, CA, USA, pp 1–5. https://doi.org/10.1109/ECRIME.2018.8376207

Hu H, Bai Q, Xu Y (2022) SCSGuard: deep scam detection for ethereum smart contracts. In: IEEE INFOCOM 2022—IEEE conference on computer communications workshops (INFOCOM WKSHPS), New York, NY, USA, pp 1–6. https://doi.org/10.1109/INFOCOMWKSHPS54753.2022.9798296

India Portal Team (2023) Tools and resources. https://guidelines.india.gov.in/tools-and-resources/

Ivanov MA, Kliuchnikova BV, Chugunkov IV, Plaksina AM (2021) Phishing attacks and protection against them. In: 2021 IEEE conference of russian young researchers in electrical and electronic engineering (ElConRus), St. Petersburg, Moscow, Russia, pp 425–428. https://doi.org/10.1109/ElConRus51938.2021.9396693

Kabla AHH, Anbar M, Manickam S, Karupayah S (2022) Eth-PSD: a machine learning-based phishing scam detection approach in ethereum. IEEE Access 10:118043–118057. https://doi.org/10.1109/ACCESS.2022.3220780

Liu Y, Zhao Q, Li MH et al (2018) Device-independent quantum random-number generation. Nature 562:548–551. https://doi.org/10.1038/s41586-018-0559-3

Martinis J, Boixo S (2019) Quantum supremacy using a programmable superconducting processor. Quantum computing theory, google AI quantum. https://ai.googleblog.com/2019/10/quantum-supremacy-using-programmable.html. Accessed 5 June 2023

Mihail and Anna (2023) Blockchain attack vectors: main vulnerabilities against blockchain technology. https://www.apriorit.com/dev-blog/578-blockchain-attack-vectors. Accessed 15 May 2023

NIC & SQTC (2023) Guidelines for Indian Government Websites (GIGW). https://cdnbbsr.s3waas.gov.in/s3c92a10324374fac681719d63979d00fe/uploads/2023/05/2023051731.pdf

NIST (2023) Selected algorithms 2022, Post-Quantum Cryptography (PQC). https://csrc.nist.gov/Projects/post-quantum-cryptography/selected-algorithms-2022. Accessed 5 June 2023

Phillips R, Wilder H (2020) Tracing cryptocurrency scams: clustering replicated advance-fee and phishing websites. In: 2020 IEEE International Conference on Blockchain and Cryptocurrency (ICBC), Toronto, ON, Canada, pp 1–8. https://doi.org/10.1109/ICBC48266.2020.9169433

Quiel S (2013) Social engineering in the context of Cialdini's psychology of persuasion and personality traits. https://tore.tuhh.de/bitstream/11420/1126/1/Social_Engineering_in_the_Context_of_Cialdinis_Psychology_of_Persuasion_and_Personality_Traits.pdf. Accessed 11 May 2023

Reserve Bank of Australia, Education, Digital currencies. https://www.rba.gov.au/education/resources/explainers/pdf/cryptocurrencies.pdf?v=2023-06-18-00-19-38. Accessed 5 May 2023

Roy KS, Kalita HK (2019) A survey on post-quantum cryptography for constrained devices. Int J Appl Eng Res 14(11):2608–2615

Seifert C, Ice phishing' on the blockchain. Microsoft 365 defender research team, Microsoft threat intelligence. https://www.microsoft.com/en-us/security/blog/2022/02/16/ice-phishing-on-the-blockchain/. Accessed: 13 May 2023

Trozze A, Kamps J, Akartuna EA et al (2022) Cryptocurrencies and future financial crime. Crime Sci 11:1. https://doi.org/10.1186/s40163-021-00163-8

Vokerla RR, et al (2019) An overview of blockchain applications and attacks. In: 2019 international conference on vision towards emerging trends in communication and networking (ViTECoN), Vellore, India, 2019, pp 1–6. https://doi.org/10.1109/ViTECoN.2019.8899450.

Wang L, Xu M, Cheng H (2023) Phishing scams detection via temporal graph attention network in Ethereum. Elsevier, Inf Process Manag 60(4) (2023)

Weber K, Schütz AE, Fertig T, Müller NH (2020) Exploiting the human factor: social engineering attacks on cryptocurrency users. In: Zaphiris P, Ioannou A (eds) Learning and collaboration technologies. human and technology ecosystems. HCII 2020. Lecture notes in computer science, vol 12206. Springer, Cham. https://doi.org/10.1007/978-3-030-50506-6_45

Wen T, Xiao Y, Wang A, Wang H (2023) A novel hybrid feature fusion model for detecting phishing scam on Ethereum using deep neural network. Elsevier, Expert Syst Appl 211 (2023)

Wu J et al (2022) Who are the phishers? Phishing scam detection on ethereum via network embedding. IEEE Trans Syst, Man, Cybern: Syst 52(2):1156–1166. https://doi.org/10.1109/TSMC.2020.3016821

Xu M, et al (2022) When quantum information technologies meet blockchain in Web 3.0. IEEE Netw. https://doi.org/10.1109/MNET.134.2200578

Zhou X, Yang W, Tian X (2023) Detecting phishing accounts on ethereum based on transaction records and EGAT. Electronics 12:993

Revolutionizing Military Technology: How the Fusion of BlockChain and Quantum Computing is Driving in Defense Application

10

Vinay Aseri, Harish Chowdhary, Naveen Kumar Chaudhary, Shivam Kumar Pandey, and Viveka Kumar

Abstract

Blockchain is a cutting-edge innovation that has reformed the manner in which society connects and exchanges. It very well may be characterized as a chain of blocks that stores data with computerized marks in a circulated and decentralized network. Many years of hacking and taking advantage of digital protection frameworks have more than once demonstrated how a decided digital assailant might think twice and regular citizen organizations. The danger of refined weapon frameworks being hurt or debilitated by non-dynamic effects has constrained militaries to foster a long haul and unmistakably savvy safeguard for military frameworks. Blockchain, and it's at this point untested military purposes, can move the security weaknesses of some digital frameworks from a weak link weakness model, in which an aggressor just has to think twice about hub to disregard the framework, to a greater part compromised weakness model, in which a noxious entertainer can't take advantage of a weak link. Quantum

V. Aseri (✉)
School of Behavioral Sciences and Forensic Investigation, Rashtriya Raksha University, Ministry of Home affairs, Government of India, Gujrat, India
e-mail: vinay4n6@gmail.com

H. Chowdhary
School of Doctoral Studies and Research, National Forensic Sciences University, Delhi, New Delhi, India

N. K. Chaudhary
School of Cyber Security and Digital Forensics, National Forensic Sciences University, Gandhinagar, Gujarat, India

S. K. Pandey
Research Scholar, Rashtriya Raksha University, Gujrat, India

V. Kumar
School of Criminal Law and Military Law, Rashtriya Raksha University, Gujrat, India

A. Kumar et al. (eds.), *Sustainable Security Practices Using Blockchain, Quantum and Post-quantum Technologies for Real Time Applications*, Contributions to Environmental Sciences & Innovative Business Technology,
https://doi.org/10.1007/978-981-97-0088-2_10

innovation interprets the standards of quantum physical science into mechanical applications. As a general rule, quantum innovation has not yet arrived at development; be that as it may, it could hold critical ramifications for the fate of military detecting, encryption, and correspondences, as well as concerning legislative oversight, approvals, and assignments.

Keywords

Blockchain • Security • Defense • Military • Quantum technology • Applications • Cybersecurity • National security

10.1 Introduction

Blockchain technology is a trend mover that has produced novel concepts for reliably transferring information and data. This innovative technology uses shared and distributed networks to establish a chain of blocks that safely keeps track of every transaction made (Rimal et al. 2009; Flynn 1966). To do this, a number of core technologies are employed, including digital signatures, distributed consensus methods, and the cryptographic hash function. Decentralized execution of every transaction eliminates the requirement for intermediaries to check and verify them (Casavant and Kuhl 1988; Yu and Buyya 2005).

Blockchain was initially proposed as an underpinning for the well-known cryptocurrency Bitcoin (Compton and Hauck 2002). But during the past few years, blockchain technology has been embraced in a number of businesses that are completely new and go beyond cryptocurrency (Singh and Chana 2016a), like the Internet of Things (IoT) (The evolution of Distributed Systems), automated transportation (Bonomi et al. 2012a), and medical care (Alrawais et al. 2017). In fact, the handling of important and delicate matters pertaining to governments and states is one of the many areas that this technology is having an impact on (Wang et al. 2015). This is due to the fact that it may increase accountability and accessibility while also saving corporation's both time and money.

The sector technological advances that predominantly emerged from the second quantum revolution are referred to as "quantum technology" (QT). Modern technologies that we are familiar with, consisting of semiconductors, magnetic resonance imaging (MRI), modern communication gadgets, or electronic cameras, and other imaging equipment, were originally introduced by the first quantum revolution. Nuclear energy and weaponry were produced by the initial quantum technology, and later the classical computer came to play a vital role. The use and testing of laser weapons have recently begun (Nakamoto 2008).

The second quantum revolution (Antonopoulos 2014) is defined as the manipulation and control of discrete quantum structures (such as atoms, ions, electrons, photons, molecules, or other quasiparticles) with the objective to reach the conventional quantum limit, or the limit of accuracy of measurement at quantum scales.

The technology of the second quantum revolution has been referred to as "quantum technology" in the current research. Instead of introducing substantially new military equipment or systems, quantum technology greatly improves the measurement, sensing, precision, computing, as well as the effectiveness of both current and upcoming military hardware.

Combined intelligence process is defined as "includ[ing] the organisations, capabilities, and processes involved in the collection, processing, exploitation, analysis, and dissemination of information or finished intelligence" by the United States' joint doctrine (Reyna, et al. 2018).

The basics of sound collective intelligence are defined by doctrine utilizing terminology like networked, decentralized, shared, distributed, protected, and secure, which appear to be the ideal match for the technology of blockchain. In order to better foresee the links across the two industries, the following subsections give a general overview of both blockchain-based systems and intelligence from the military.

The United States' joint doctrine describes the integrated intelligence process which "include[s] the organizations, capabilities, and processes involved in the collection, processing, exploitation, analysis, and dissemination of information or finished intelligence" (Kosba et al. 2016).

By employing words like networked, decentralized, shared, distributed, protected, and safe, which appear to be the perfect fit with blockchain technology, doctrine defines the basics of strong collective intelligence. To better anticipate the connections between the two companies, the following subsections provide a comprehensive overview of both the security intelligence and the blockchain technology (Tuli e al. 2019).

10.2 Fundamentals of Blockchain and Quantum Computing in Military Industry

Blockchain is a kind of disseminated data set that makes it conceivable to keep up with records in a protected, open, and extremely durable way. The tactical business can utilize blockchain innovation in different ways, like further developing coordinated operations, improving network protection, and empowering secure correspondence (Stanciu 2017; Samaniego and Deters 2016).

One way blockchain innovation can be utilized in the tactical business is by further developing strategies. The tactical depends vigorously on supply chains to furnish troops with vital hardware and assets. Be that as it may, supply chains can be helpless against digital assaults or extortion. Each exchange that happens inside the store network might have a carefully designed record made utilizing blockchain innovation, guaranteeing that all gear and assets are represented and diminishing the gamble of extortion (Li et al. 2017).

Another way blockchain technology can be used in the military industry is by enhancing cybersecurity. The military is a prime target for cyberattacks, and its computer networks must be protected at all times. Networks may be built using blockchain technology that is safe, decentralized, and challenging to hack. By using a blockchain-based system, the military can ensure that its sensitive data is protected from unauthorized access (Singh and Chana 2015).

Finally, blockchain-based technology is helpful in enabling safe communication between military personnel. Communication is critical in military operations, but traditional communication channels can be vulnerable to interception and eavesdropping (Gill et al. 2019a).

10.3 Blockchain for Military and Defense

Military leaders who adopted cyber technology in the 1990s and early 2000s are now working to address the severe problems produced by digital technologies (Glikson et al. 2017; Gubbi et al. 2013; Tordera et al. 2016; Bonomi et al. 2012b; Lin et al. 2017). The ability of a motivated hacker to infiltrate both military and civilian networks has been repeatedly shown during the course of decades-long attempts to hack and misuse computer security measures. It has been required by military forces to develop a long time and, ideally, affordable protection for weapons and equipment due to the risk that advanced weapon systems will be damaged or rendered inoperable by non-kinetic impacts (Lin et al. 2017).

Thanks to blockchain and its as-of-yet-untested military applications due to which a single point-of-failure threat model may be replaced with a majority-compromised risk model, where an attacker cannot exploit a single point of failure and only has to compromise one node to compromise the system. The following issues may be covered by the military's deployment of blockchain: A few examples are encrypted communications, supply chain management, combat management, infrastructural monitoring, detection of intrusions, and UAV management.

10.4 Quantum Computing and Its Uses in the Military Industry

Entanglement, quantum bits (qubits), and superposition are only a few of the fundamental concepts that are essential to quantum applications. Superposition describes a quantum system's capacity to exist in two or more states simultaneously. A qubit is a particular kind of computer component that encrypts data by employing the concept of superposition. (In contrast to a conventional computer, which stores data in bits that are limited to identifying binary states of either 0 or 1, a quantum computer stores data in qubits, which can individually represent 0, 1, or a simultaneous combination of 0 and 1. The results show that each qubit adds exponentially to the power of a quantum computer.) (Singh et al. 2017a; Shirazi et al. 2017; Iorga et al. 2018).

According to the National Academy of Sciences (NAS), entanglement is characterized by "two or more quantum objects in a system can be intrinsically linked such that measurement of one dictates the possible measurement outcomes for another, regardless of how far apart the two objects are." Entanglement supports a variety of possible military applications for quantum technology. However, due to the brittleness of quantum states, which can be disrupted by extremely small motions, temperature changes, or additional outside variables, both superposition and entanglement are challenging to maintain (Rahmani et al. 2018; Fang et al. 2011).

10.5 Military Applications of Quantum Technology

The Department of Defence (DOD) is mainly interested in three quantum technology applications: quantum sensing, quantum computing, and quantum networking, according to the Defence Science Council (DSB), an independent council of scientific advisors. The DSB came to the conclusion that quantum radar, which was thought capable of determining the object's performance traits, particularly low observable or stealth aircraft (e.g., cross-section, speed, and radar), "will not provide upgraded capability to DOD."

10.6 Quantum Sensing

A sensor and quantum physics are used in quantum sensing. According to the DSB, it represents the most advanced military application of quantum technology and it is currently "poised for mission use." Numerous improved military capabilities could be provided by quantum sensing. For instance, it might offer alternate location, timing alternatives, and navigation, which could enable the military forces to carry on with their normal operations in GPS-denied or -degraded circumstances.

Quantum sensors may also be utilized for ISR (intelligence, surveillance, and reconnaissance) purposes. Such sensors' successful creation and use might significantly enhance the ability to detect submarines, which would jeopardize the ability of sea-based nuclear deterrents to remain operational. Quantum sensors might possibly be useful to armed forces because of their anticipated "extreme sensitivity for ecological disturbances" to find nuclear or subterranean structures. Similar to how electromagnetic emissions might potentially be detected by military forces using quantum sensors, this could potentially improve electronic warfare capabilities and help find covert enemy forces (Fig. 10.1).

10.7 Quantum Communications

The field of quantum communications is still in its infancy, with the exception of quantum key distribution (QKD). Theoretically, Quantum communications may enable secure networking of quantum warfare sensors, computers, and various other systems, enhancing performance compared to a single quantum system or conventional communications channel. As a result, the possible use scenarios (i.e.,

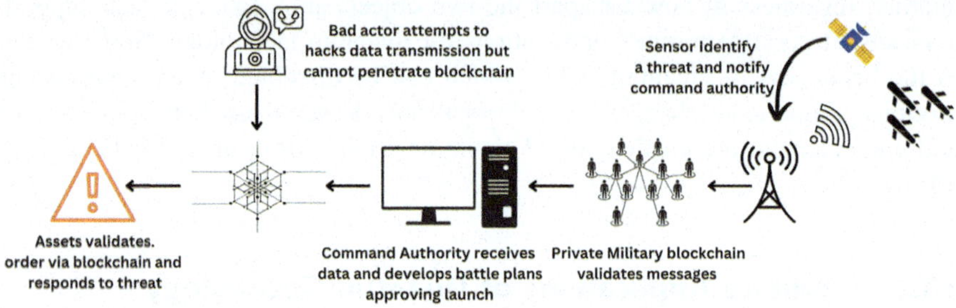

Fig. 10.1 Cyberwarfare mechanism process of blockchain and computing

environments other than the laboratory conditions generally required to maintain fragile quantum states) might be expanded. Networking could also make such systems more resilient at a distance. This might considerably increase the utility of quantum communications in the military (Madsen et al. 2013).

10.8 Benefits of Quantum Computing

Integrating blockchain and quantum computing technologies can enhance cybersecurity and create more secure communication channels in several ways.

With the ability to detect the insensible, revolutionize cybersecurity, and give us access to previously unsolvable issues, quantum technologies have the potential to give humans radically new powers. Two applications will have particularly substantial effects on the defense and security environment in the short- to medium-term (Lin et al. 2016; Michie et al. 1994; Chen et al. 2017).

Starting with quantum sensing. There are some intriguing military uses for quantum sensors. Position, Navigation, and Timing (PNT) applications might benefit from quantum sensors for detecting stealth aircraft and submarines, for instance. These "quantum PNT devices" may be used as reliable inertial navigation devices, enabling travel without the aid of external markers like GPS technology. This will be considered as a revolutionary advancement in underwater navigation for submarines, for example, in the event of GPS signal loss, a backup navigating system is used for platforms above water (Jordan and Mitchell 2015; Jia et al. 2013; Carbonneau et al. 2008).

The most sophisticated type of sensing, communication, and computation is quantum computing, which has the first commercially feasible quantum sensors in operation. Furthermore, given the enormous potential value that quantum communications and computation have for the civil industry, the civilian sector is likely to push breakthroughs in these fields. For the military, however, quantum sensing applications like quantum PNT and quantum radar hold considerable appeal. Therefore, in order to make these prospective applications a reality, the military should promote, support, and oversee this area's research and development (Steane 1998).

Second, there is a "quantum threat" possessed by quantum computing. The preceding section indicates that the problem the quantum computers are particularly good at solving is the factorization of integers. Cryptographic methods are used to encrypt the majority of our digital infrastructure and virtually all of our online activities, such as video conferencing, emailing, and accessing our online bank accounts (like the RSA algorithm), that are based on how challenging it is to solve these types of integer factorization problems. Shor's algorithm, developed in 1994 and awaiting a computer with quantum technology that can execute it, is a quantum algorithm that can solve these challenges and decode our digital communication while practically usable quantum computers are still being constructed (Leuenberger and Loss 2001).

This offers an immediate danger to society in general and also primarily to the military given the requirement for safe and confidential information for defense and security. We would require to totally update all of our protected digital systems employing "quantum-resistant" encryption that is safe against both quantum and conventional computers in order to prevent this menace. While waiting for quantum communication (QKD or quantum teleportation) to develop, one strategy would be to use this quantum technology to counteract the other quantum technology (Han and Kim 2002).

By giving end-user military operators access to validation and testing infrastructure (test centers), allied armies might significantly add value to existing initiatives in business and academics. Early testing of these technologies assists in both their development and eventual acceptability by allowing the military to get familiar with them and their capabilities. The military's understanding of the possible risks posed by quantum technology, notably in the field of cyber, is also increased through active participation in the quantum ecosystem (Schuld et al. 2015).

10.9 Benefits of Blockchain in Defense

The application of the technology known as blockchain is gaining popularity among every sector. The government of the United States has taken notice and is looking into how the technology may be used for its benefit (Lloyd et al. 2013; Singh and Singh 2017).

Agencies are investigating the use of blockchain technology, which might have ramifications in a variety of fields, including identity management, patent, trademark, and supply chain management, as well as copyrights. The Department of Health and Human Services is at the forefront of the pack with significant blockchain efforts in the pipeline. For instance, HHS Accelerate aims to save the organization money and time by expanding access to safe, standard data and significantly cutting down on paperwork handling time (Singh and Singh 2016).

The Army and Treasury Department are creating blockchain-based systems for tracking devices. While the Army is looking into healthcare device tracking, the Treasury is concentrating on mobile device tracking in order to better organize its resources and get rid of outdated or unused equipment (Singh et al. 2017b).

10.10 The Secure Data Storage and Management

Integrating blockchain and quantum computing technologies can also lead to the development of sophisticated data analytics tools that can provide enhanced insights into large datasets (Singh et al. 2017b).

The order data framework is a significant strategy and hardware to acknowledge ongoing, exact, and programmed war zone order at war and help base camps at all levels to execute logical and proficient administration of subordinate units and weapons at common times (Gill et al. 2019b). As a natural "human–machine" framework, accepting leaders as the center, and PCs, and involving PCs and other data innovation supplies as the essential and material assurance, the framework naturally joins different order and-control strategies and administrators to empower undeniable level computerization in the assortment, transmission, handling and utilization of military order data. Along these lines, it guarantees information uprightness, accessibility, and security of the order data framework, and essentially ensures the improvement of battle adequacy and the progress in seizing (Gill et al. 2019c; Teerapittayanon et al. 2017; Nguyen et al. 2018; Eshratifar et al. 2019).

10.11 Military Management

The conventional military administration structure is a hierarchical "order and control" framework, which brings along such issues as swollen establishments, high administration costs, muddled meaning of liabilities, wasteful administration, repetitive administration, unfortunate data transmission, and power gathering at the high level while the lower levels having exceptionally restricted independence and in this way restricted development potential (Osia et al. 2017; Singh and Chana 2016b; Singh et al. 2017).

Blockchain + Delicate information to the board utilizing the "untamperable" component of blockchain innovation, we can give an answer for the issue of "difficult to keep up with proof" in delicate information the executives in military review and oversight, HR and clinical and medical care. "Honest record" of all data can be acknowledged through "entire organization witness," subsequently staying away from archive fakes, document missing, and data altering (Eshratifar and Pedram 2018; Singh and Chana 2013).

10.12 Conclusion and Future Outlook

The safeguard area is fundamental to keeping up with public safety and protecting residents from different risks. The safeguard business should keep steady over the latest advancements in innovation and investigate new ones that could work on their capacities. In this part, we talked about how blockchain innovation and quantum figuring could alter the protection area, including how they may be utilized for cryptography, store networks on the board, information from the executives, knowledge and observation, and network safety. We additionally accentuated the meaning of continuous examination in fields like network safety, expanded and computer-generated reality, and AI. The guard business can increment efficiency, support security, and empower better critical thinking abilities by embracing new innovations and burning through cash on innovative work. To remain on the ball and ensure we are prepared to confront the dangers of today and tomorrow, it is urgent that we continue to research new innovations and their applications as we advance.

References

Alrawais A, et al. Fog computing for the internet of things: Security and privacy issues. IEEE Internet Comput 21(2):34–42

Antonopoulos AM (2014) Mastering Bitcoin: "Unlocking Digital Cryptocurrencies" O'Reilly Media, Inc.

Bonomi F, Milito R, Zhu J, Addepalli S (2012) Fog computing and its role in the internet of things. In: Proceedings of the first edition of the MCC workshop on Mobile cloud computing. ACM, pp 13–16

Bonomi F, Milito R, Zhu J, Addepalli S (2012) Fog computing and its role in the internet of things characterization of fog computing. In: Proc. first Ed. MCC Work. Mob. cloud Comput, pp 13–15

Carbonneau R, Laframboise K, Vahidov R (2008) Application of machine learning techniques for supply chain demand forecasting. Eur J Oper Res 184(3):1140–1154

Casavant TL, Kuhl JG (1988) A taxonomy of scheduling in general-purpose distributed computing systems. IEEE Trans Software Eng 14(2):141–154

Chen M, Hao Y, Hwang K, Wang L, Wang L (2017) Disease prediction by machine learning over big data from healthcare communities. IEEE Access 5:8869–8879

Compton K, Hauck S (2002) Reconfigurable computing: a survey of systems and software. ACM Comput Survey 34(2):171–210

Eshratifar AE, Pedram M (2018) Energy and performance efficient computation offloading for deep neural networks in a mobile cloud computing environment. In: Proceedings of the 2018 on Great Lakes Symposium on VLSI. ACM, pp 111–116

Eshratifar AE, Esmaili A, Pedram M (2019) BottleNet: a deep learning architecture for intelligent mobile cloud computing services. arXiv preprint arXiv:1902.01000

Fang X, Member S, Misra S, Xue G, Yang D (2011) Smart grid – the new and improved power grid, pp 1–37

Flynn MJ (1966) Very high-speed computing systems. Proc IEEE 54(12):1901–1909

Gill SS, Chana I, Singh M, Buyya R (2019) RADAR: self-configuring and self-healing in resource management for enhancing quality of cloud services. Concurr Comput Pract Exp 31(1):e4834

Gill SS, Garraghan P, Buyya R (2019) ROUTER: Fog enabled cloud based intelligent resource management approach for smart home IoT devices. J Syst Soft

Gill SS, Garraghan P, Stankovski V, Casale G, Thulasiram RK, Ghosh SK, Ramamohanarao K, Buyya R (2019) Holistic resource management for sustainable and reliable cloud computing: an innovative solution to global challenge. J Syst Software

Glikson A, Nastic S, Dustdar S (2017) Deviceless edge computing: extending serverless computing to the edge of the network. In: Proceedings of the 10th ACM international systems and storage conference. ACM, p 28

Gubbi J, Buyya R, Marusic S, Palaniswami M (2013) Internet of Things (IoT): A vision, architectural elements, and future directions. Futur Gener Comput Syst 29(7):1645–1660

Han KH, Kim JH (2002) Quantum-inspired evolutionary algorithm for a class of combinatorial optimization. IEEE Trans Evol Comput 6(6):580–593

Iorga M, Feldman L, Barton R, Martin MJ, Goren N, Mahmoudi C (2018) Fog computing conceptual model: recommendations of the national institute of standards and technology. NIST Spec Publ, pp 500–325

Jia L, Zhao Q, Tong L (2013) Retail pricing for stochastic demand with unknown parameters: an online machine learning approach. In: 2013 51st Annual Allerton conference on communication, control, and computing (Allerton). IEEE, pp 1353–1358

Jordan MI, Mitchell TM (2015) Machine learning: trends, perspectives, and prospects. Science 349(6245):255–260, 29

Kosba A, Miller A, Shi E, Wen Z (2016) Papamanthou C. Hawk: the blockchain model of cryptography and privacy-preserving smart contracts. Security and Privacy (SP). In: 2016 IEEE Symposium on, San Jose, CA, USA, IEEE, pp 839–858

Leuenberger MN, Loss D (2001) Quantum computing in molecular magnets. Nature 410(6830):789

Li X, Jiang P, Chen T, Luo X, Wen Q (2017) A survey on the security of blockchain systems. Futur Gener Comput Syst. https://doi.org/10.1016/j.future.2017.08.020

Lin J, Yu W, Yang X (2016) Towards multistep electricity prices in smart grid electricity markets. IEEE Trans Parallel Distrib Syst 27(1):286–302

Lin J, Yu W, Zhang N, Yang X, Zhang H, Zhao W (2017) A survey on internet of things: architecture, enabling technologies, security and privacy, and applications. IEEE Internet Things J 4(5):1125–1142

Lloyd S, Mohseni M, Rebentrost P (2013) Quantum algorithms for supervised and unsupervised machine learning. arXiv preprint arXiv:1307.0411

Madsen H, Albeanu G, Burtschy B, Popentiu-Vladicescu F (2013) Reliability in the utility computing era: Towards reliable fog computing. Int Conf Syst Signals, Image Process, pp 43–46

Michie D, Spiegelhalter DJ, Taylor CC (1994) Machine learning. Neural Stat Class 13

Nakamoto S (2008) Bitcoin: a peer-to-peer electronic cash system

Nguyen KK, Hoang DT, Niyato D, Wang P, Nguyen D, Dutkiewicz E (2018) Cyberattack detection in mobile cloud computing: A deep learning approach. In: 2018 IEEE wireless communications and networking conference (WCNC). IEEE, pp 1–6

Osia SA, Shamsabadi AS, Taheri A, Katevas K, Sajadmanesh S, Rabiee HR, Lane ND, Haddadi H (2017) A hybrid deep learning architecture for privacy-preserving mobile analytics. arXiv preprint arXiv:1703.02952

Rahmani AM et al (2018) Exploiting smart e-Health gateways at the edge of healthcare Internet-of-Things: a fog computing approach. Futur Gener Comput Syst 78:641–658

Reyna A, et al. (2018) On blockchain and its integration with IoT. Challenges and opportunities. Futur Gener Comput Syst 88:173–190

Rimal BP, Choi E, Lumb I (2009) A taxonomy and survey of cloud computing systems. In: INC, IMS and IDC, 2009. NCM'09. Fifth International Joint Conference on IEEE, pp 44–51

Samaniego M, Deters R (2016) Blockchain as a Service for IoT. In: 2016 IEEE International Conference on Internet of Things (iThings) and IEEE Green Computing and Communications (GreenCom) and IEEE Cyber, Physical and Social Computing (CPSCom) and IEEE Smart Data (SmartData). IEEE

Schuld M, Sinayskiy I, Petruccione F (2015) An introduction to quantum machine learning. Contemp Phys 56(2):172–185

Shirazi SN, Gouglidis A, Farshad A, Hutchison D (2017) The extended cloud: review and analysis of mobile edge computing and fog from a security and resilience perspective. IEEE J Sel Areas Commun 35(11):2586–2595

Singh S, Chana I, Buyya R (2017) STAR: SLA-aware autonomic management of cloud resources. IEEE Trans Cloud Comput

Singh S, Chana I (2015) Q-aware: quality of service based cloud resource provisioning. Comput Electr Eng 47:138–160

Singh S, Chana I (2016a) QoS-aware autonomic resource management in cloud computing: a systematic review. ACM Comput Survey 48(3):42
Singh S, Chana I (2016b) A survey on resource scheduling in cloud computing: Issues and challenges. J Grid Comput 14(2):217–264
Singh I, Singh S (2016) Model for targeting customers based on analytics in telecom domain. Int J Modern Educ Comput Sci 8(11):43
Singh I, Singh S (2017) Framework for targeting high value customers and potential churn customers in telecom using Big Data Analytics. Int J Educ Manag Eng 7(1):36–45
Singh S, Chana I, Singh M (2017a) The journey of QoS-aware autonomic cloud computing. IT Professional 19(2):42–49
Singh S, Chana I (2013) Consistency verification and quality assurance (CVQA) traceability framework for SaaS. In: 2013 3rd IEEE international advance computing conference (IACC). IEEE, pp 1–6
Singh I, Singh KV, Singh S (2017) Big data analytics based recommender system for value added services (VAS). In: Proceedings of sixth international conference on soft computing for problem solving. Springer, Singapore, pp 142–150
Stanciu A (2017) Blockchain based distributed control system for edge computing. In: 2017 21st international conference on control systems and computer science (CSCS). IEEE
Steane A (1998) Quantum computing. Rep Prog Phys 61(2):117
Teerapittayanon S, McDanel B, Kung HT (2017) Distributed deep neural networks over the cloud, the edge and end devices. In: 2017 IEEE 37th international conference on distributed computing systems (ICDCS). IEEE, pp 328–339
The evolution of Distributed Systems. Available Online: https://medium.com/microservices-learning/the-evolution-of-distributedsystems-fec4d35beffd
Tordera EM, et al. (2016) What is a fog node a tutorial on current concepts towards a common definition
Tuli, Shreshth, Redowan Mahmud, Shikhar Tuli, and Rajkumar Buyya. "FogBus: A blockchain-based lightweight framework for edge and fog computing." Journal of Systems and Software (2019). Volume 154, August 2019, Pages 22–36
Wang Y, Uehara T, Sasaki R (2015) Fog computing: Issues and challenges in security and forensics. In: 2015 IEEE 39th annual computer software and applications conference, Vol 3. IEEE
Yu J, Buyya R (2005) A taxonomy of workflow management systems for grid computing. J Grid Comput 3(3–4):171–2005

Building Resilient Digital Forensic Frameworks for NoSQL Database: Harnessing the Blockchain and Quantum Technology

11

Rizwan Ur Rahman, Kuldeep Singh, Deepak Singh Tomar, and Rabia Musheer

Abstract

Digital forensics is the process of gathering, examining, and presenting digital evidence from devices like computers, smart phones, and cameras with supporting documentation. Investigating cybercrimes, retrieving deleted data, confirming the veracity of documents and photos, and locating the source and location of digital information are just a few of the many uses for digital forensics. The process of investigating logs of database systems and metadata to search for clues and signs of criminal activity or security breaches is called database forensics. Forensic experts could recover lost or damaged data with tools of database forensics and determine the origin, nature, and scope of an attack. Two cutting-edge technologies that can be employed in digital forensics are blockchain and quantum computing. While quantum computing can be used to find patterns and encrypt data, blockchain can be used to guarantee the accuracy of logs and monitor data flow. These technologies offer verifiable audit trails, an unbreakable chain of custody, and secure storage of forensic artifacts for forensic investigators. Digital forensic professionals must develop quantum-safe digital evidence preservation techniques and deal with other issues brought

R. U. Rahman (✉) · K. Singh
School of Computing Science and Engineering, VIT Bhopal University, Bhopal 466114, India
e-mail: rizwan.rahman12@gmail.com; rizwan.ur@vitbhopal.ac.in

K. Singh
e-mail: kuldeepsingh2022@vitbhopal.ac.in

D. S. Tomar
Maulana Azad National Institute of Technology, Bhopal, India
e-mail: deepaktomar@manit.ac.in

R. Musheer
School of Advanced Sciences and Language, VIT Bhopal University, Bhopal 466114, India

A. Kumar et al. (eds.), *Sustainable Security Practices Using Blockchain, Quantum and Post-quantum Technologies for Real Time Applications*, Contributions to Environmental Sciences & Innovative Business Technology,
https://doi.org/10.1007/978-981-97-0088-2_11

on by quantum computing. However, with careful planning and preparation, these difficulties are surmountable. For the wide-column store NoSQL database, we provide a six-phase forensic investigation framework in this chapter. Our system includes the full process of forensic examination, from preparation to reporting, in contrast to other studies that concentrated on particular parts of NoSQL forensics, such as transaction log analysis or deleted data recovery. We also discuss the difficulties associated with finding and evaluating distributed evidence in a NoSQL setting, which is distinct from a relational DBMS. Our system may be used with a variety of documents and wide-column NoSQL DBMSs including MongoDB, CouchDB, and Cassandra. We do a case study using Cassandra as an illustration to show its efficacy.

Keywords

Blockchain • Digital forensic • Evidence preservation • No SQL • Quantum technology

11.1 Introduction

In the last few years, data volume is increasing very rapidly in almost all fields. There are basically two types of databases that can be used to handle such huge data. These are relational databases and non-relational databases. Relational databases are the traditional database systems that have been used since very old times. These database systems follow ACID (Atomicity, Consistency, Isolation and Durability) properties. They have a predefined schema and follow a rigid approach, thus resulting in degrading the performance. So, now a days, non-relational databases are being widely used for handling such large amounts of data. These non-relational databases should not be considered as a replacement of relational database, instead, they can be a more suitable and easier option for dealing with these large databases (Han et al. 2011).

There are various non-relational databases, which are also known as NoSQL databases. The term NoSQL is an acronym for Not Only SQL. These databases do not rely only on a single language, i.e., structured query language, but also make use of other languages for querying the data.

In contrast with relational database systems that follow ACID properties, NoSQL database systems support for BASE (basically available, soft state, Eventual Consistency) property. Basically, available means that the system does insure availability of data. It also guarantees to be in a responding condition. Although that response can even be success or failure of the request. "Soft" state refers to a system, which can change its state over time even if there is no input. NoSQL database systems have a property of eventual consistency. It means that the system does not check for consistency after every transaction but will become eventually consistent once it stops receiving input (Chandra 2015).

NoSQL database systems support schema-less modeling, i.e., they do not have a fixed and predefined schema. The database structure can vary as per the requirement. Thus, NoSQL system provides flexibility as well as scalability to users. Scalability refers that the system performance and efficiency will be increased even when the data load increases. Scaling can be classified in two forms—horizontal scalability and vertical scalability. In horizontal scalability, more systems are added together to improve the performance whenever workload increases. On the other hand, in the case of vertical scalability more memory, processor, and hardware units will be added in a single system to improve the performance. NoSQL database system supports horizontal scalability resulting in better and improved performance.

Moreover, to improve performance, NoSQL database systems use the sharding mechanism, which is similar to horizontal scaling. Sharding refers to the process of distributing the load over multiple servers. Thus, the load is divided into multiple systems such as a distributed environment. In this distributed environment NoSQL database system supports a property known as CAP theorem. This theorem was first proposed by Professor Eric Brewer in 2000. The main tenet of this theorem is that a system must satisfy three essential requirements which are consistency, availability, and partition tolerance for its proper functioning. Consistency means that the system must always be in a consistent state. This is as same as we learn in ACID properties. Availability means that the system must be available whenever required. Partition tolerance means that failure of a single node should not impact on complete system performance. The rest of the system should be in operating condition.

The CAP theorem states that a distributed system cannot meet all the above three properties simultaneously. Kit can only support any two properties at a time. Thus, the NoSQL database system can be classified into three classes that are concerned with consistency and availability (CA), concerned with consistency and partition tolerance (CP), and concerned with availability and partition tolerance (AP).

Due to all these features, NoSQL systems are being used in multimedia, word processing, emails, and social media in various other fields. There are various companies that are now using NoSQL database systems. MongoDB, Cassandra, Redis, Hbase, and Neo4j are the top five NoSQL database systems, that are being widely used by big companies.

There are various data storage models that are used in NoSQL databases. Different NoSQL databases uses different models. These models are

1. Key-value store
2. Document store
3. Column store
4. Object store
5. Graph databases

The key-value store model does not support the ACID property. It is a kind of hashing technique. In this model, a unique key value is associated with each data object. This key can be auto-generated or synthetic and it always contains a string type of value. The problem with this type of model is that as the data volume increases, it becomes very difficult to maintain unique key values. Redis, OrientDB, MUMPS, Aerospike, etc., are examples of key-value data store models.

Document store model is similar to a key-value store model. But, in this model, a complex data structure known as a document is mapped to a unique key value. The document is itself a structure and each document has its own schema. The document can itself be a grouping of key-value pairs or string, numeric, Boolean arrays. MongoDB, RavenDB, CouchDB, BaseX, eXist, etc., are examples of document store data models.

In a column store database, a group of columns are stored together rather than as a group of rows. These cells are grouped together and stored in continuous locations in database. Due to this search becomes faster and access time gets reduced. Cassandra, Vertica, and HBase are examples of column store databases.

Object data stores can be considered as a hybrid approach which is a combination of relational database management and object-oriented procedure. In this information is stored in the form of objects. This model supports all the features of object-oriented programming such as data encapsulation, inheritance, and polymorphism.

Graph store model stores data in the form of nodes and edges. Nodes are connected with edges showing the interconnectivity between them. The queries are executed by traversing these edges. These models are generally used to contain information about the networks. OrientDB, Allegro, Stardog, etc., are examples of graph store models.

Since NoSQL database systems focus more on improving performance and faster access, there exist some security issues with them. Security was never considered a priority for NoSQL. There are various vulnerabilities in these database systems that can be used to perform various attacks. These vulnerabilities arise due to huge databases, large-scale infrastructures, different sources of data and varying formats, etc. Attackers can also perform NoSQL injection by exploiting these vulnerabilities in the system. Thus, various researches are still going on to make these NoSQL database systems more secure.

11.2 Introduction to Cassandra Database

Facebook created the distributed, highly scalable NoSQL database technology known as Cassandra. While ensuring high availability and fault tolerance, it is made to handle massive amounts of data across numerous commodity servers. Many major businesses utilize Cassandra, which is ideal for applications requiring great scalability and real-time data processing (Cassandra 2014).

11.2.1 Need for Cassandra

Earlier, it was developed for Facebook inbox search feature. After that, Facebook made it as an open source and then many versions were released. To overcome big data workloads across multiple nodes without any failure, Cassandra was designed. Apache Cassandra is a distributed, decentralized, scalable, and highly available NoSQL database.

Cassandra is a distributed database that can run on multiple machines. It is decentralized, meaning that there is no master–slave relationship between nodes. All nodes are identical and there is no single point of failure. This makes Cassandra highly available.

Cassandra is also scalable. We can add or remove nodes at any time without impacting performance. There is no need for manual intervention or rebalancing. The data is automatically redistributed across the nodes in the background.

11.2.2 When to Use Cassandra?

- Large and growing data.
- High availability is important.
- Changing data (Non-static data).
- Consistency is not a concern.
  [Cassandra is an eventually consistent database by default, it means if we write something and read something out immediately, it is possible that the data is inconsistent]
  [Cassandra has very high throughput for Faster writes]

11.2.3 Data Model of Cassandra

- Columns: A name-value pair is represented here.
- Row: This is a container for columns that the primary key refers to. Rows are contained in a table.
- Tables that span one or more nodes are contained in a key space. [Identical to Database in RDBMS] Cluster: The key space's container. In RDBMS, we have joins, foreign keys, and there are flexible queries but in Cassandra, we cannot join two tables (Raut 2017).
- We cannot refer from one table to another table; i.e., no foreign keys in this database, as these two operations are very expensive in terms of distributed databases (Fig. 11.1).

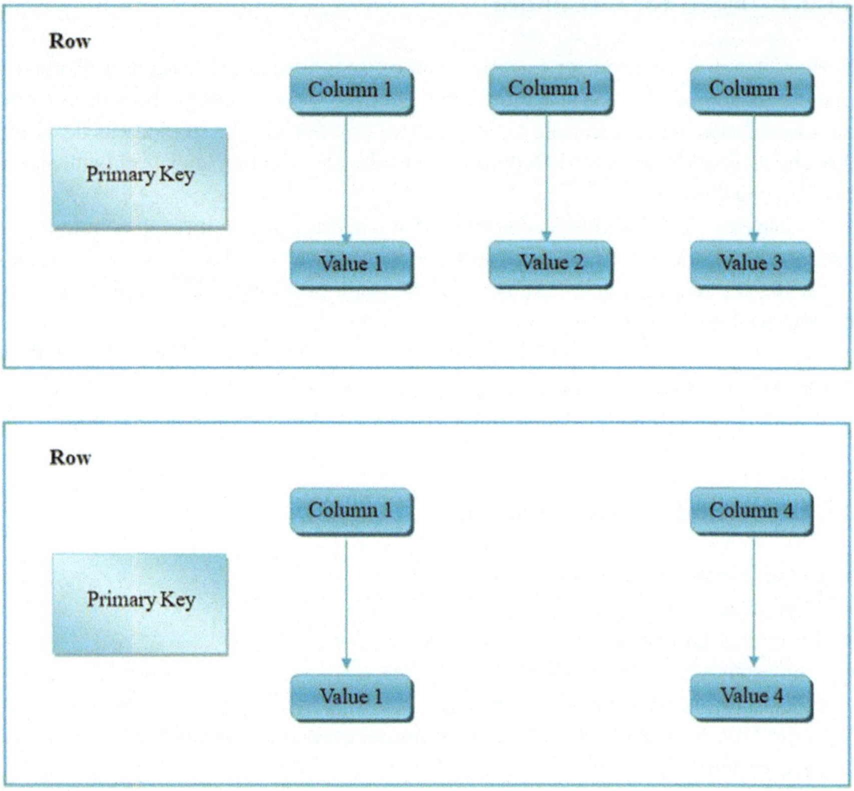

Fig. 11.1 Data models in Cassandra

11.2.4 Applications of Cassandra

Apache Cassandra is a highly scalable distributed NoSQL database that is built to handle enormous volumes of data across many commodity machines. Cassandra can be used for a variety of purposes, including:

- Big Data: Cassandra is often used for big data applications because it can handle massive amounts of data in real time (Fan et al. 2014). It is used by companies like Apple, Netflix, and eBay for storing and processing data.
- Internet of Things (IoT): Cassandra can handle the large volume of data generated by IoT devices and can store and process that data in real time. Companies like Cisco and Verizon use Cassandra for their IoT applications.
- Chatting Applications: It is capable of managing large volumes of data generated by messaging and chat applications in real-time data. Companies like Twitter and Facebook use Cassandra for their chatting applications.

- Financial Services: It is capable of handling large volumes of financial data and the financial industry uses it for trading and risk management systems.
- Healthcare: It is capable of storing healthcare data, such as medical reports and images. Companies like Philips Healthcare use it to store data.
- Gaming: It stores and processes data of gaming applications, such as game state, user data, and player stats. Companies like Zynga use it to store gaming applications data.

11.2.4.1 Cassandra Real-Time Applications

- Netflix: uses to store 95% of its data, including customer data and viewing history. CassandraDB allows them to manage a high volume of data and its distribution of data to nodes ensures data reliability and availability (https://www.infoq.com/news/2023/02/netflix-annotations-cassandra/; https://www.infoworld.com/article/2078771/big-movies--big-data--netflix-embraces-nosql-in-the-cloud.html).
- Apple: uses to store and manage data in Siri, iCloud, and other services. This Database has features like high availability and fault tolerance which make it an ideal option for critical services (https://www.techrepublic.com/article/apples-secret-nosql-sauce-includes-a-hefty-dose-ofcassandra/; https://cassandra.apache.org/_/case-studies.html).
- eBay: eBay effectively stores user profiles, transaction history, and product catalogue data by employing the robust capabilities of Cassandra. The capability of Cassandra to handle extensive volumes of data in a timely and effective manner serves as the critical foundation for eBay's capacity to offer an uninterrupted shopping experience for their client (https://cassandra.apache.org/_/case-studies.html).
- Twitter: Twitter effectively stores and manages all their tweets, user data, and activity streams. The capability of Cassandra's performance and high scalability allows Twitter to handle millions of tweets per second and provide a real-time user experience.
- Airbnb: Airbnb uses to manage its booking system, property listings, and user data. The capabilities of Cassandra to handle high traffic and large volumes of data enable Airbnb to offer a dependable and scalable platform to its users.
- Uber: Uber uses to store and manage their driver and rider data, trip history, and payment transactions. The real-time data processing and high scalability of Cassandra make it a highly suitable choice for Uber's requirements (https://cassandra.apache.org/_/case-studies.html).

Cassandra is a widely accepted option for enterprises that handle massive data sets and necessitate a database solution that is both exceptionally accessible and scalable.

11.2.4.2 Cassandra in Banking System

1. Transactional Data Storage: Cassandra database is used to store transactional data for banking systems such as account balances, transaction histories, and other financial data. With its features like low latency and high write throughput, Cassandra can ensure that transactional data is stored accurately and quickly.
2. Fraud Detection: Cassandra database can be used to store and analyze vast amounts of data to detect patterns of fraud in real-time. By using ML (Machine Learning) algorithms, Cassandra database can analyze transactional data to detect fraudulent activity and alert banking systems to potential security breaches.
3. Customer Analytics: Customer information like transaction history, demographics, and social media interaction can be stored and analyzed using the Cassandra database. These data can help improve customers' experiences, increase customer engagement, and optimize target marketing efforts.
4. Compliance Reporting: Banking systems are subject to strict laws and compliance requirements. Cassandra database can be used to store data for compliance reporting purposes, for example, account balances, transaction histories, and audit logs.
5. Operational Analytics: Operational Analytics: Cassandra database can be used to store and analyze operational data such as server logs, network traffic, and system performance. This information is very useful to improve system performance, improve overall system efficiency, and identify and resolve issues.

11.3 Security of Cassandra

The widely used distributed NoSQL (Not Only Structured Query Language) database management system Cassandra is free and open source. That is capable of managing enormous data volumes across several servers. The most important component of every database is security (Okman et al. 2011).

11.3.1 Architecture of Cassandra

Node: All data are stored in the node. This is the most basic component of Cassandra.

Data Center: The collection of nodes is called the data center. Data centers can be physical as well as virtual, depending upon workload.

Cluster: Cluster comprises one or more data centers. Clusters usually span different physical locations (Fig. 11.2; Table 11.1).

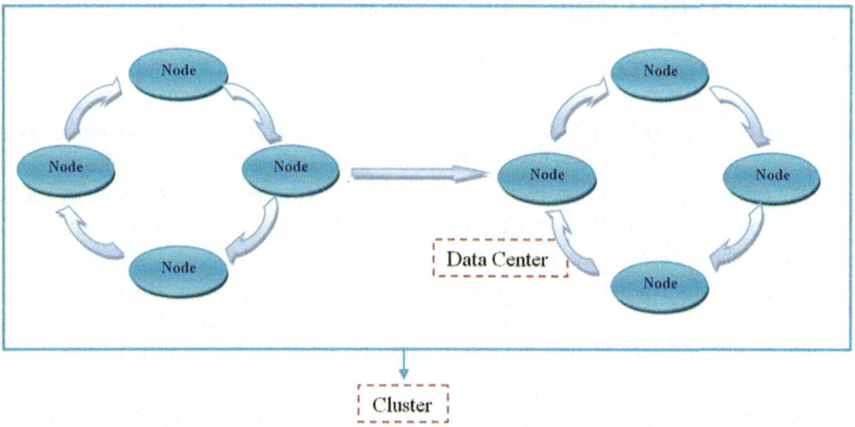

Fig. 11.2 Architecture of Cassandra

Table 11.1 Versions of Cassandra

S. No.	Version	Original release date	Features added
1	0.6	2010-04-12	Secondary indexes, counters, atomic batch operations, Time-to-Live (TTL)
2	0.7	2011-01-10	Online schema changes
3	0.8	2011-06-03	Self-tuning memtables, the Cassandra Query Language (CQL), and support for zero-downtime upgrades
4	1	2011-10-18	Improved read performance, even compaction, and integrated compression
5	1.1	2012-04-24	Row-level isolation, self-tuning caches, and assistance with mixed SSD/spinning disc deployments
6	1.2	2013-01-02	Request tracing, atomic batches, inter-node communication, and clustering across virtual nodes
7	2	2013-09-03	Lightweight transactions, triggers, improved compactions
8	2.1	2014-09-16	User-Defined Types (UDTs), collections in materialized views, CQL binary protocol v2
9	2.2	2015-07-20	JSON data type, SASI (SSTable Attached Secondary Indexes), CQL binary protocol v3, CQL tracing, virtual tables
10	3	2015-11-09	Triggers, Improved compaction, performance and stability
11	3.11	2017-06-23	Java 11, CQL binary protocol v5
12	4	2021-07-26	Simple Authentication and Security Layer (SASL) authentication, Improved compaction
13	4.1	2022-06-17	SASI (SSTable Attached Secondary Index)

11.3.2 Security Components of Cassandra

There are mainly three types of components for security in Cassandra:

1. **TLS/SSL Encryption**: Cassandra offers secure communication between a client system and a database cluster (Rahman and Tomar 2020a) thanks to TLS/SSL encryption. You can enable SSL/TLS encryption to secure data in transit between nodes and clients. Cassandra also supports encryption of data stored on disk using Transparent Data Encryption (TDE).
2. **Client Authentication**: Cassandra supports authentication through various authentication providers, including Password Authenticator, LDAP Authenticator, and Kerberos Authenticator. Cassandra uses the Password Authenticator to authenticate users by default.
3. **Authorization**: Cassandra uses role-based access control (RBAC) to provide authorization.

Other security components are:

1. **Firewall**: Firewall rules can be set up to restrict access to Cassandra nodes. This helps to prevent unauthorized access to data (Ioannidis et al. 2000). Cassandra does not have a built-in firewall. Accordingly, you can configure the firewall running on nodes in your Cassandra cluster.
2. **Datastax Enterprise (DSE)**: Datastax Enterprise is a commercial version of Cassandra that includes additional security features (Ivanov et al. 2015). These features include data masking, dynamic data masking, and row-level access control.
3. **Audit logging**: Cassandra provides audit logging to track database activity. You can configure logging to record events such as successful and failed login attempts, schema changes, and more.

11.3.3 Attacks on Cassandra

Any attempt to gain unauthorized access, making an unauthorized modification or unauthorized destruction of data stored in the Cassandra database is known as Attack on Cassandra. Since the concerned database is a distributed NoSQL database, i.e., it is spread across multiple nodes which in return makes it a more complex and challenging target as compared to traditional relational database. This property also helps to make the Cassandra database strong enough to withstand an attack and bounce back quickly after an attack.

11.3.3.1 Apache Cassandra's Lightweight Directory Access Protocol (LDAP) Authentication Mechanism Vulnerability

In February 2021, Cassandra's LDAP authentication process was found to be security-susceptible, affecting the widely used distributed database management system and LDAP protocol for user authentication and authorization in enterprise environments (Yeong et al. 1995).

Damage Done

The susceptibility found was CVE-2021-27807, which allowed an attacker with valid login details to bypass LDAP authentication and take unauthorized access to the Cassandra cluster. This could lead to the exposure of sensitive data and the compromise of the entire database. The susceptibility was due to the way Cassandra's LDAP authenticator processed certain LDAP search queries by sending specially crafted search queries an attacker could manipulate the LDAP server's response and bypass authentication checks.

Precaution Taken

The Cassandra project freed a spot for the susceptibility in March 2021, along with suggestions for users to elevate to the fresh version of the Cassandra database. In inclusion, users were also suggested to analyze their LDAP configuration and consider executing additional security measures, such as enabling SSL/TLS encryption and implementing strong password policies.

Overall, this susceptibility brings out the significance of constancy updating and securing enterprise software as well as implementing suitable practices for user authentication and access control.

11.3.3.2 Vulnerability in Cassandra's Built-In Authentication Mechanism

In 2020, a security researcher detected a susceptibility in Cassandra's built-in authentication procedures. Cassandra is popular for its open-source as well as distributed database management system which is used by many businesses for its high availability, scalability, and fault tolerance (Becker et al. 2004).

Damage Done

The susceptibility found as CVE-2020-13,922, allowed an attacker to bypass authentication checks and gain unauthorized access to the Cassandra cluster. The flaw was caused by the way Cassandra's internal authentication system validated credentials, which could be exploited by an attacker to inject malicious code and execute capricious commands on the system. The said vulnerability affected all versions of Apache Cassandra before version 3.11.9 and 4.0-beta4 in a drastic way and considered a high-severity attack by Common Vulnerability Scoring System (CVSS).

Precaution Taken

Upon discovering the susceptibility, the Cassandra project brings out patches for affected versions and recommended that users upgrade to the fresh version of Cassandra to minimize the risk of exploitation. In inclusion, users were advised to review their authentication configuration and implement best practices for securing the database cluster, such as using strong passwords and enabling SSL/TLS encryption.

This susceptibility shows the significance of regular software updates and security evaluates for enterprise systems, as well as the need to execute robust authentication and access control procedures to protect from unauthorized access and data breaches.

11.3.3.3 Cryptojacking

In 2019, A small group called "Cryptojacking" targeted unsecured Apache Cassandra databases to mine cryptocurrency. Cryptojacking is the unauthorized use of computer without the permission of the owner to mine cryptocurrency This is done by installing malware on their computers (Eskandari et al. 2018).

Damage Done

The attackers targeted unsecured Cassandra databases that were exposed to the internet, exploiting a vulnerability in the default configuration of the database that allowed unauthorized access. Once inside, the intruders installed a crypto-mining program that utilized the compromised systems' computational power to mine cryptocurrency for the intruders.

A large rise in crypto-mining attacks in 2019 was brought on by the hack, which affected thousands of unprotected Cassandra databases. In addition to causing financial damages for the owners of the infected systems, the attack also resulted in performance problems and system outages.

Precaution Taken

It is crucial to safeguard Apache Cassandra databases by adhering to recommended practices for access control and network security in order to thwart such assaults. Implementing secure network setups, such as firewalls and VPNs, and securing the database with robust authentication and authorization processes are examples of what this entails.

11.3.3.4 Cassandra Namer Ransomware Campaign

In 2016, a ransomware campaign named "Cassandra" targeted unsecured Apache Cassandra databases, encrypting the data stored in the database and demanding payment in exchange for the decryption key (Conti et al. 2018).

Damage Done

The attack affected numerous unsecured Cassandra databases that were exposed to the internet, exploiting a loophole in the default configuration of the database that allowed access. After getting access, the attackers encrypt the data with an unknown key and demanded payment in Bitcoin or other cryptocurrency for the decryption key.

Precaution Taken

The Cassandra ransomware assault made it clear that businesses must follow best practices for network security and secure their databases with robust authentication and authorization systems. Securing the database with strong passwords, encryption, implementing safe network setup (such firewalls and VPNs), and ensuring that the software is current with the most recent security patches are some techniques that can be used. Business organizations must also ensure that they have effective backup and recovery protocols that can be used to quickly and securely restore data in the event of a ransom attack.

11.4 Introduction to Blockchain and Quantum Computing

11.4.1 Introduction to Blockchain

A type of distributed database that keeps track of all the blocks is known as blockchain. Blocks are nothing but a continuously growing list of ordered data and connecting these blocks results in the making of a blockchain. Each block in blockchain has a timestamp and reference to the previous block. Following are algorithmic steps to save log data in blockchain:

1. The very first step is to pre-process the log data, which involves tasks like data cleaning, deleting personally identifiable information (PII), and converting the log into a format that could be stored in a blockchain.
2. Block Chain Storage: After preprocessing is done, the log could be stored in blockchain by parting into numerous blocks or saved in one single block.
3. Linking of Blocks: Using the hash function these blocks can be connected together and doing this guarantees that log cannot be tampered without any change in the hash of the block.
4. Distribution of Blockchain: After completing all of these steps, then the blockchain is distributed to a network of computers. With these steps, we ensure that the log is protected and no single entity can have complete control (Fig. 11.3).

Fig. 11.3 Illustration of the process of saving log's preprocessed data in blockchain

The above diagram illustrates the three-block chain. A different set of log's preprocessed data is represented by each block of data in blockchain (Zheng et al. 2018). All the blocks are linked with each other by use of the hash function. It works in a way where the hash of one block is the hash of its next block, i.e., the hash of the first block is the hash of the second block and the hash of the second block is the hash of the third block and it goes on in the same manner. The blockchain is then distributed to a network of computers. This ensures that the data is secure and that no single entity can control it.

Blockchain could be used to enhance the security and veracity of log preprocessing since it gives a tamper-proof record of the data transformation procedure. This is a primary requirement in forensic analysis since log data can include secluded information and can be used to identify security issues.

11.4.2 Introduction to Quantum Computing

Complex calculations are performed utilizing concepts from quantum physics in the area of computing known as quantum computing. By using bits, which can only be either 0 or 1, classical computers represent and process data. On the other hand, quantum bits, or qubits, are utilized in quantum computers and can exist in several states simultaneously due to a phenomenon called superposition (Steane 1998). Because of this, quantum computers are able to perform parallel computations and solve particular issues faster than conventional computers.

11.4.2.1 Applications of Quantum Computing

1. Quantum computers are more effective at solving optimization issues, such as improving supply networks, complex financial portfolios, or logistics.
2. Simulation: With the help of quantum computers, scientists may explore events in physics, chemistry, and materials science.
3. Machine Learning: Clustering, pattern identification, and optimization in data analysis are just a few examples of how quantum algorithms might improve machine learning tasks.
4. Cryptography: Cryptography is affected by quantum computing. While it has the ability to defeat some traditional encryption algorithms, it also makes it possible to create quantum-resistant cryptographic techniques to guarantee safe communication in the post-quantum future.

11.4.2.2 Use in Security and Cryptography

Traditional cryptography methods, which rely on the complexity of certain mathematical problems for security, are challenged by quantum computing (Keplinger 2018). For instance, the commonly used public-key encryption techniques RSA and elliptic curve cryptography can both be broken using Shor's algorithm, a quantum algorithm that is capable of swiftly factoring enormous numbers.

Table 11.2 Traditional cryptography and post-quantum cryptography algorithms

Traditional cryptography algorithms	Post-quantum cryptography algorithms
• RSA: performance is critical, but vulnerable to quantum attacks	• CRYSTALS-Kyber: secure against quantum computers but slower than RSA
• ECDSA: based on the NIST P-256 curve, slower but widely used	• NTRU Prime: based on a custom curve, more efficient, faster, and resilient to quantum attacks
• Diffie-Hellman: based on the difficulty of factoring large integers. Faster and well-established in the real world	• SIKE: based on the difficulty of solving isogeny problems. Slower but resilient to quantum attack but newer

1. **Post-Quantum Cryptography**

Post-quantum cryptography is the term used to describe cryptographic methods created to withstand assaults from quantum computers (Rahman and Tomar 2018). In response to the danger that quantum computers pose to established cryptography systems, several methods are being developed (Table 11.2).

In this table, traditional cryptography algorithms are represented on the left side, which are vulnerable to attacks from quantum computers. On the right side, we have post-quantum cryptography algorithms, which are designed to be secure against quantum attacks. These algorithms are used for encryption and decryption processes to ensure the confidentiality and integrity of data.

2. **Quantum Key Distribution (QKD)**

In order to distribute cryptographic keys between two parties, quantum key distribution (QKD) makes use of the laws of quantum mechanics (Scarani et al. 2009). Quantum physics laws serve as the foundation for the security of QKD and offer a way to identify any efforts at interception or tampering.

Through a quantum channel, two parties—often referred to as Alice and Bob—converse in QKD. The principle is demonstrated in the diagram below (Fig. 11.4).

During the QKD process, Alice and Bob exchange quantum states (often photons) through the quantum channel. These states represent the cryptographic keys. Any attempt to intercept or measure these states would disturb them due to the principles of quantum mechanics, alerting Alice and Bob to the presence of an eavesdropper.

Quantum computing has the potential to revolutionize various fields, including optimization, simulation, and machine learning. While it poses challenges to traditional cryptography, ongoing research in post-quantum cryptography and the development of quantum-resistant algorithms aim to ensure secure communication in the future.

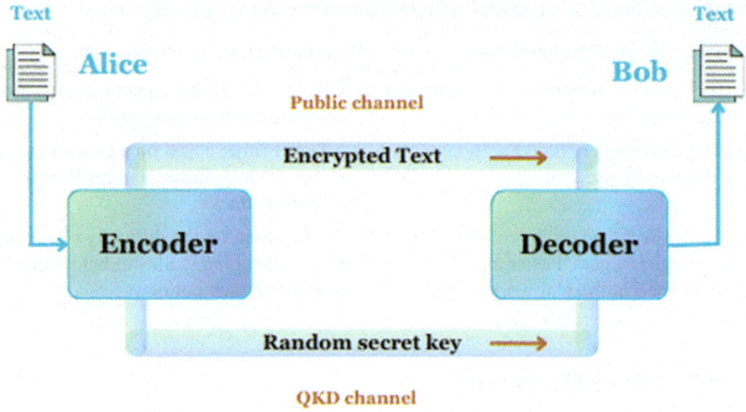

Fig. 11.4 Diagram represents the QKD process

11.5 Proposed Forensic Framework for Cassandra NoSQL Database

By keeping the above-mentioned attacks in perspective, we propose a six-phase forensic investigation framework for the wide-column store NoSQL DBMS (Fig. 11.5).

11.5.1 Phase I: Log Collection

Log collection is an essential step in digital forensics investigations (Ieong 2006) and it involves gathering evidence from different sources of system activity, such as files, processes, network connections, and user actions. Logs can provide valuable information about what happened on a system, when, and by whom.

Cassandra is a distributed database system that supports high availability and scalability. It also generates various types of logs that can be useful for forensic analysis. These are System logs such as records (Fig. 11.6).

Cassandra's startup, shutdown, configuration changes, errors, and warnings.

Debug log: records detailed information about the Cassandra operations, such as compaction, repair, streaming, and gossip (Fig. 11.7).

Commit log: Cassandra uses commit logs to store all mutations that occur on a node. A commit log is an append-only log that ensures durability in case of a crash. When Cassandra restarts, it replays the commit log to update the memtables. The commit log is essential for Cassandra's performance (Fig. 11.8).

Audit Log: Authentication and Authorization events, like login attempts, role changes and permission grants are recorded in this log.

For example, a typical audit log message might look like this (Fig. 11.9).

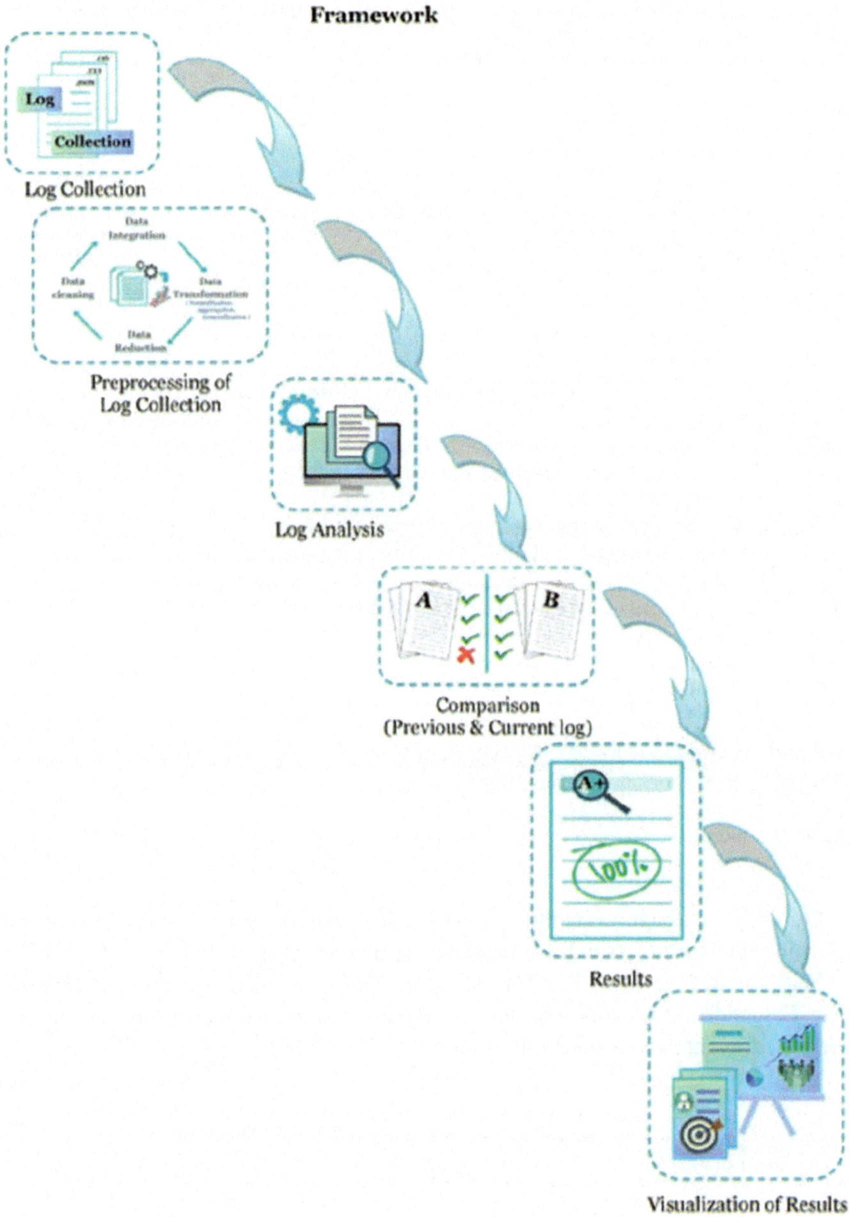

Fig. 11.5 Phases of framework

```
INFO  [StorageServiceShutdownHook] 2023-04-30 10:30:08,504 HintsService.java:235 - Paused hints dispatch
INFO  [StorageServiceShutdownHook] 2023-04-30 10:30:08,667 Server.java:176 - Stop listening for CQL clients
INFO  [StorageServiceShutdownHook] 2023-04-30 10:30:08,679 Gossiper.java:2095 - Announcing shutdown
INFO  [StorageServiceShutdownHook] 2023-04-30 10:30:08,690 StorageService.java:2977 - Node localhost/127.0.0.1:
INFO  [StorageServiceShutdownHook] 2023-04-30 10:30:08,720 StorageService.java:2977 - Node localhost/127.0.0.1:
```

Fig. 11.6 System log

```
DEBUG [StorageServiceShutdownHook] 2023-04-30 10:30:08,413 StorageService.java:1797 - DRAINING:
INFO  [StorageServiceShutdownHook] 2023-04-30 10:30:08,504 HintsService.java:235 - Paused hints
INFO  [StorageServiceShutdownHook] 2023-04-30 10:30:08,667 Server.java:176 - Stop listening for
```

Fig. 11.7 Debug log

```
C  ,G¯   GEü/bïä;è1♠)‰cCœ¨# system  compaction_history   22 üU<||
-3702326138260541532           $ roles-5517614638419036830-47
nt   rolesprimary  $ rolesprimary-14914418895014140609856266301 24
rimary66617047229763796267506081370467541384           $ roles
local_primary  $ roles
local_primary-149144188950141406098562663012484577          $
local_primary-2458806132579949558-1491441889501414060      
local_primary-3702326138260541532-2458806132579949558      
local_primary-4781698945708143189-3702326138260541532      
```

Fig. 11.8 Commit log

2023-04-20T06:10:15.123Z|10.0.0.1|192.168.0.100|alice|N/A|ATTEMPTED|SELECT * FROM users WHERE id = '123'|CQL|my_keyspace|users

Fig. 11.9 Audit log

GC Log: Garbage collection activity and memory usage of the Java Virtual Machine that runs Cassandra is recorded in this log (Fig. 11.10).

With the help of these logs, investigators can reconstruct the state and behavior of a Cassandra cluster as well as identify anomalies and malicious actions, and correlated events across multiple sources.

```
-XX:-UseBiasedLocking -XX:+UseCMSInitiatingOccupancyOnly -XX:+UseConcMarkSweepGC -XX:+UseGCLogFileRotation -XX:-UseNUMA -XX:+U
2023-04-30T11:11:44.020+0530: 5.325: Total time for which application threads were stopped: 0.0092950 seconds, Stopping thread
2023-04-30T11:11:43.830+0530: 5.938: Total time for which application threads were stopped: 0.0004068 seconds, Stopping thread
2023-04-30T11:11:44.231+0530: 6.340: Total time for which application threads were stopped: 0.0003517 seconds, Stopping thread
```

Fig. 11.10 GC log

11.5.2 Phase II: Log Preprocessing

The process of preparing and cleaning raw data before it could be used for forensic analysis is known as data preprocessing. Data preprocessing is the most crucial and important phase of analysis. It involves phases such as data cleaning, data transformation, data integration, and data reduction (Khan and Ullah 2017). The initial stage of data preprocessing involves data cleaning, locating and fixing flaws and inconsistencies in data, eliminating duplicates, dealing with missing values, and fixing inaccurate data. Data transformation, which comes after data cleansing, is putting data in a format that can be used for analysis with ease. Scaling, normalization, and data discretization are a few more examples. The act of combining multiple datasets is referred to as data integration. Data reduction, which condenses the size of the dataset while retaining its integrity and usefulness. Data reduction is a complex phase since data is arriving from different sources, that may come in varied forms and structure.

11.5.2.1 Data Cleaning

Data cleaning is an essential component of data preparation for forensic analysis. It involves the process of identifying and correcting or erasing flaws, inconsistencies, or inaccuracies in a dataset before it is analyzed for the investigation. Cassandra's log data can be cleaned in a variety of ways, including the following.

Deleting redundant entries: Duplicate log entries, which may be formed by glitches like network problems or system failures, which might influence the accuracy of forensic analysis and should be removed.

Removing useless information: The entire log data could not be used since it might not be relevant to our forensic analysis. For example, some log entries may contain details about commonplace events that may not be significant for forensic investigation.

Standardizing data format: The log data is generated by different systems and applications and it may appear in different date formats or naming conventions. Altering it into standard format will benefit to make data more consistent and makes it easier to examine.

Handling missing data: Some log entries may be missing important information or may contain null values. Handling missing data can involve either predicting missing values or discarding the entries altogether, depending on the type of the analysis.

Correcting errors: There might be times when log Entries contain errors like incorrect timestamps or values. These errors can be corrected manually to ensure the accuracy of the analysis.

```
DEBUG [PendingRangeCalculator:1] 2023-04-30 10:30:08,754 TokenMetadata.java:881 - Pending range calculation for system_traces completed (took: 0ms)
DEBUG [StorageServiceShutdownHook] 2023-04-30 10:30:10,782 MessagingService.java:517 - Shutting down: timeout=60s, gracefully=true, shutdownExecutors=true
INFO  [StorageServiceShutdownHook] 2023-04-30 10:30:10,782 MessagingService.java:525 - Waiting for messaging service to quiesce
INFO  [StorageServiceShutdownHook] 2023-04-30 10:30:13,182 ColumnFamilyStore.java:1009 - Enqueuing flush of system.compaction_history, Reason: DRAIN, Usage: 4.391KiB (0%) on-heap, 0.000KiB
(0%) off-heap
DEBUG [StorageServiceShutdownHook] 2023-04-30 10:30:13,195 CompactionStrategyManager.java:519 - Recreating compaction strategy - disk boundaries are out of date for
system.compaction_history.
```

Fig. 11.11 Delete log

For example, we have done data cleaning on log guests and further analyzed it (Fig. 11.11).

Deleted log at the Timestamp: 2023-04-30 10:30:10,782 and 2023-04-30 10:30:13,182 From the given log, we can identify the following fields:

Log level: INFO

Timestamp: 2023-04-30 10:30:10,782 and 2023-04-30 10:30:13,182

Log message: Waiting for messaging service to quiesce and Enqueuing flush of system.compaction_history, Reason: DRAIN, Usage: 4.391KiB (0%) on-heap, 0.000KiB (0%) off-heap

To perform data cleaning on this log, we can consider the following steps:

Remove any irrelevant fields: Since this log contains only one message, all fields are relevant. However, in other logs, we may need to remove any irrelevant fields.

Standardize the timestamp: The timestamp format in this log is already standardized. Nevertheless, it is essential to acknowledge that under specific circumstances, the configuration of the temporal record in the records may exhibit dissimilarities. Consequently, there may be a requisite to convert it into a standardized format.

Extraction of Useful Information: We can extract a lot of useful information from the log message of this log as it contains all the required information about the operations that are being performed and stores this extracted information in a separate field. For example: We can extract the following information (Fig. 11.12).

Finally the cleaned data can be correctly preserved in a systematic arrangement, like a database or a CSV file for further analysis (Fig. 11.13).

Analyzation using LOG VIEWER

The log shows two pieces of information related to the shutdown of a storage service:

The first line indicates that the messaging service is waiting for all pending messages to be processed before the service is shut down. This is a methodology,

operation: "Waiting for messaging service to quiesce" and "Enqueuing flush of system.compaction_history"reason: "DRAIN"usage_on_heap: "4.391KiB (0%)"usage_off_heap:"0.000KiB (0%)"

Fig. 11.12 Info

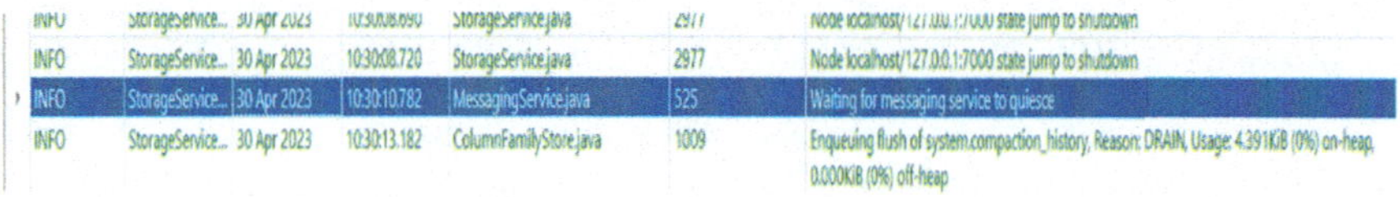

Fig. 11.13 Insert log at the Timestamp: 2023-04-30 10:30:10,782 and 2023-04-30 10:30:13,182

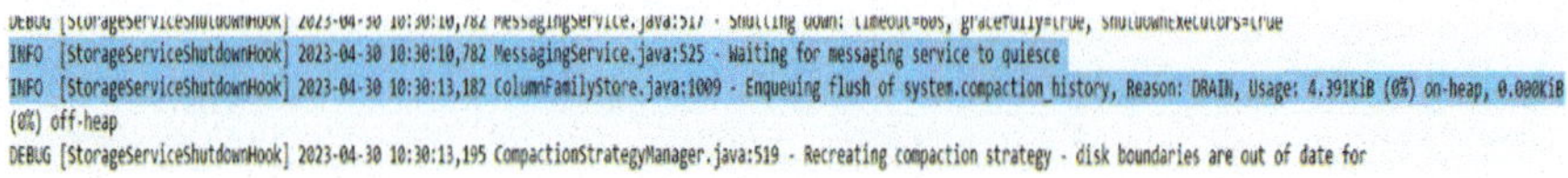

Fig. 11.14 System log

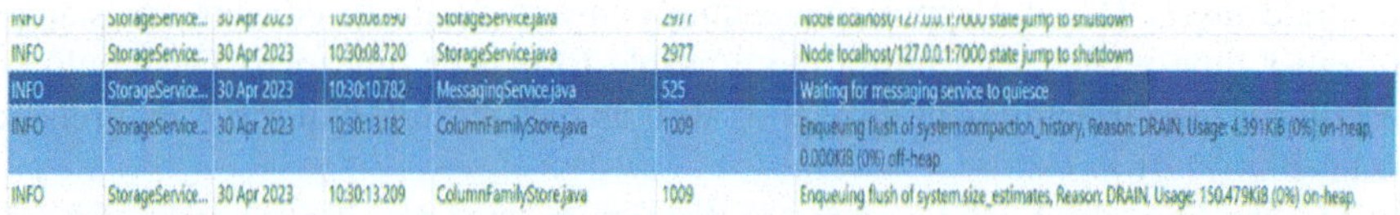

Fig. 11.15 System log

which ensures the proper processing of all messages and the prevention of any loss of data during the shutdown (Fig. 11.14).

The second line shows that a flush of the "system compaction_history column family store has been enqueued with a reason of DRAIN." This indicates that the storage service is flushing all data in the column family store before shutting down. The usage of on-heap and off-heap memory for this column family store is also shown.

Overall, this log provides useful information to ensure that data is not lost during the shutdown process, and it also shows that the system is properly managing its memory usage. No data cleaning is needed for this log (Fig. 11.15).

11.5.2.2 Data Transformation

Data preparation, often known as the process of turning raw data into a format that is more suitable for data analysis, includes data transformation as a crucial stage (Selamat et al. 2008). To increase the utility and quality of data, a variety of strategies and methodologies are used, such as:

Example: Done data transformation on log guest and analyzed it.

Deleted Log at

The system appears to be scheduling two-time conversion and checking-related jobs, based on the log messages from MonotonicClock.java. The first task is planned with a precision of 2 ms, and the second one with an interval of 10,000 ms (10 s). Data transformation or analysis utilizing time-related data may be relevant

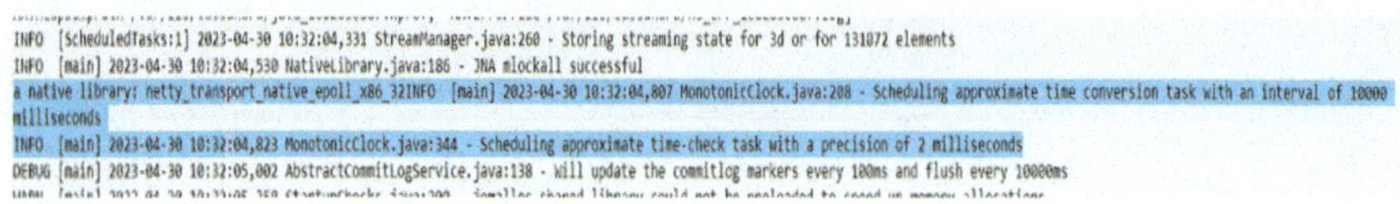

Fig. 11.16 System log for deleted record

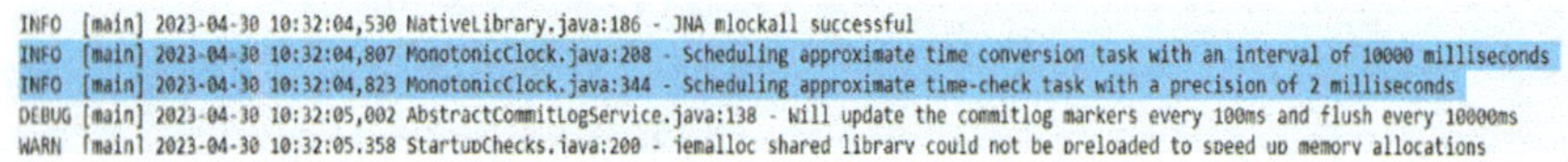

Fig. 11.17 System log for inserted record

to these tasks. Use the approximate time conversion job to convert timestamps or other time-related data into a more usable format or timezone. This could be useful for user-friendly data display as well as calculations requiring time-related data (Fig. 11.16).

Use the time-check job to guarantee that time-related data is accurately captured or to verify the precision of the system clock. Applications that need precise timing data, such real-time systems or financial applications, may depend on this.

Inserted Log at

There appear to be two jobs linked to time that are being planned with precise intervals and timing, according to the log messages from MonotonicClock.java.

The first task, "approximate time conversion task," is set to run once every 10 s (10,000 ms interval). This task could involve changing the time zone or format of timestamps or other time-related data, which is frequently required for data transformation operations (Fig. 11.17).

The second assignment, called "approximate time-check task," is slated to run with a 2 ms accuracy. This task could be used to verify that time-related data is accurately captured or to check the system clock's correctness. Additionally, it can be used to synchronize clocks between several systems, which is frequently required in contexts with distributed computing.

Overall, these tasks indicate that time-related data transformation and analysis are important in the system and that the system is taking measures to ensure the accuracy and consistency of the time-related data (Fig. 11.18).

Example: Done data transformation on log guest and analyzed it.

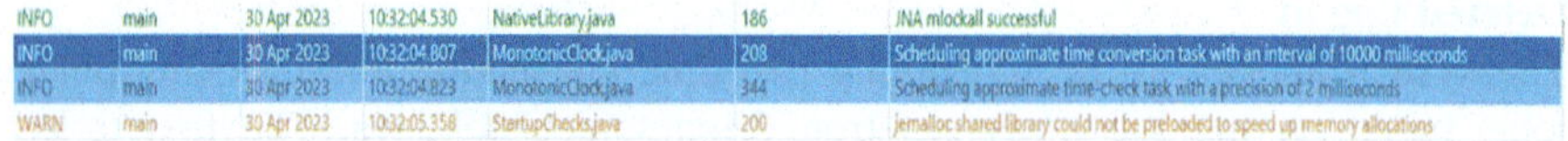

Fig. 11.18 System log after updating record

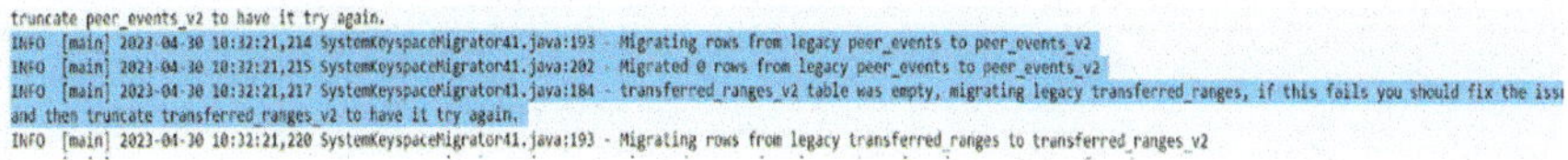

Fig. 11.19 Deleted log

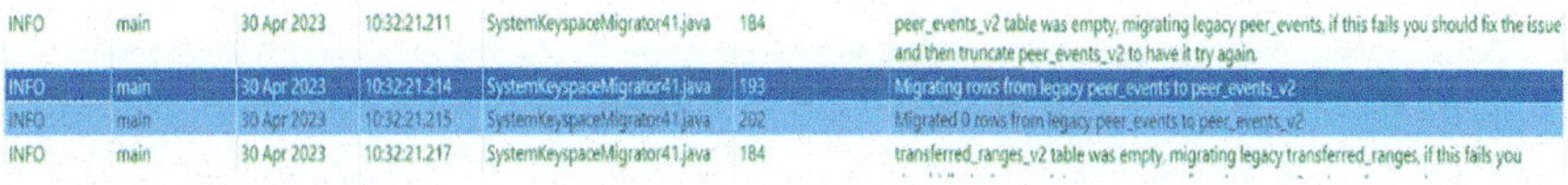

Fig. 11.20 Updated log

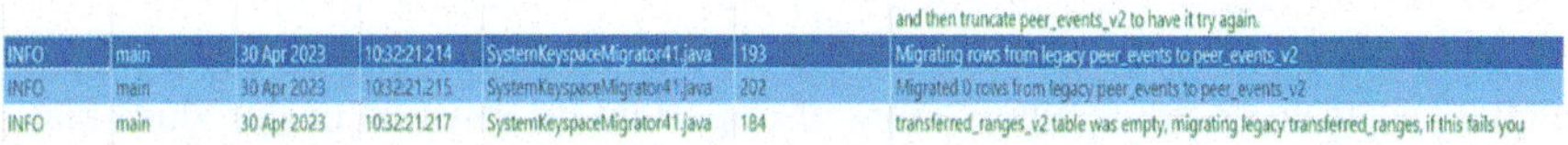

Fig. 11.21 Inserted log

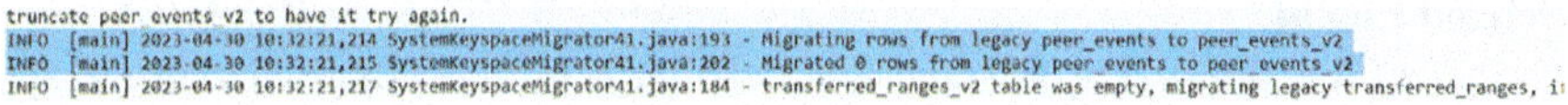

Fig. 11.22 Updated log

Deleted Log at

These log messages indicate that the key space "system" is undergoing a migration operation. In particular, data is being transferred from a historical table called "peer_events" to a brand-new table called "peer_events_v2." The message "Migrated 0 rows from legacy peer_events to peer_events_v2" indicates that no rows have yet been moved as part of the migration process. Additionally, a new table called "transferred_ranges_v2" is being created from data that was previously in a legacy table called "transferred_ranges," but it is now empty. "transferred_ranges_v2," but the latter table is currently empty (Figs. 11.19 and 11.20).

Inserted Log at

According to this log entry, the system is transferring rows from the old table peer_events to the table's new version, peer_events_v2. The notification indicates that "Migrated 0 rows" in this specific instance, meaning that no rows were migrated (Figs. 11.21 and 11.22).

11.5.2.3 Data Reduction

By reducing the amount of data, data reduction can improve the performance of data analysis algorithms, reduce the storage space required, and improve the interpretability of the results. However, it is important to ensure that the data reduction

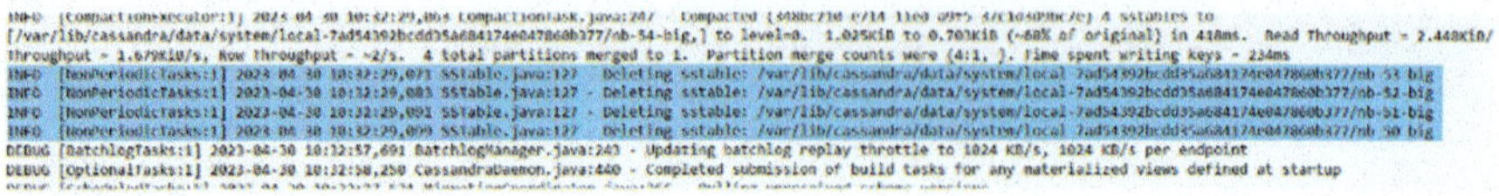

Fig. 11.23 Deleted log

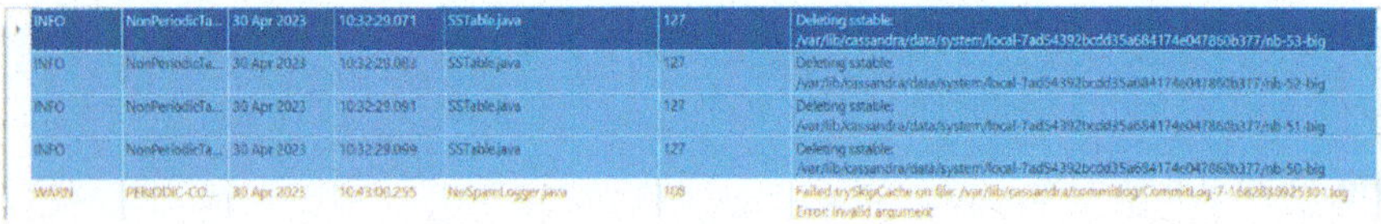

Fig. 11.24 Updated log

techniques used do not result in a loss of important information or introduce biases in the analysis.

Example: Done data reduction on log guest and analyzed it.

Deleted Log at

The given log shows that Cassandra is deleting several SSTable files in the system /local-7ad54392bcdd35a684174e047860b377 directory. SSTable files are the primary storage format used by Cassandra to store data on disk (Fig. 11.23).

The deletion of these files could indicate that Cassandra is performing data reduction by removing old or unnecessary data from the system. This could be due to a variety of reasons such as reclaiming disk space or removing outdated data that is no longer needed (Fig. 11.24).

Inserted Log at

This log contains information about the deletion of SSTables in a Cassandra database. Each line starts with the log level (in this case, "INFO"), followed by the thread that generated the message ("NonPeriodicTasks:1"), the timestamp ("2023-04-30 10:32:29,071"), the class that generated the message ("SSTable.java"), and the message itself ("Deleting sstable: ...") (Fig. 11.25).

Based on this log, it appears that several SSTables are being deleted from the Cassandra system's "local" keyspace. Specifically, the files being deleted are

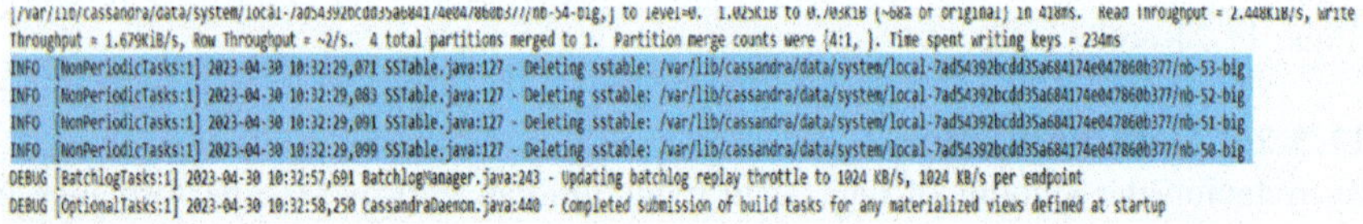

Fig. 11.25 Inserted log

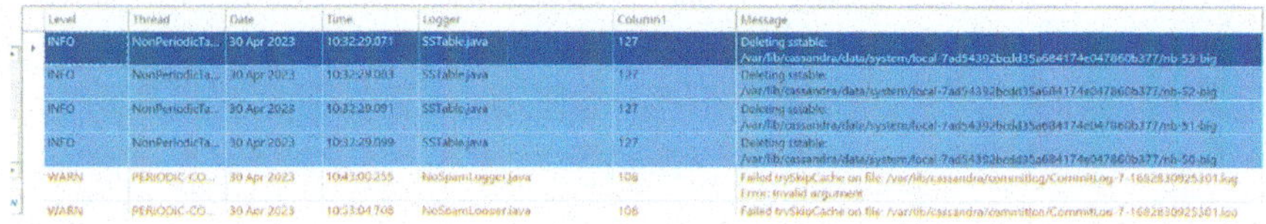

Fig. 11.26 Updated log

named "nb-53-big," "nb-52-big," "nb-51-big," and "nb-50-big." Without more context, it's difficult to say why these files are being deleted or what their contents were. However, it's possible that they were no longer needed or were being deleted as part of routine maintenance or cleanup (Fig. 11.26).

11.5.2.4 Log Preprocessing Through Blockchain and Quantum Computing

Quantum computing can be used to improve the efficiency of log preprocessing in CassandraDB by using quantum algorithms to identify patterns in data, classify log data, and compress log data. Blockchain can be used to secure the log data that is processed by the quantum algorithm.

The following are the phases of saving log preprocessing data of Cassandra DB in the blockchain using quantum computing and quantum algorithms:

1. Data Preprocessing: The log data at the start must be preprocessed. This necessitates transforming the raw log data into a quantum algorithm-processable format. The data might need to be compressed, cleared of extraneous information, and arranged in a way that is beneficial to quantum computing.
2. Quantum Algorithm: The next process is to originate a quantum algorithm that is applicable to pinpoint patterns in the log data. This algorithm could use the principles of quantum mechanics to find patterns that would be strenuous to find with traditional methods.
3. Blockchain: The third step is to safeguard the log data that is processed by the quantum algorithm. This might be potentially done by stockpiling the data in a blockchain, which is an allocated ledger technology that is secure and tamper-proof.
4. Results: This last step is to verify the results of the quantum algorithm. This involves patterns identification of the data, arranging the data, or contracting the data.

Diagram of Phases

Diagram of the phases of saving log preprocessing data of Cassandra DB in the blockchain using quantum computing and quantum algorithms (Fig. 11.27).

Interdisciplinary Solutions

The Interdisciplinary solutions that can be used to improve the efficiency of log preprocessing in Cassandra using quantum computing and blockchain include:

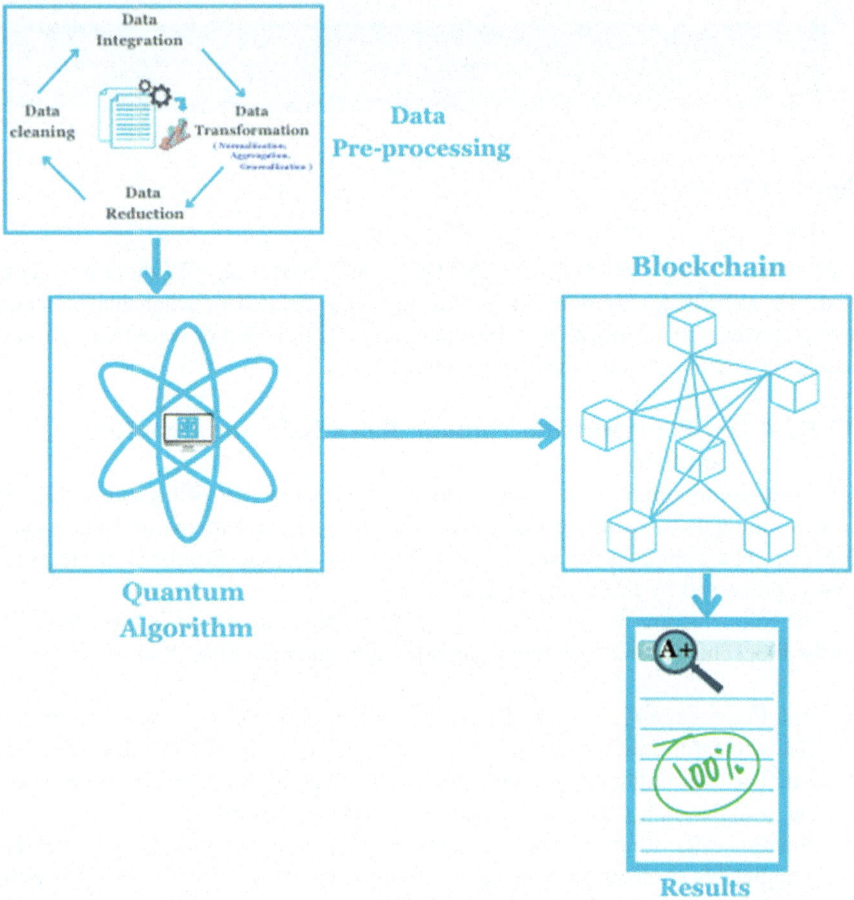

Fig. 11.27 Data preprocessing ⟹ Quantum algorithm ⟹ Blockchain ⟹ Results

- Identify patterns in data: Quantum algorithms can be used to find patterns in data that would be difficult to find with traditional methods. This purpose of using quantum algorithm is that it helps us to identify security threats, improve the performance of applications, and make it easier to analyze log data.
- Secure log data: Blockchain can be used to store log data, which makes it more secure and tamper-proof blockchain is a type of database which a more secure and reliable approach for storing log data, which will prevent unauthorized access and make it temperproof.
- Quantum computing and blockchain together: Both technologies can be used together to create new and innovative ways to store and process data. The application of quantum computing in log data compression offers a viable solution for data storage and transmission. The use of blockchain technology in

securing the compressed log data presents an effective approach to preventing unauthorized access and tampering by malicious parties.

11.5.3 Phase III: Log Analysis

Log analysis is a process of identifying bugs, error security threats, and other risks in log data.

Grouping of similar data points together based on their characteristics and attributes can done through a data mining technique known as clustering analysis (Oliner et al. 2012).

To understand how implementing clustering analysis will help in log analysis we can look into the following case where the system administrator wants to identify activity that are associated with a particular type of error message. To make sure there isn't any unauthorized activity, i.e., leading to error message. The administrator will use said data mining known clustering analysis to group log entries that contain that error message and based on attributes such as the time of day, the user account, the source IP address, and the type of request that triggered the error. Then administrator will use log analysis techniques to further analyze the patterns within each cluster of log Entries gained from clustering analysis. This process involves examining the timing of events and frequency within each cluster, comparing the behavior of different clusters, or identifying any external factors that might be influencing the behavior of the system. System Administrator and developers can troubleshoot problems, optimize system performance, and identify opportunities for improvement more effectively and efficiently using the insights drawn from log and cluster analysis.

11.5.4 Phase IV: Comparison

The previous and current logs of the Cassandra database are compared to identify missing data, users, access, operations performed, and errors for investigation.

This phase uses a log analysis tool to extract and compare relevant information from both logs. The findings are summarized, and recommendations for further actions are provided.

We can do this in the following way:

Step 1: We will go ahead and extract our logs manually.

Step 2: Then we will apply conditional statements to the data extracted and filter the data on the bases of operations performed into subsets.

Step 3: We can then create a live line graph using Google Charts for each operation performed and based on the irregularity of the line graph we can easily trace when and how our database has been tampered with.

A normal operation on our database will result in a line graph with steady spikes whereas when it's being tempered the spike might be too frequent compared to what it is on normal days.

For example, Let's suppose that deletion is not an operation that is performed frequently on our database, which means that the line graph of deletion will have a rare spike. But, on a particular day, the line graph suddenly has multiple spikes, which in turn alerts the admin and makes them aware of when and which change took place.

One better way to understand this is by taking the real-life example of the stock market news, where fluctuations in stock price are constantly shown by a line graph.

11.5.4.1 Comparison of Linear and Logistic Regression

Linear Regression:

This approach uses supervised machine learning to predict continuous values (Montgomery et al. 2021). The relationship between dependent and independent variables is linear, as the name would imply.

Process:

Step 1: Identify the dependent and independent variables. Write the function for dependent variable. Use random values for slope and intercept and generate output.

Step 2: As we have our calculated output, now check for residual error, i.e., what was the actual output vs what we calculated for a value. We use MSE (mean squared error) as loss function.

Step 3: We need to find the line of best fit so that we can plot all the points with minimal residual errors. To minimize the error.

Gradient Descent:

- MSE means it is in second order and if we plot a graph for this, it would be a parabolic curve. Means we are plotting loss function for given weights (slope and intercept).
- Our motive is to minimize the loss function and for that we should reach bottom of the curve.
- Take the first order derivative of the loss function, subtract the result of the derivative from the initial weight multiplying with a learning rate (α). Repeat this until we reach a minimum value. We set the threshold value near to 0 else we will never reach to the bottom of the curve.

Step 4: Now we get the line for best fit and now we can predict output.

Logistic Regression:

It is a classification algorithm. We use it mostly when dependent variable has binary solution (Menard 2002). It classifies elements of a set into two groups (binary classification) by calculating the probability of each element of the set.

Process:

We decide on a probability threshold upon which we categorize them into respective classes.

Step 1: First follow linear regression to find the line of best fit.

Step 2: The regression line is highly susceptible to outliers (random data or noise) and then convert the predicted values into probability using sigmoid function.

Step 3: Finally the output value of sigmoid function gets converted into 0 or 1 according to the threshold value.

To understand the comparison in a better way we use a dataset that contrains the height and weight of a group of people.

11.5.4.2 Comparison of Log Analysis

There are different tools and approaches used for log analysis.

1. **CQL**: query logs and perform various operations such as filtering, sorting, and aggregation.
2. **Cassandra Log4j Appender**: It is a plugin, that allows applications to send events to Apache Cassandra Repositories. With this, we can easily configure the level of logs and send them to Log4j for analysis.
3. **Elasticsearch**: A popular search and analytics engine. We can store logs in a centralized location and perform complex searches, filtering, and aggregation.
4. **Grafana**: It is another popular open-source platform in which we can monitor and visualize data. We can create an interactive dashboard to visualize logs in real time.

When choosing the approach, it really depends on the use case and requirements.

If you need to perform complex searches, and filtering on your logs—Elasticsearch may be a better choice.

Each approach has its own strength and weaknesses, some factors to consider are:

- Real-time-analysis
- Search and filtering
- Visualization
- Integration

11.5.5 Phase V: Result Analysis

Lubuntu is a lightweight Linux-based operating system that is designed to run on older and lower specification hardware. In this lab, we will investigate a case where a user's images with the names test1.jpeg, test2.png, and test3.webp have been deleted from the Lubuntu OS (Glanz 2016). In addition, we will analyze the logs of the Cassandra database to determine all the performed operations like database insertion, updation, and deletion.

To investigate this case, we acquired a forensic image of the Lubuntu OS disk using a forensically sound tool to ensure the integrity of the evidence. The acquired image was analyzed using the Autopsy tool, log view (Rahman and Tomar 2020b). A search was performed to locate the deleted images by searching for files with the names test1.jpeg, test2.png, and test3.webp; in order to identify all performed operation log files of Cassandra database were thoroughly studied and analyzed.

Through the analysis of these logs, we were able to recover the deleted images from our records named test1.jpg, test2.png, and test3.webp. The evidence of the same has been shown (Fig. 11.28).

The logs of the Cassandra database were also analyzed, and we found that all the performed operations were logged in detail. The evidence screenshot shows a sample of the logs (Breeuwsma et al. 2007). We were able to identify all the insertions, updates, and deletions in the database (Fig. 11.29).

After analysis, we can conclude that images named test1.jpeg, test2.png, and test3.webp were deleted from Lubuntu OS but, however, there were no suspicious activities were noted. By analysis of Cassandra database logs it was revealed that all the performed operations like insertions, updates, and deletions were logged in detail.

Recommendations

From our investigation, we recommend the following measures that can be implemented to improve security of the Lubuntu OS and Cassandra database:

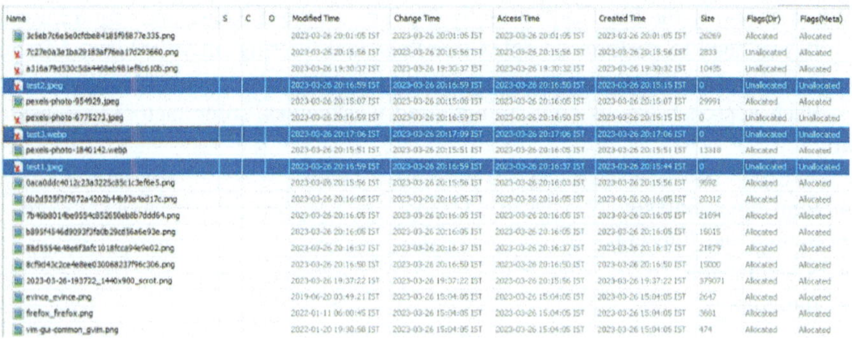

Fig. 11.28 Autopsy view

```
0x000004d0: 73 65 72 69  64 20 3D 31  30 36 3B 0A  53 45 4C 45   serid =106;.SELE
0x000004e0: 43 54 20 2A  20 46 52 4F  4D 20 75 73  65 72 20 3B   CT * FROM user ;
0x000004f0: 0A 44 45 4C  45 54 45 20  66 72 6F 6D  20 75 73 65   .DELETE from use
0x00000500: 72 20 57 48  45 52 45 20  75 73 65 72  69 64 20 3D   r WHERE userid =
0x00000510: 31 30 35 3B  0A 44 45 4C  45 54 45 20  66 72 6F 6D   105;.DELETE from
0x00000520: 20 75 73 65  72 20 57 48  45 52 45 20  70 68 5F 6E    user WHERE ph_n
0x00000530: 6F 20 3D 20  39 34 33 39  38 3B 0A 44  45 4C 45 54   o = 94398;.DELET
0x00000540: 45 20 66 72  6F 6D 20 75  73 65 72 20  57 48 45 52   E from user WHER
0x00000550: 45 20 65 6D  61 3B 0A 44  45 4C 45 54  45 20 66 72   E ema;.DELETE fr
0x00000560: 6F 6D 20 75  73 65 72 20  57 48 45 52  45 20 75 73   om user WHERE us
0x00000570: 65 72 69 64  20 3D 31 30  33 3B 0A 53  45 4C 45 43   erid =103;.SELEC
0x00000580: 54 20 2A 20  46 52 4F 4D  20 75 73 65  72 20 3B 0A   T * FROM user ;.
0x00000590: 65 78 69 74  0A                                      exit.
```

Fig. 11.29 Hex view of CQLsh_history.log

1. Regular backups of the Lubuntu OS and the Cassandra database should be performed to ensure that data can be recovered in the event of a security breach or data loss.
2. Access controls should be implemented to restrict access to the Lubuntu OS and the Cassandra database to authorized users only.
3. Logging and monitoring should be implemented to detect any suspicious activity or unauthorized access attempts.
4. Regular security awareness training should be provided to all users to educate them on the importance of security and how to protect their data.

We give measures to ensure, the security of the Lubuntu OS and the Cassandra database can be significantly improved, and the risk of unauthorized access and data loss can be minimized (Reith et al. 2002).

11.5.6 Phase VI: Visualization of Result

Log data visualization can help understand trends, patterns, and anomalies in system behavior. This can be done by creating graphs, charts, and dashboards that display the log data in a user-friendly format. Examples of visualization tools that can be used for log analysis include Elasticsearch, Kibana, Grafana, and Splunk (Wijk 2005). These tools allow users to create custom dashboards and visualizations to monitor and analyze real-time log data. By visualizing log data, system administrators can quickly identify issues and take appropriate actions to improve system performance and security (Fig. 11.30).

Distribution of different types of operations that are performed in the log file is depicted by the above pie chart. The distribution of operations is as follows it shows that 30% of operations are of insertion, 20% of deletion, 15% warning, 10% error, and finally 25% are informational. Records in log file stating new records or table being created are insertion records.

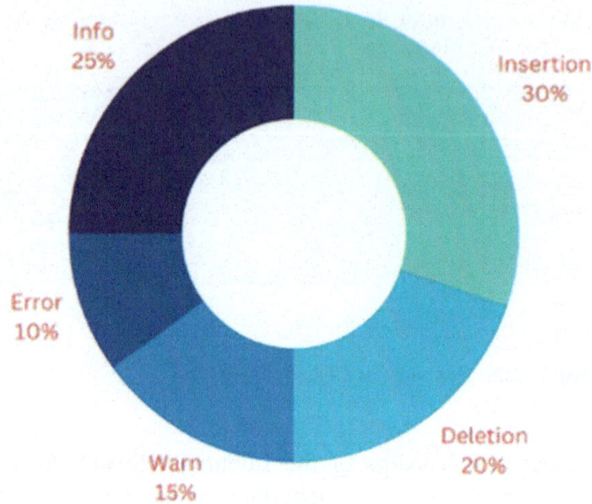

Fig. 11.30 Operation and Info

The process of adding new data to the Cassandra database is called an insertion operation. Records in the log file indicating that a record, table, or expired data is being erased are referred to as deletion records.

The process of deleting data from the database is known as a deletion operation. Up until this point, all of the operations we have seen have been fundamental activities; now, let's understand the final three operations, which may be unfamiliar to certain users.

Warning; working of this operation is pretty self-explanatory from its name, this operation sends messages to users that indicate potential issues that can affect the performance or stability of the system (Haerder and Reuter 1983). In the log file, it can indicate issues such as low disk space, high memory usage, or network connectivity issues.

Errors are messages that come into play when something goes wrong with the system. In the log file, error could indicate issues like data corruption, database crash, or failed operation.

In the log file, these messages indicate details about the system's memory usage, disk space, or performance of the network.

Visual representation of the types of operations that are being performed in the Cassandra database is provided to us by the above pie chart. This pie chart helps to identify patterns or issues that might be affecting the system's performance or stability.

11.6 Conclusion

The six phases of log analysis for a Cassandra cluster are crucial for ensuring its security, performance, and reliability. In the first phase, logs are collected from various sources of system activity. We also see how one can use blockchain and quantum computing technology to process and secure log data. Integrating both of these technologies results in the enhancement of security and efficiency of log management systems. By using quantum-resistant cryptographic algorithms one can secure blockchain transactions and prevent unauthorized access to log data, even in the presence of quantum computers. In the second phase, collected logs are preprocessed to extract relevant information and transform it into a unified format for analysis. The third phase involves the analysis of the logs to troubleshoot issues, optimize performance, and enhance security. The fourth phase compares previous and current logs to identify missing data, users, access, operations performed, and errors for investigation. The fifth phase involves the preparation of a comprehensive report that includes recommendations for improving the Cassandra cluster. Finally, in the sixth phase, visualizations are created to help stakeholders understand the state and behavior of the Cassandra cluster. Overall, the six phases provide a structured approach to analyze and monitor logs for a Cassandra cluster and help ensure its efficient functioning.

References

Becker MY, Sewell P (2004) Cassandra: flexible trust management, applied to electronic health records. In: Proceedings 17th IEEE computer security foundations workshop. IEEE, pp 139–154

Breeuwsma M, De Jongh M, Klaver C, Van Der Knijff R, Roeloffs M (2007) Forensic data recovery from flash memory. Small Scale Digit Device Forens J 1(1):1–17

Cassandra A (2014) Apache Cassandra, 13. Website. http://planetcassandra.org/what-is-apache-cassandra

Chandra DG (2015) BASE analysis of NoSQL database. Futur Gener Comput Syst 52:13–21

Conti M, Gangwal A, Ruj S (2018) On the economic significance of ransomware campaigns: a Bitcoin transactions perspective. Comput Secur 79:162–189

Eskandari S, Leoutsarakos A, Mursch T, Clark J (2018) A first look at browser-based cryptojacking. In: 2018 IEEE European symposium on security and privacy workshops (EuroS&PW). IEEE, pp 58–66

Fan J, Han F, Liu H (2014) Challenges of big data analysis. Natl Sci Rev 1(2):293–314

Glanz O (2016) ETCBC-vm v1. 62. Lubuntu 16.04 64bit

Haerder T, Reuter A (1983) Principles of transaction-oriented database recovery. ACM Comput Surv (CSUR) 15(4):287–317

Han J, Haihong E, Le G, Du J (2011) Survey on NoSQL database. In: 2011 6th International conference on pervasive computing and applications. IEEE, pp 363–366

https://cassandra.apache.org/_/case-studies.html

https://www.infoq.com/news/2023/02/netflix-annotations-cassandra/

https://www.infoworld.com/article/2078771/big-movies--big-data--netflix-embraces-nosql-in-the-cloud.html

https://www.techrepublic.com/article/apples-secret-nosql-sauce-includes-a-hefty-dose-of-cassandra/

Ieong RS (2006) FORZA–Digital forensics investigation framework that incorporate legal issues. Digit Investig 3:29–36

Ioannidis S, Keromytis AD, Bellovin SM, Smith JM (2000) Implementing a distributed firewall. In: Proceedings of the 7th ACM conference on computer and communications security, pp 190–199

Ivanov T, Niemann R, Izberovic S, Rosselli M, Tolle K, Zicari RV (2015) Performance evaluation of enterprise big data platforms with HiBench. In: 2015 IEEE Trustcom/BigDataSE/ISPA, vol 2. IEEE, pp 120–127

Keplinger K (2018) Is quantum computing becoming relevant to cyber-security? Netw Secur 2018(9):16–19

Khan MNA, Ullah S (2017) A log aggregation forensic analysis framework for cloud computing environments. Comput Fraud Secur 2017(7):11–16

Khan SN, Loukil F, Ghedira-Guegan C, Benkhelifa E, Bani-Hani A (2021) Blockchain smart contracts: applications, challenges, and future trends. Peer-to-Peer Netw Appl 14:2901–2925

Menard S (2002) Applied logistic regression analysis (No. 106). Sage

Montgomery DC, Peck EA, Vining GG (2021) Introduction to linear regression analysis. Wiley

Okman L, Gal-Oz N, Gonen Y, Gudes E, Abramov J (2011) Security issues in NoSQL databases. In: 2011 IEEE 10th International conference on trust, security and privacy in computing and communications. IEEE, pp 541–547

Oliner A, Ganapathi A, Xu W (2012) Advances and challenges in log analysis. Commun ACM 55(2):55–61

Rahman RU, Tomar DS (2018) Security attacks on wireless networks and their detection techniques. In: Emerging wireless communication and network technologies: principle, paradigm and performance, pp 241–270

Rahman RU, Tomar DS (2020a) New biostatistics features for detecting web bot activity on web applications. Comput Secur 97:102001

Rahman RU, Tomar DS (2020b) A new web forensic framework for bot crime investigation. Forens Sci Int Digit Investig 33:300943

Raut ABPD (2017) NOSQL database and its comparison with RDBMS. Int J Comput Intell Res 13(7):1645–1651

Reith M, Carr C, Gunsch G (2002) An examination of digital forensic models. Int J Digit Evid 1(3):1–12

Scarani V, Bechmann-Pasquinucci H, Cerf NJ, Dušek M, Lütkenhaus N, Peev M (2009) The security of practical quantum key distribution. Rev Mod Phys 81(3):1301

Selamat SR, Yusof R, Sahib S (2008) Mapping process of digital forensic investigation framework. Int J Comput Sci Netw Secur 8(10):163–169

Steane A (1998) Quantum computing. Rep Prog Phys 61(2):117

Van Wijk JJ (2005) The value of visualization. In: VIS 05. IEEE visualization. IEEE, pp 79–86

Yeong W, Howes T, Kille S (1995) Lightweight directory access protocol (No. rfc1777)

Zheng Z, Xie S, Dai HN, Chen X, Wang H (2018) Blockchain challenges and opportunities: a survey. Int J Web Grid Serv 14(4):352–375

Revolutionizing Gait Analysis with Blockchain Technology: Enhancing the Privacy and Security

12

Pratibha Tokas, Vijay Bhaskar Semwal, Sweta Jain, and Deepak Singh Tomar

Abstract

Gait analysis is essential for evaluating the gait patterns of individuals and for diagnosing and treating conditions such as multiple sclerosis and Parkinson's disease. However, traditional gait analysis techniques have their own limitations. They can be time-consuming, costly, and detrimental to patient data security. Blockchain technology therefore offers a probable solution. It provides stakeholders with an efficient and transparent system for storing and exchanging patient information. Blockchain promotes the collection of patient data from wearable devices such as sensors, smartwatches, and smartphones by utilizing a decentralized framework. This data is encrypted and securely stored on the blockchain to ensure its immutability and prevent any manipulation attempts. The integration of blockchain technology with gait analysis provides many advantages including enhanced efficiency, traceability, and a reduction in the risks associated with human errors and expenditures. Additionally, the transparent nature of the blockchain ledger system provides healthcare professionals with comprehensive and accurate records of each patient's gait analysis over time. However, there are challenges such as substantial investment for required infrastructure, complex integration, and ethical concerns surrounding data ownership and patient privacy. The rest of the chapter is organized as follows. It

P. Tokas (✉) · V. B. Semwal · S. Jain · D. S. Tomar
Department of CSE, MANIT, Bhopal, Madhya Pradesh, India
e-mail: 213112001@stu.manit.ac.in

V. B. Semwal
e-mail: vsemwal@manit.ac.in

S. Jain
e-mail: swetajain@manit.ac.in

D. S. Tomar
e-mail: deepaktomar@manit.ac.in

A. Kumar et al. (eds.), *Sustainable Security Practices Using Blockchain, Quantum and Post-quantum Technologies for Real Time Applications*, Contributions to Environmental Sciences & Innovative Business Technology,
https://doi.org/10.1007/978-981-97-0088-2_12

provides a comprehensive discussion on gait, covering the different phases and the integration of blockchain technology. Additionally, the chapter explores the limitations and advantages of incorporating blockchain technology in gait analysis. Furthermore, some case studies are presented. Later, the framework for the Decentralized Gait Analysis and Rehabilitation Platform is presented and finally, the conclusion is followed by references.

Keywords

Blockchain technology • Gait analysis • Healthcare • Privacy • Security • Wearable sensors

12.1 Introduction

12.1.1 Gait Analysis

The movement pattern involved in walking or running is referred to as gait (Callisaya et al. 2017). In other words, Gait is the coordinated steps that result in forward movement of the body while maintaining the balance of the body (Ataullah and Jesus 2023). The gait cycle consists of two major phases, swing and stance phase. The alternate occurrences of swing and stance phase is gait. The swing phase comprises of five sub-phases, namely, initial, contact, loading response, mid-stance, terminal-stance, and finally pre-swing. Furthermore, initial swing, mid-swing, and terminal swing are the sub-phases associated with the swing phase. The various sub-phases associated with a gait cycle are illustrated in Fig. 12.1.

Gait analysis involves analyzing the walking pattern of an individual and measuring various factors like stride length, joint angles related to knee, hip and ankle, etc.

Gait analysis is very important for diagnosing and monitoring various medical conditions like neurological disorders (e.g., Parkinson's disease) that can have an

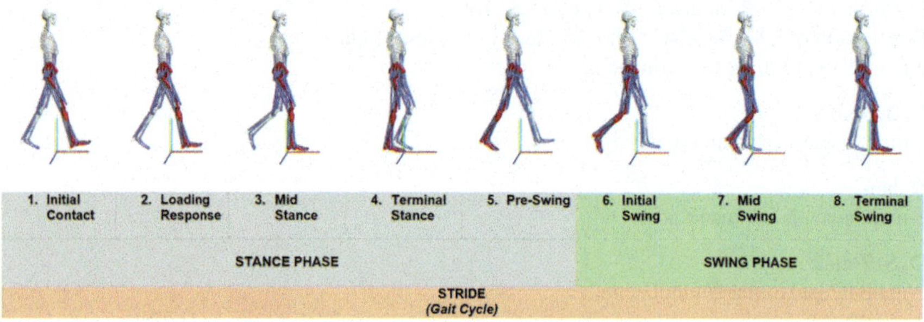

Fig. 12.1 Different phases and sub-phases of the gait cycle

impact on individual's gait (Muniz et al. 2010; Mariani et al. 2013; Arcolin et al. 2019; Yoneyama et al. 2013; Handojoseno et al. 2012; Prakash et al. 2018).

Additionally, gait analysis proves useful in evaluating the effects of orthopedic disorders such as arthritis or lower limb surgery on an individual's walking pattern.

Various techniques can be employed to gather data for analyzing gait. These include the use of pressure sensors (Tian et al. 2021; Li et al. 2022; Martínez-Martí et al. 2014), video-based analysis (Zhao et al. 2023; Sabo et al. 2022; Wang et al. 2022), force plates (Miyazaki and Iwakura 1978), thrust plates (Angin et al. 2007), and IMU sensors (Eikerling et al. 2018). Such information obtained from the analysis proves invaluable in guiding treatment decisions and tracking progress over time.

Phases of Gait Analysis

Gait analysis typically consists of several phases: preparation, data collection, analysis, and interpretation. Let's delve into each phase briefly:

1. **Preparation**: In this phase, the patient undergoes necessary steps before gait analysis. These include obtaining medical history, conducting physical examination, and arranging required equipment or sensors. Data acquisition involves utilizing various sensors such as IMU sensors (Eikerling et al. 2018; Semwal et al. 2022; Bijalwan et al. 2021), Microsoft Kinect sensors (Huang and Onnela 2020), smartphones, or smartwatches (Bijalwan et al. 2023).
2. **Analysis**: Once the data is acquired, it needs to undergo preprocessing to eliminate any null values or noise. This step also involves extracting various features such as time domain, frequency domain, or statistical features based on the specific application of the data. Subsequently, machine learning or deep learning models can be trained to analyze these extracted features.
3. **Interpretation**: The interpretation phase represents the final stage in gait analysis. During this phase, the obtained results are carefully analyzed. By comparing the observed gait pattern before and after treatment, it becomes possible to assess the improvement in an individual's movement. This assessment plays a crucial role in tailoring a personalized treatment plan.

Gait analysis holds significant importance for several reasons. Firstly, it plays a crucial role in identifying the underlying causes of medical conditions that lead to difficulties in walking (Eldeeb and Abdelraheem 2021; Jain et al. 2021; Dua et al. 2021). This identification allows for early detection and timely treatment, leading to improved outcomes and quality of life. Additionally, gait analysis facilitates the monitoring of treatment progress and enables necessary adjustments to the approach as needed. Lastly, this analysis assists in creating personalized rehabilitation strategies for individuals with injuries by pinpointing areas of weakness or dysfunction in their gait patterns. By addressing these concerns specifically, rehabilitation plans can promote a more effective and efficient recovery.

12.1.2 Blockchain Technology

Blockchain technology is a decentralized digital ledger that logs and verifies transactions across computer networks (Sable et al. 2022; Modi 2023). By employing encryption, it ensures data transparency (Majeed et al. 2021; Guo et al. 2023), immutability (Abdul Hadi and Au 2021), and trustworthiness (Esmaili et al. 2020). Each transaction, or block, is linked to the previous one to form a secure data chain. This eliminates the need for intermediaries, reduces fraud, and improves data integrity. Figure 12.2 depicts the basic architecture of blockchain technology for Healthcare.

Blockchain technology is not just limited to Bitcoin; it holds immense potential in various applications across industries like supply chain management (Chugh et al. 2020), healthcare (Kaushik and Kumar 2022; Kumar et al. 2020; Kumar and Jain 2021) [45], and the banking sector. By creating a secure and decentralized platform for transactions and data sharing, blockchain revolutionizes these sectors by ensuring tamper resistance.

12.2 Integration of Blockchain Technology with Gait Analysis

Blockchain technology can be applied in various ways to analyze gait. Below, we provide several examples of its utilization:

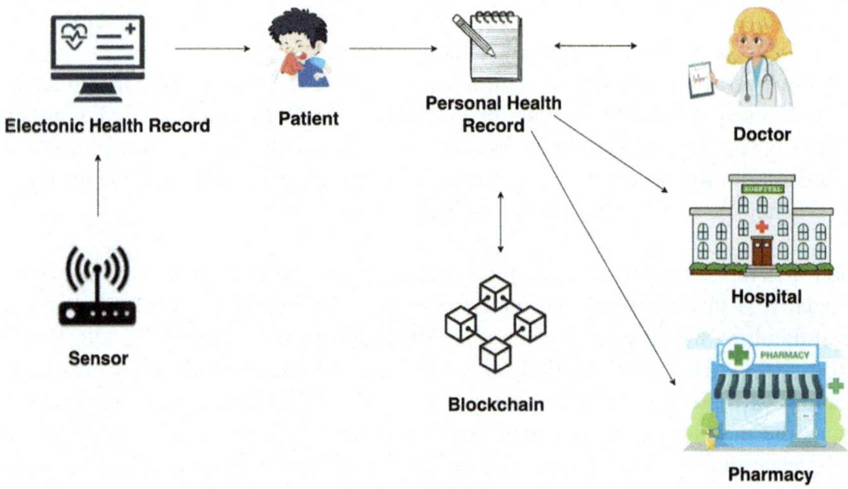

Fig. 12.2 Basic architecture of blockchain technology for healthcare

1. **Record gait analysis data**: A blockchain-based system offers a secure and immutable solution to record gait analysis data. With each gait analysis, the system captures and logs essential information such as data, time stamp, and other relevant details onto the blockchain.
2. **Enhancing Data Security**: In healthcare, safeguarding sensitive patient information is of utmost importance. The decentralized and immutable nature of blockchain presents a robust solution for protecting gait analysis and rehabilitation data. By storing this data across a dispersed network of computers, blockchain ensures transparency while greatly reducing the risk of unauthorized access or tampering.

When clinical data, evaluation results, and treatment plans are involved, this particular feature becomes invaluable. It instills a sense of confidence among healthcare providers, patients, and researchers alike.

3. **Ensure data privacy**: Gait analysis data holds significant sensitivity, warranting its privacy. To address this concern, a viable solution lies in employing blockchain technology. By utilizing a blockchain-based system, patients gain the authority to manage and safeguard their own data, ensuring its accessibility only to authorized individuals.
4. **Enhancement in the accuracy of Gait Analysis**: Blockchain technology can improve the accuracy of gait analysis by securely recording patient data. This ensures that the information remains untampered and guarantees reliable and precise analysis results.
5. **Interoperability and Data Interchange**: Interoperability and data interchange play a vital role in gait analysis and rehabilitation. To ensure effective communication and seamless data sharing among healthcare providers, the use of standardized protocols and smart contracts in blockchain technology can be beneficial. This facilitates secure connections between multiple systems, enabling authorized parties to access and update gait analysis data, rehabilitation progress, and treatment plans. This well-coordinated approach not only enhances the personalized patient care but also ensures utmost security. The integration of blockchain technology with gait analysis can secure data sharing in healthcare, enhance the quality of services provided to the patient and also aid in providing optimal treatment.
6. **Facilitate Research and Collaboration Opportunities**: Blockchain technology has the potential to advance the research and collaborations in the fields of gait rehabilitation. The making use of blockchain, the researchers can securely access the aggregated gait data. This enables the researchers to dig deeper and get data-driven insights. Moreover, blockchain technology eliminates the intermediaries and promotes seamless collaboration between various organization. The free, yet secured flow of information among stakeholders expedites the progress in gait rehabilitation.

7. **Remote Monitoring and Real-Time Feedback**: Blockchain technology is a potential solution that can revolutionize the gait analysis and lower limb rehabilitation. By making use of the wearable devices, critical data on stride length, cadence, and joint motions can be collected remotely. This information is securely stored on the blockchain, enabling healthcare providers to effectively monitor their patients' progress from a distance. Furthermore, smart contracts come into play by automating feedback systems, delivering timely instructions and encouragement to ensure adherence to rehabilitation programs and facilitating necessary adjustments.
8. **Patient Engagement and Incentivization Mechanisms**: To enhance patient outcomes, active participation in gait analysis and rehabilitation programs is encouraged. Blockchain-based incentivization methods, including token awards and gamification elements, can be introduced to engage and motivate patients. By rewarding adherence to treatment regimens, monitoring progress, and actively participating in their recovery journey, patients can earn blockchain-based tokens that signify their achievements or advancements.

Blockchain technology offers several advantages. It enhances security and privacy by safeguarding the patient's gait data, ensuring encryption, and granting access solely to authorized individuals. Furthermore, it enables secure data sharing (Tripathi et al. 2020). However, it is imperative to carefully evaluate the potential drawbacks that may arise before integrating a blockchain-based system.

12.2.1 Advantages of Using Blockchain Technology for Gait Analysis

There are several possible advantages to using blockchain for gait analysis instead of conventional methods:

1. **Data Protection**: Blockchain offers one of the most crucial advantages: data protection. Through encryption and decentralized storage, blockchain technology ensures that manipulation or theft of data becomes exceedingly challenging. This aspect holds immense significance in gait analysis, where sensitive patient information such as medical history and personal details are involved (Floridi 2018; Lopes and Oliveira 2018).
2. **Immutability of data**: Data recorded on a blockchain is immutable, prohibiting any alterations or deletions. This inherent feature instills trust in its integrity and reliability. When it comes to gait analysis, safeguarding the data becomes crucial as any form of tampering or removal can significantly compromise the accuracy and dependability of the results.
3. **Data Transparency**: Blockchain technology provides transparency by granting all network participants access to the data. By storing information using blockchain, data consistency is ensured and authorized individuals can effortlessly obtain the same data. This not only fosters trust but also mitigates

potential disputes, making it particularly valuable in scenarios involving gait analysis where multiple stakeholders like healthcare professionals, patients, and insurance companies are involved.

4. **Data ownership**: Data ownership can be explicitly established and controlled through blockchain technology. This means that patients have the ability to own and manage access to their gait analysis data, which is particularly important when it comes to their medical information, where data privacy holds significant importance.

Blockchain adoption in gait analysis offers improvements in security, data immutability, transparency, and data ownership. However, implementing blockchain-based systems can present challenges and expenses. The necessity for such implementation depends on the specific use case at hand.

12.2.2 Limitations of Blockchain Technology for Gait Analysis

Blockchain technology holds immense potential to transform numerous industries, including healthcare. While its advantages in gait analysis and rehabilitation have been extensively documented, it is crucial to acknowledge and understand the associated limitations and challenges that come with its implementation.

1. **Complexity**: Developing a secure, reliable, and user-friendly blockchain-based solution that is also robust can pose significant challenges. It requires a certain level of technical expertise to accomplish this task successfully.
2. **Cost**: Implementing blockchain technology in gait analysis and rehabilitation can be expensive for healthcare organizations. The upfront costs include infrastructure construction, network maintenance, and data storage. For smaller or resource-limited organizations, adopting blockchain-based solutions can be financially burdensome. Before implementing on a larger scale, it is essential to assess the cost-to-benefit ratio and long-term viability of these solutions.
3. **Scalability**: Scalability is a critical concern when it comes to blockchain technology. The more transactions there are, the slower the processing times and the greater the computational requirements for the blockchain network. When considering gait analysis and rehabilitation data, incorporating blockchain could impact real-time analysis and timely interventions needed for accurate evaluations. Although efforts are underway to address these scalability issues, ongoing research and development are still needed to find suitable solutions.
4. **Regulatory compliance**: Regulatory compliance in the healthcare industry becomes a challenge when dealing with sensitive patient information during gait analysis and rehabilitation. While blockchain encryption and decentralized storage offer data security, privacy concerns arise. The transparency and immutability of blockchain may conflict with data protection regulations like the General Data Protection Regulation (GDPR), potentially triggering legal and ethical issues such as the right to be forgotten or data deletion. Balancing

data security, privacy, and legislative requirements is crucial when implementing a blockchain-based system in healthcare. Ensuring regulatory compliance can be difficult due to the stringent legal and regulatory standards governing patient privacy and data security.

5. **Integration and Adoption Issues**: Integrating blockchain technology into existing healthcare systems and workflows can pose challenges in terms of difficulty and time consumption. The compatibility between blockchain and legacy systems, as well as other healthcare technologies, may introduce technological hurdles. Moreover, incorporating blockchain in gait analysis and rehabilitation requires a significant shift in mindset along with training investments for healthcare professionals. Additionally, user adoption can be demanding. Both healthcare providers and patients might approach new technologies cautiously, requiring time and effort to ensure their comfort and familiarity with the system. Overcoming these integration and adoption obstacles necessitates meticulous planning, coordination, and adequate stakeholder support.
6. **Energy Consumption**: Blockchain networks, especially those employing proof-of-work consensus techniques, consume substantial processing power and energy. Consequently, concerns have arisen regarding the environmental impact of this extensive energy usage in blockchain technology. The energy requirements associated with blockchain networks can further pose challenges in gait analysis and rehabilitation, where continuous monitoring and data processing are vital. Therefore, it is crucial to undertake further research into exploring energy-efficient alternatives or optimizing blockchain protocols for healthcare applications.

Blockchain technology offers significant advantages for gait analysis. However, before implementing a blockchain-based system, it is crucial to thoroughly analyze its obstacles and limitations.

12.3 Case Studies

12.3.1 Case Study 1: Blockchain Based Fall Prediction Model for Aged Care (Callisaya et al. 2017; Rupasingh et al. 2019)

This research article introduces a groundbreaking approach to predicting falls in eldercare facilities by leveraging blockchain technology. The authors proposed a blockchain-based fall prediction model that utilizes smart contracts and the FHIR (Fast Healthcare Interoperability Resources) standard in order to identify elderly individuals with a higher risk of falling.

Traditional fall prediction models have their limitations, making a blockchain-based system an appealing solution due to enhanced security and confidentiality benefits. The proposed approach comprises three critical components: a data collector, a prediction engine, and a blockchain network. The data collector acquires

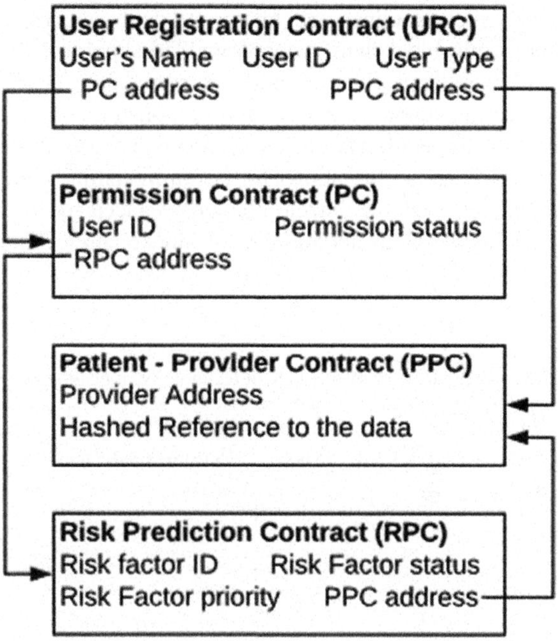

Fig. 12.3 Smart contracts (Rupasingh et al. 2019)

information from various sources such as wearable devices and medical records. Subsequently, the prediction engine analyzes this data using machine learning methods to forecast the likelihood of falls. To ensure privacy, the blockchain network secures and tamper-proofs these predictions while enabling individuals to manage their own data access. To evaluate its efficacy, the authors conducted experiments using fall datasets obtained from a nursing home setting. Their finding show that the solution for blockchain-enabled fall predictions outperforms the conventional methods. In conclusion, this study suggests that employing a blockchain-based fall prediction model holds great potential for enhancing fall prevention among the elderly population. However, it is still at a very nascent stage and a lot more in depth research is needed to assess its implementability (Fig. 12.3).

12.3.2 Case Study 2: PatientDataChain: A Blockchain-Based Approach to Integrate Personal Health Records (Dua et al. 2021; Cernian et al. 2020)

It is a novel technique that uses a distributed healthcare infrastructure. It incorporates a reliability layer in the value chain for healthcare, ensuring secure access to patients' medical history records and enhancing privacy and data confidentiality. By leveraging wearable devices and sensors, PatientDataChain acquires patient data and seamlessly integrates it into a unified medical record system.

The article outlines the blueprint, deployment, and validation of a PHR system using blockchain technology. This approach facilitates distributed storage of healthcare data while promoting secure data exchange between providers. Through validation research, it has been demonstrated how this method enables seamless sharing of health records among healthcare professionals. Furthermore, PatientDataChain's scalability allows for accommodating a large number of users effectively. This addresses the current challenges faced by the healthcare system. Table 12.1 illustrates various applications of blockchain technology in healthcare.

Table 12.1 Review of blockchain technology in healthcare

Name	Type of system	Description
MedBlocks (https://secure.health/)	Electronic medical records (EMR)	Distributed ledger technology is used to create an immutable P2P filesystem for storing medical records Strength: An enhanced consensus process that prevents network overload
Secure health chain (https://medicalchain.com/en/)	EMR	It assures that only persons with permission to access the data can access electronic patient's medical records stored on a blockchain
Medical chain (https://doc.ai/)	EMR	Medical data can be stored, shared, and used on a distributed network. It also enables the users to grant access to their personal medical records to healthcare providers
Doc.AI (Roehrs et al. 2017)	Generate insights from medical data	Enterprise AI platform that mines medical data for insights using computer vision, blockchain, and natural language processing
OmniPHR (Zhou et al. 2018)	PHR	A mechanism for distributing data storage for patient health. It maintains various patient datasets in various chain blocks
MIStore (Rupasingh et al. 2019)	Medical insurance storage system	A system based on blockchain for storage of health insurance

12.4 Proposed Approach for Decentralized Gait Analysis and Rehabilitation Platform

To develop a decentralized platform that integrates gait analysis with blockchain technology in order to empower individuals with gait impairments, facilitate secure data sharing among healthcare providers, and incentivize data contribution for research and rehabilitation purposes.

12.4.1 Algorithm

Step 1: Registration of Users

(1.1) User supplies the required personal data and initiates the registration procedure.
(1.2) The platform generates a unique wallet address for each user and associates it with their profile.
(1.3) The user's private information and wallet address are securely stored on the blockchain.

Step 2: Analysis of Gait and Data Collection

(2.1) The user wears sensor for gait analysis, such as an IMU or sEMG, and engages in daily activities such as walking, jogging, ascending stairs, etc.
(2.2) The sensor captures data such as cadence, walking speed, and stride length in real time.
(2.3) The collected data is encrypted and stored on the wallet address on the blockchain that corresponds to the particular user.

Step 3: Authorization for Data Sharing

(3.1) The user has access to their gait data via their pocketbook and can grant permission to a specific healthcare provider or researcher.
(3.2) The user specifies the criteria for granting access, such as the duration or purpose, and initiates a transaction involving a smart contract.
(3.3) The smart contract keeps a record of the data access authorization and its condition on the blockchain.

Step 4: Incentive Mechanism

(4.1) Users who volunteer their gait data for scientific research are eligible for rewards.
(4.2) The platform determines the amount of compensation based on the quantity and quality of the shared data.

(4.3) As a reward, tokens or cryptocurrency are issued to the user's wallet.

Step 5: AI-Based Gait Analysis

(5.1) Artificial intelligence models for locomotion analysis are deployed on the blockchain or made available via secure APIs.
(5.2) When the blockchain receives new locomotion data, the AI models are triggered for analysis.
(5.3) Artificial intelligence analyzes gait data and generates personal insights, such as gait irregularities and risk factors.

Step 6: Individualized Rehabilitation Plans

(6.1) The platforms generate personalized rehabilitation programs for users based on AI analysis.
(6.2) Rehabilitation programs comprise of gait-improving exercises and recommendations.
(6.3) These programs are stored in a secure manner on the blockchain and made available to the user.

Step 7: The Data Market

(7.1) Data marketplace access is granted to aggregated and anonymized gait analysis information.
(7.2) Through smart contracts, researchers and institutions can request access to specific datasets.
(7.3) Data access is granted if the requested criteria match the user-defined conditions.

12.4.2 Advantages of Decentralized Gait Analysis and Rehabilitation Platform

The benefits of the above approach is listed below.

(1) Users with gait impairments gain insight into their gait patterns and have access to individualized rehabilitation programs, empowering them to take charge of their health.
(2) Researchers and healthcare professionals have access to diverse datasets for analyzing gait patterns, designing treatments, and advancing rehabilitation innovations.
(3) The incentive mechanism encourages users to contribute their gait data, promoting a collaborative environment that is advantageous to both individuals and researchers.

(4) The use of blockchain and encryption mitigates privacy and security concerns by ensuring the secure storage and sharing of sensitive gait analysis data.

12.5 Conclusion and Future Scope

Blockchain technology has the potential to completely revolutionize gait analysis and rehabilitation. Through improved data security, enhanced interoperability, remote monitoring capabilities, incentivized patient engagement, and strengthened research collaborations, it can transform the way we approach patient care and outcomes. Technology advancements pave the way for further research and applications of blockchain in gait analysis and rehabilitation. This progress is set to enhance patient experiences, improve therapy effectiveness, and elevate healthcare delivery as a whole.

Blockchain technology offers a secure and efficient system for gait analysis, enhancing the storage and sharing of patient data. By leveraging advanced encryption, this system ensures the utmost security and privacy. Moreover, through automation and transparency, it drives efficiency and fosters traceability. Although there are challenges like scalability, integration complexity, regulatory considerations, and energy consumption to tackle, overcoming these hurdles can yield remarkable benefits in terms of improved patient experiences, more effective medicines, and enhanced healthcare delivery. The potential advantages for patients and the healthcare system are enticing; furthermore, further research and development in blockchain applications for gait analysis and rehabilitation procedures present exciting possibilities.

In the proposed approach for Decentralized Gait Analysis and Rehabilitation Platform, the data marketplace can foster collaborations between researchers and healthcare providers, leading to groundbreaking discoveries and innovative treatments for gait impairments. The platform's incentivization mechanism can be further refined to motivate more users to share their data, thereby creating a more comprehensive and diverse dataset for research. The future holds promising opportunities for patients, healthcare providers, and researchers alike, making this innovative integration an exciting prospect for the healthcare industry as a whole.

References

Abdul Hadi ZL, Au TW (2021) Blockchain for the authentication and immutability of academic credentials issued in Brunei Darussalam. In: Suhaili WSH, Siau NZ, Omar S, Phon-Amuaisuk S (eds) Computational intelligence in information systems, CIIS 2021. Advances in intelligent systems and computing, vol 1321. Springer, Cham. https://doi.org/10.1007/978-3-030-68133-3_8

Angin S, Karatosun V, Unver B et al (2007) Gait assessment in patients with thrust plate prosthesis and intramedullary stemmed prosthesis implanted to each hip. Arch Orthop Trauma Surg 127:91–96. https://doi.org/10.1007/s00402-006-0281-8

Arcolin I, Corna S, Giardini M et al (2019) Correction to: proposal of a new conceptual gait model for patients with Parkinson's disease based on factor analysis. BioMed Eng OnLine 18:80. https://doi.org/10.1186/s12938-019-0698-2

Ataullah AHM, De Jesus O (2023) Gait disturbances. In: StatPearls. StatPearls Publishing

Bijalwan V, Semwal VB, Singh G et al (2023) HDL-PSR: modelling spatio-temporal features using hybrid deep learning approach for post-stroke rehabilitation. Neural Process Lett 55:279–298. https://doi.org/10.1007/s11063-022-10744-6

Bijalwan V, Semwal VB, Mandal TK (2021) Fusion of multi-sensor-based biomechanical gait analysis using vision and wearable sensor. IEEE Sens J 21(13):14213–14220. https://doi.org/10.1109/JSEN.2021.3066473

Callisaya ML, Launay CP, Srikanth VK et al (2017) Cognitive status, fast walking speed and walking speed reserve—the gait and Alzheimer interactions tracking (GAIT) study. GeroScience 39:231–239. https://doi.org/10.1007/s11357-017-9973-y

Cernian A, Tiganoaia B, Sacala I, Pavel A, Iftemi A (2020) PatientDataChain: a blockchain-based approach to integrate personal health records. Sensors 20(22):6538. MDPI AG. https://doi.org/10.3390/s20226538

Chugh N, Sharma D, Singhal R, Jain S, Pulipeti S, Kumar A, Agarrwal A (2020) Blockchain-based decentralized application (DApp) design, implementation, and analysis with healthcare 4.0 trends. Basic Clin Pharmacol Toxicol 126:139–140. https://doi.org/10.1111/bcpt.13405

Doc.AI. https://doc.ai/

Dua N, Singh SN, Semwal VB (2021) Multi-input CNN-GRU based human activity recognition using wearable sensors. Computing 103:1461–1478. https://doi.org/10.1007/s00607-021-00928-8

Eikerling HJ, Uelschen M, Stuntebeck B (2018) Efficient techniques for gait-analysis: comparing marker-less and IMU-based tracking systems for monitoring rehabilitation processes. In: Eskola H, Väisänen O, Viik J, Hyttinen J (eds) EMBEC & NBC 2017. EMBEC NBC 2017 2017. IFMBE proceedings, vol 65. Springer, Singapore. https://doi.org/10.1007/978-981-10-5122-7_215

Eldeeb HM, Abdelraheem HS (2021) Functional gait assessment in early and advanced Parkinson's disease. Egypt J Neurol Psychiatry Neurosurg 57:145. https://doi.org/10.1186/s41983-021-00399-w

Esmaili M, Javidan R (2020) A distributed blockchain-based video sharing system with copyright, integrity, and immutability. In: 2020 8th Iranian joint congress on fuzzy and intelligent systems (CFIS). Mashhad, Iran, pp 86–92. https://doi.org/10.1109/CFIS49607.2020.9238685. MedBlocks. https://medblocks.org/

Floridi L (2018) Soft ethics: its application to the general data protection regulation and its dual advantage. Philos Technol 31:163–167. https://doi.org/10.1007/s13347-018-0315-5

Guo S, Sun X, Lam HKS (2023) Applications of blockchain technology in sustainable fashion supply chains: operational transparency and environmental efforts. IEEE Trans Eng Manag 70(4):1312–1328. https://doi.org/10.1109/TEM.2020.3034216

Handojoseno AMA, Shine JM, Nguyen TN, Tran Y, Lewis SJG, Nguyen HT (2012) The detection of freezing of gait in Parkinson's disease patients using EEG signals based on wavelet decomposition. In: 2012 Annual international conference of the IEEE engineering in medicine and biology society. San Diego, CA, USA, pp 69–72. https://doi.org/10.1109/EMBC.2012.6345873

Huang EJ, Onnela J-P (2020) Augmented movelet method for activity classification using smartphone gyroscope and accelerometer data. Sensors 20:3706. https://doi.org/10.3390/s20133706

Jain S et al (2021) Blockchain and autonomous vehicles: recent advances and future directions. IEEE Access 9:130264–130328. https://doi.org/10.1109/ACCESS.2021.3113649

Kaushik K, Kumar A (2022) Demystifying quantum blockchain for healthcare. Secur Privacy 6. https://doi.org/10.1002/spy2.284

Kumar A, Krishnamurthi R, Nayyar A, Sharma K, Grover V, Hossain E (2020) A novel smart healthcare design, simulation, and implementation using healthcare 4.0 processes. IEEE Access 8:118433–118471. https://doi.org/10.1109/ACCESS.2020.3004790

Kumar A, Jain S (2021) Proof of game (PoG): a proof of work (PoW)'s extended consensus algorithm for healthcare application. In: Gupta D, Khanna A, Bhattacharyya S, Hassanien AE, Anand S, Jaiswal A (eds) International conference on innovative computing and communications. Advances in intelligent systems and computing, vol 1165. Springer, Singapore. https://doi.org/10.1007/978-981-15-5113-0_2

Li W et al (2022) Wearable gait recognition systems based on MEMS pressure and inertial sensors: a review. IEEE Sens J 22(2):1092–1104. https://doi.org/10.1109/JSEN.2021.3131582

Lopes IM, Oliveira P (2018) Implementation of the general data protection regulation: a survey in health clinics. In: 2018 13th Iberian conference on information systems and technologies (CISTI). Caceres, Spain, pp 1–6. https://doi.org/10.23919/CISTI.2018.8399156

Majeed U, Khan LU, Yousafzai A, Han Z, Park BJ, Hong CS (2021) ST-BFL: a structured transparency empowered cross-silo federated learning on the blockchain framework. IEEE Access 9:155634–155650. https://doi.org/10.1109/ACCESS.2021.3128622

Mariani B, Jiménez MC, Vingerhoets FJG, Aminian K (2013) On-shoe wearable sensors for gait and turning assessment of patients with Parkinson's disease. IEEE Trans Biomed Eng 60(1):155–158. https://doi.org/10.1109/TBME.2012.2227317

Martínez-Martí F, Martínez-García MS, García-Díaz SG et al (2014) Embedded sensor insole for wireless measurement of gait parameters. Australas Phys Eng Sci Med 37:25–35. https://doi.org/10.1007/s13246-013-0236-7

MedicalChain. https://medicalchain.com/en/

Miyazaki S, Iwakura H (1978) Foot-force measuring device for clinical assessment of pathological gait. Med Biol Eng Comput 16:429–436. https://doi.org/10.1007/BF02442661

Modi N (2023) Comparative analysis of different block chain technology to improve the security in social network. In: Balas VE, Semwal VB, Khandare A (eds) Intelligent computing and networking. lecture notes in networks and systems, vol 632. Springer, Singapore. https://doi.org/10.1007/978-981-99-0071-8_7

Muniz AMS, Liu W, Liu H, Lyons KE, Pahwa R, Nadal J (2010) Gait initiation evaluation after deep brain stimulation for Parkinson's disease: a 7-year follow-up. In: 2010 Annual international conference of the IEEE engineering in medicine and biology. Buenos Aires, Argentina, pp 3650–3653. https://doi.org/10.1109/IEMBS.2010.5627419

Prakash C, Kumar R, Mittal N (2018) Recent developments in human gait research: parameters, approaches, applications, machine learning techniques, datasets and challenges. Artif Intell Rev 49:1–40. https://doi.org/10.1007/s10462-016-9514-6

Roehrs A, da Costa CA, da Rosa Righi R (2017) OmniPHR: a distributed architecture model to integrate personal health records. J Biomed Inform 71:70–81. https://doi.org/10.1016/j.jbi.2017.05.012

Rupasinghe T, Burstein F, Rudolph C, Strange S (2019) Towards a blockchain based fall prediction model for aged care. In: Wang Y, Patterson D (eds) ACSW'19—Proceedings of the Australasian computer science week multiconference. Association for Computing Machinery (ACM), Sydney, NSW, Australia. https://doi.org/10.1145/3290688.3290736

Sable NP, Rathod VU, Sable R, Shinde GR (2022) The secure e-wallet powered by blockchain and distributed ledger technology. In: 2022 IEEE Pune section international conference (PuneCon). Pune, India, pp 1–5. https://doi.org/10.1109/PuneCon55413.2022.10014893

Sabo A, Mehdizadeh S, Iaboni A, Taati B (2022) Estimating Parkinsonism severity in natural gait videos of older adults with dementia. IEEE J Biomed Health Inform 26(5):2288–2298. https://doi.org/10.1109/JBHI.2022.3144917

Secure Health Chain. https://secure.health/

Semwal VB, Gaud N, Lalwani P et al (2022) Pattern identification of different human joints for different human walking styles using inertial measurement unit (IMU) sensor. Artif Intell Rev 55:1149–1169. https://doi.org/10.1007/s10462-021-09979-x

Tian X et al (2021) Pressure sensor array with low-power near-sensor CMOS chip for human gait monitoring. IEEE Sens Lett 5(2):1–4, Art no 6000304. https://doi.org/10.1109/LSENS.2021.3053963

Tripathi G, Ahad MA, Paiva S (2020) S2HS—A blockchain based approach for smart healthcare system. Healthcare (Amsterdam, Netherlands) 8(1):100391. https://doi.org/10.1016/j.hjdsi.2019.100391

Wang Y et al (2022) SAIL: a deep-learning-based system for automatic gait assessment from TUG videos. IEEE Trans Hum-Mach Syst 52(1):110–122. https://doi.org/10.1109/THMS.2021.3123232

Yoneyama M, Kurihara Y, Watanabe K, Mitoma H (2013) Accelerometry-based gait analysis and its application to Parkinson's disease assessment—Part 2: a new measure for quantifying walking behavior. IEEE Trans Neural Syst Rehabil Eng 21(6):999–1005. https://doi.org/10.1109/TNSRE.2013.2268251

Zhao Y, Wang X, Yu X, Liu C, Gao Y (2023) Gait-assisted video person retrieval. IEEE Trans Circuits Syst Video Technol 33(2):897–908. https://doi.org/10.1109/TCSVT.2022.3202531

Zhou L, Wang L, Sun Y (2018) MIStore: a blockchain-based medical insurance storage system. J Med Syst 42(8):149. https://doi.org/10.1007/s10916-018-0996-4

A Blockchain-Enabled Framework to Implement Supply Chains in Digital Land Registry in the Kingdom of Saudi Arabia

13

Omar Alsukaiti and Mohammed Alojail

Abstract

With the introduction of the Land Administration Domain Model (LADM), countries started adapting this new standard into their real estate systems. This research describes a sustainable supply chain framework model to implement a digital land registry in the Kingdom of Saudi Arabia. The proposed research evaluates and analyzes the existing land registries and their digital aspects. It also conforms to capture the best practices for designing new processes and procedures by inheriting and satisfying current laws and regulations of the country. It discussed how blockchain can be a useful asset in the supply chain digital land registry model. The work also aims to design a framework for such implantation and fitting it to the model such as the Software Development Life Cycle (SDLC). The research followed an exploratory research approach, which is more suitable for qualitative-based researches. The data used in the research was collected through interviews with key individuals in organizations that are related to real estate. The results discussed the viability of the framework, its advantages, and shortcomings with the detailed discussion of the future directions.

Keywords

Land registry • Land Administration Domain Model • Supply chain • Blockchain • Decision making

O. Alsukaiti · M. Alojail (✉)
Department of Management Information Systems, College of Business Administration, King Saud University, Riyadh 11587, Saudi Arabia
e-mail: malojail@ksu.edu.sa

A. Kumar et al. (eds.), *Sustainable Security Practices Using Blockchain, Quantum and Post-quantum Technologies for Real Time Applications*, Contributions to Environmental Sciences & Innovative Business Technology,
https://doi.org/10.1007/978-981-97-0088-2_13

13.1 Introduction

All countries need to maintain their land ownership, boundaries, rights, and restrictions in some form or another; it is the reason land registries have been developed in different ways since the beginning of time (Kalantari et al. 2015; Beck and Moss n.d.). To satisfy the need to standardize such a practice and to have better governance and measurable results, the International Standardization Organization (ISO) came up with the Land Administration Domain Modern (LADM), which contains the governing guidelines and data models required to establish a land registry that fulfils the needs of any country based on their profile (Oosterom et al. 2013; Kalantari et al. 2015; Kalogianni et al. 2021; Lemmen et al. 2015; The World Bank n.d.).

With the rise of technology and automation, as in the case of most services provided by governments or the private sector land registries went digital through multiple frameworks and using different technologies, allowing responsible parties to have better visibility and enhanced overall governance to the land registry (Kalantari et al. 2015; Deininger and Goyal 2012). Nonetheless, this came with different types of challenges as technology in this case is mandated to absorb old systems, some of which are hundreds of years old. Bring into the equation the LADM and its implementation, a conclusion can be reached that the technology is not the biggest challenge but the actual process of authenticating, standardizing, and transforming such land registries is (Beck n.d.; Kalogianni et al. 2021). Many countries have gone through the implementation of LADM-based land registries usually in a digital format and with that came different types of steps, usually the development of a country profile (Kalogianni et al. 2021). Different steps will follow and they change based on the country but the best is described by Dr. Kalantari through his study to establish a road map to adapt LADM in cadastral information systems which follow six stages:

Organizational Motivation
Institutional Arrangements
Information Interpretation
Data Organization
Governance and Engagement
Capacity Building

All above stages are broken down to further steps establishing a framework in which LADM can be adapted into an information system (Kalantari et al. 2015).

This research is expected to face multiple challenges. First, the huge differences between approaches taken to implement a land registry, maintain rights, and acquire geospatial data to link to those rights. Papers have described the approach taken by different parties and how they differ greatly, bringing technology into the equation and this will amplify the challenge. Many recent researches have talked about the implementation of blockchain, 3D models, and advanced AI implementation in a land registry. To incorporate such technologies is such a significant feat for any land registry and to design a framework to suit them is no less of an accomplishment. Another challenge rises when the framework suggested needs to be adaptive enough to handle any future changes in laws, regulations, and technology. This will require the framework to give enough details that it can guide the implementation process without being blocked by minor changes in the mentioned aspects. A different challenge is that the framework needs to find the middle ground between the Saudi regulation and the international standard (LADM) and how to make sure to follow the regulations while keeping the standard guidelines implied unless conflicted with the regulations.

The proposed research is about designing a framework to implement a digital land registry in the Kingdom of Saudi Arabia, and will attempt to fit this framework to satisfy all the needs presented in the Saudi laws, regulations, and requirements. What is known is there are many issues in the current land registry system in the kingdom; although, the kingdom is relatively advanced in its government's E-Services. However, there are multiple situations where the issue is not technology, but process and procedure. A better overall land registry system should take place where all rights on the land are preserved and authenticated to become absolute truth. Such a land registry system has already been established around the world in countries such as the UK, Australia, and the USA; however, the kingdom offers some unique and challenging scenarios that have not been addressed previously. Most prominently is the existence of two holy cities that are thousands of years old in which land has been held so dearly. Also, digitization may not have reached the desired levels for which creating a special treatment for any transaction might happen in the land registry. A solution for the problem facing the kingdom will be proposed in the research as a framework to implement a digital land registry that will satisfy the special requirements of the kingdom, in addition to applying the international standards that are related to land registry and technology. Hopefully, this framework will provide the satisfactory solution that is required by the kingdom to make the change it needs for its real estate sector.

Motivation and Contributions

It is evident that an organized and well-standardized land registry is a must in any country that wishes to enhance its economic power and increase the value of its real estate market for its own citizens and foreign investors. In the above, the research defines a land registry, the LADM standard, and the stages required for any country to implement it, and as our research will talk about a framework to implement a digital land registry in the Kingdom of Saudi Arabia then it was essential to fully understand how an undertaking should be approached. This research aims to establish a framework for implementing a digital land registry in the Kingdom of Saudi Arabia. This will assist in future advancements in the real estate sector allowing better operations and advancements with regard to process, standard, and technology.

The land registry of any country needs to be run smoothly with a very small margin of error. When handling one of the most expensive assets a human can have, errors and mistakes become very intolerable. Technology comes to play a key role in modern land registries, identity access management systems, double factor authentications, biometrics, escrow accounts, blockchains and so many other aspects that can secure and enhance the quality of both information and user experience of a land registry. The challenge is to absorb the old land registry, provide a cadastral map, and integrate all related rights, restrictions, and responsibilities into these parcels. Technology can offer amazing solutions but without the right processes and correct data models, the implementation of such a registry is a difficult undertaking. So, the aim can be broken down into the following objectives:

- To evaluate and analyze the existing land registries and the digital aspects of it.
- To capture the best practices for designing new processes and procedures and make sure they satisfy laws and regulations.
- To design a framework for such implantation and fitting it to an SDLC.

13.2 Literature Review

The following section will contain the literature review of the existing studies related to the topic of land registry and Land Administration Domain Model "LADM."

13.2.1 Land Registration General Concepts, Roles, and Frameworks

In the following section, the general concepts of land registry will be elaborated, which the current literature has discussed in length; most researchers have mentioned these concepts in their papers in some form.

13.2.2 Land Administration Domain Model (LADM)

Land Administration Domain Model (LADM) is defined as "*a conceptual model which supports the modelling of social relations with land articulated through rights. There are three principal concepts within LADM: the party (the who) that has rights relationship (the what) with a plot of land (the where)*". LADM is a part of the land administration that is considered rights, responsibilities, and restrictions (Lemmen et al. 2015). Those rights responsibilities and restrictions describe who holds a right and this right is considered a restriction on whom and whether they have a responsibility on the land or not (Beck n.d.).

LADM has several goals and motivations, some of which are to battle poverty, offer sustainable housing, help with climate change response, and provide guidelines for governance (Oosterom et al. 2013; Lemmen et al. 2015; The World Bank n.d.). Those goals in the end will go a long way to serve as a solid foundation in any real estate market (Lemmen et al. 2015). You can see in Fig. 13.1 a diagram

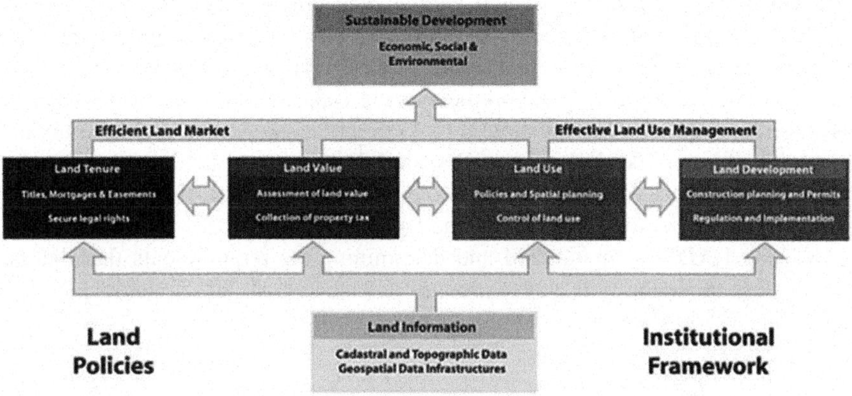

Fig. 13.1 A diagram used in the study done by the World Bank for innovations in land rights recognition, administration, and governance. *Source* An innovations in land rights recognition on, administration, and governance

used by the World Bank to show how the components of sustainable development of any land administration system should interact (The World Bank n.d.).

LADM is designed to be flexible to suit any country profile yet any standard without continuous improvement and monitoring will be outdated and fail (Lemmen et al. 2015; The World Bank n.d.). New technologies are being introduced to land registries such as blockchains (Aborujilah et al. 2021; Barbieri 2017), new geospatial capabilities such as 3D modeling for strata (Nasorudin et al. 2016). All those new technologies and frameworks are to be introduced and governed by the LADM for it to survive.

13.2.3 Country Profile

A country must develop a profile to ensure that the implementation of a land registry will meet its requirements; country profiles can be described in six characteristics (Kalogianni et al. 2021; Mithöfer 2008):

Profile scope
Stakeholder's involvement
Status existing LAS
Profile development stage
3D LA
Future LADM scope of a country profile

The above characteristics describe the LADM-based country profile. Profile scoping strives to understand the country's existing situation of the Land Administration System (LAS) to better model future LADM (Kalogianni et al. 2021; Beck and Moss n.d.). Stakeholders' involvement describes the main involvement of academic intuitions, government entities, and other influencing parties to the LADM. The status of the existing LAS describes the maturity of the existing land administration system. Profile development stage will describe the modeling process and whether it will depend on the existing cadastral database or not based on its maturity. 3D Land Administration (LA) describes the extent to which the new country profile will depend on the 3D spatial description of the land register. And finally, the future LADM scope of a country profile. This is the most important and usually the first to come in any undertaking. It is setting the scope of the future LAS using LADM as a standard and determining the requirements that the data and process modeling will evolve around (Kalantari et al. 2015; Kalogianni et al. 2021; The World Bank n.d.).

13.2.4 Rights, Restrictions, and Responsibilities on Land

Rights are an interesting concept on property, as a right does not describe the relationship between an owner and the land (Beck n.d.). Rather, it is the relationship between the owner and others concerning the land. Imagine the land is a door and the owner is the doorkeeper and he chooses who can get in and what he must do in return for access (Beck n.d.; Lazăr n.d.). So, what is the relationship between rights, restrictions, and responsibilities? This question was discussed by Dr. Beck and he stated, "*This duality of owners and others is also present in restrictions and responsibilities affecting landowners and users,*" "*Each restriction/responsibility involves a duality that imposes obligations on owners in relation to the land for the benefit of others.*" So, what is considered a right for a party can be a restriction for another. If the property is mortgaged to a lender, the lender will become a right holder to the property and a restriction will be in place on the owner in relation to the land (Beck n.d.; Beck and Moss n.d.; Lazăr n.d.).

13.2.5 Implementation of a Land Register

The following section will discuss the literature, which talks about the actual implantation of a land registry as a general roadmap, the technology used, and geospatial capabilities needed to actually establish an effective and efficient land registry (Fig. 13.2).

13.2.5.1 Current Status of the Land Registry in KSA

This process goes through the following steps

A party meets the qualification of a new grant

The royal court based on their internal process issues an order to grant a property according to certain specifications to the party

The party receives the royal court's order and goes to the Amanah or municipality where the land is allocated

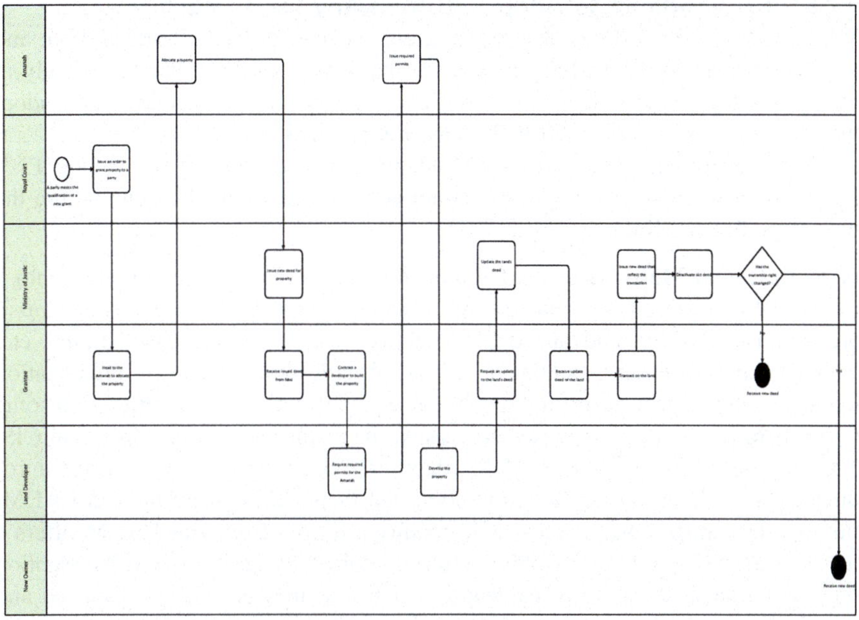

Fig. 13.2 The current high-level process flow of land distribution

The Amanah or municipality allocates the land of the party in adherence to the specification suggested in the royal court's order

A letter is sent "using mail or through the grantee" to the Ministry of Justice to issue an ownership deed for the grantee

Grantee receives an ownership deed with which she/he/it can move to develop the land by contracting a real estate developer

The real estate developer will head to the responsible Amanah or municipality to request any required permits

Amanah or municipality will issue required permits in adherence to their regulations

The real estate developer develops the property

The grantee requests an updated deed that reflects the new property's status from the Ministry of Justice

The Ministry of Justice issues a new deed including the new status of the property with any new rights, restrictions, or responsibilities or any new description

The owner receives the new deed and can choose to transact on it

If a transaction happens, the old deed is deactivated and a new deed is issued

If a transaction requires ownership change, the new owner(s) will receive the updated deed.

13.2.5.2 A Roadmap to Adopt LADM in Any Land Registry

When LADM was introduced, it was a game changer in land administration, and a standard that took more than 10 years of hard work to formulize was finally in effect. Researchers were keen to establish a clear roadmap and framework to adopt the standard in any land registry (Kalantari et al. 2015).

There are 6 main stages in LADM adaption, which you can see in Fig. 13.3. Each one of these stages describes certain activities that must be done before the next stage begins (Kalantari et al. 2015).

Organizational Motivation: land administration systems can be classified into 3 categories based on their cadastral information system (CIS) advancement. First, are countries that have advanced CIS such as Australia, France, Netherland, etc. This CIS and all processes related to it are usually built around an organization mandate, which is driven from the public needs. Secondly, are countries with some CIS capabilities but not as advanced. Lastly, the countries with no functional CIS. So, what does this mean for LADM adaption? For the advanced countries introducing the LADM, giving the assumption that their CIS was not built in LADM, this will be a huge change and it will require a lot of effort. But LADM offers a lot of value, so it might be worth the effort to make the transition. LADM adaption will get easier as it moves to categories 2 and 3 as they are not as advanced and their process can be somewhat easier to change (Kalantari et al. 2015).

Institutional Arrangements: describes the agreements to suit the existing processes and workflows with LADM; the dependencies lessen along with the complication in implementing the LADM (Kalantari et al. 2015).

Information Interpretation: describes terminology and general concepts and their relation to stakeholders. These are critical to define as best as possible in alignment with the LADM, which can be flexible to a certain extent with such terms. In

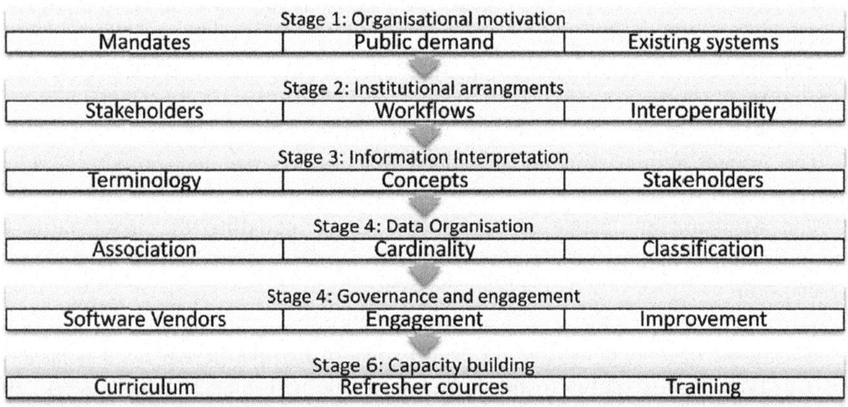

Fig. 13.3 A model that describes the stages for LADM adaption in any land registry. *Source* A roadmap to adopt the Land Administration Domain Model in cadastral information systems

LADM terminology, the word "property," which is used very widely among land administration systems around the world does not exist (Kalantari et al. 2015).

Data Organization: Data organization happens at 3 levels, which are conceptual where the entities are described and their relationships to one another, logical where the software-specific design, and physical where all attributes are shown and their metadata is described in detail (Kalantari et al. 2015; Steudler et al. 2004).

Governance and Engagement: LADM is an international standard, which was designed to serve as a foundation to build effective land registries upon, but this does not mean it is set in stone. Feedback is critical and the engagement of the communities that work with LADM in a country and government agencies involved. Committees must be formed to follow up and make sure the implementation is going in the right direction (Kalantari et al. 2015; Simsion and Witt 2005).

Capacity Building: It does not matter how well the LADM is understood in a country unless the right people are trained and given the right equipment and capabilities. They cannot offer anything; for example, the gap between land registry officers and cadastral surveying officers in technology can be hazardous. Land registry officers can have a digital platform from which they carry out their operation and the cadastral surveying officers may still use paper-based surveying reports, thus increasing the human interaction and the error margin (Kalantari et al. 2009, 2015).

The framework to adopt LADM standards in any approach encompasses the organization, people, processes, and technology.

13.2.5.3 Software Design for a Land Registry

Land registry software usually follows the same architecture as any normal software except for the separation of geospatial components (Ali Mohammed et al. 2021). Usually, a web server handles interaction with the user, and this server in turn is integrated with a web mapping server (Geoserver), which captures and reflects changes in a geospatial database (see Fig. 13.4) (Kommana n.d.; Oukes et al. 2021). All the software capabilities should be oriented to streamline processes and procedures and to make the lives of all beneficiaries of the land registry easier. It is critical that any organization that develops a land registry should not be swept away by new technologies, as it might hinder the implantation of the land register and complicate the process, procedure, and maintenance (Deininger and Goyal 2012; Kommana n.d.).

Any successful land registry should include geospatial (GIS) software that captures the spatial data and can be linked to the rights and restrictions on the land (Mithöfer 2008; Nasorudin et al. 2016). The idea of a title system is to generate a unique ID for each parcel that is described spatially and link this ID to the owner and rights holder and all restrictions and responsibilities related to the parcel (Beck n.d.).

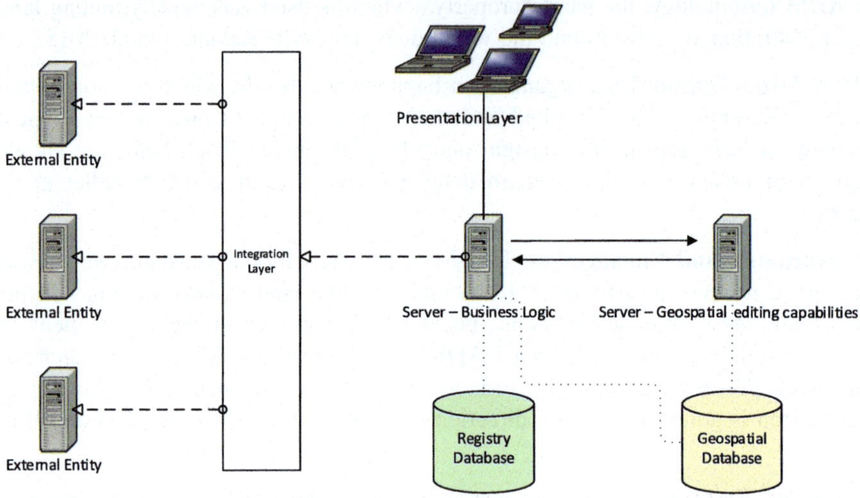

Fig. 13.4 A high-level design of land registry software architecture integrated with geospatial editing tools

Authentication and Reliability in Land Registry Using Blockchain

Blockchain is a rising technology that offers certain important aspects to a land registry. A land registry needs to be transparent, secure, and reliable. Blockchains offer this extra layer of authenticity and transparency. Blockchain is defined as "A distributed ledger technology that maintains a continuously growing list of records, called blocks, which are linked and secured using cryptography. Each block contains a timestamp and a link to a previous block. The data in a blockchain is duplicated and distributed across the entire network of computer systems on the blockchain. This makes it very difficult to tamper with the data, as any changes would have to be made to every copy of the blockchain" (Hayes 2023; Wu et al. 2023).

Blockchains depend on different ways of authentication, one of the most used is the roof of work (PoW) is a consensus mechanism used by blockchains to validate transactions and add new blocks to the chain. In PoW, miners compete to solve a computationally difficult puzzle. The first miner to solve the puzzle is rewarded with a cryptocurrency, and their block is added to the chain. This process helps to ensure that the blockchain is secure and tamper-proof (Wu et al. 2023).

There has been a different study about the implementation of blockchains in land registries. One main example is Malaysia as their land registry is facing the following challenges (Wu et al. 2023):

Indefeasibility: The current land registration system in Malaysia is based on the Torrens system, which provides for indefeasibility of title. This grant title registered absolute ownership even if the title was registered through unlawful means.

This has created some issues where titles were obtained illegally and remained registered in the land register.

Ownership: Owners' sequence isn't very well recorded in the current land register in Malaysia, which has created certain problems where disputes have been recorded about transactions, which include multiple parties.

Fraud: The current land registry of Malaysia is exposed to fraud. Manual process dominates the land registry enabling human mistakes and manipulation. This has created less of an optimal situation in regard with the authenticity provided by the land registry.

Cost: Malaysia's land register is costing a lot more than required due to manual processes. Other related government agencies are poorly integrated with the land registry, and almost all are manually integrated which skyrockets the possibility of mistakes and manipulation, as well as significantly increasing the cost.

Accuracy: Another point, which is hit hard by manual processes and dependency on human input is accuracy. Mistakes are often made due to a lack of automation and quality checks.

Blockchains can play a role to help in all the above-mentioned factors:

Reduce Fraud: Blockchains are a distributed ledger in which any transaction that needs to be recorded will need peer verification. This will help to greatly reduce fraud and maintain a sequence of records for any title, solving both the fraud issue and the ownership issue at the same time.

Increased transparency: Blockchains are a public ledger for all to see, this is true also for a land registry. This will make corruption detection highly efficient as whistleblowers will have access to the required information. Keeping in mind the privacy and confidentiality of some of the datasets blockchains can help significantly in making the land registry more transparent.

Faster process and verifications: Enabling a blockchain-based land registry will help improve the process streamlining. Enabling peer verification in a public or private blockchain will enhance the process speed as well as its quality.

Reduced cost: Blockchains can eliminate some land registry costs due to their integration with its processes. Some extra layers such as middlemen and lawyers may not be required as the blockchain will do all required verifications as well as enabling access to the data.

13.2.5.4 Development Approach

Development approaches and methodologies differ in framework and practice. Yet, two remain dominant to this day, Waterfall and Agile. Both have their advantages and disadvantages, and each has a situation where it would be perfect for it (McCormick 2012; Tilley 2020).

Waterfall excels at undertakings where the plan is clear and the requirements are very well understood. It depends on heavy engagement in the start to gather requirements and provide extensive documentation, but it lacks the required communication and error correction that might happen after development starts, which makes it suitable for projects that have a very low chance of changing or low error occurrence. On the other hand, Agile is on the other side of the spectrum where documentation is very light, enjoyment is across all phases, and it's all about finding errors and correcting them as the project goes, which makes it perfect for projects with always changing requirements and a fast-moving target (McCormick 2012; Tilley 2020).

In recent years, a hybrid methodology appeared where both Waterfall and Agile frameworks were used. Iterative Waterfall or Agile Waterfall provides some of the flexibility of Agile without ignoring the planning and deep analysis of the Waterfall across the lifecycle that represents a Waterfall phase that is broken into Agile iterations (Prenner et al. 2020).

13.2.5.5 LADM-Based Land Registries Cases

Domain-Driven Design implementation in the Netherlands

The Netherlands ran a deed-based land registry system, which has many shortcomings (Oukes et al. 2021). The system was designed in the 1980s and by the new century, the need for a new "Enhanced" system was apparent. The Netherlands took a domain-driven design "DDD" to implement a new deed system that complies with LADM (Oukes et al. 2021). DDD structures the design of a land registry into core domains and sub-domains. This is illustrated in Fig. 13.5 (Oukes et al. 2021).

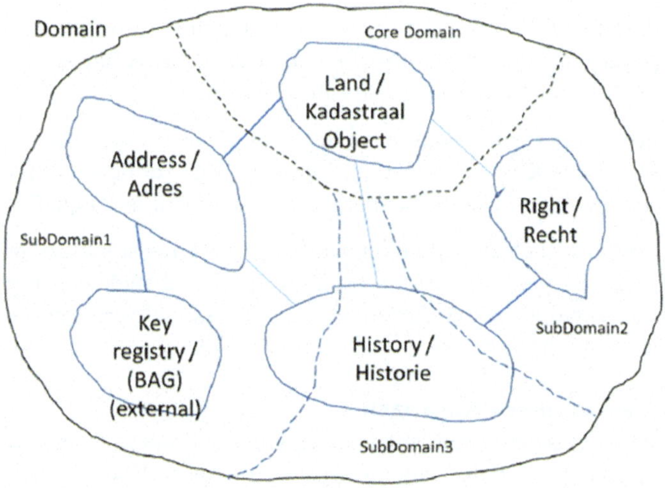

Fig. 13.5 Domain structure for the Netherlands land registry. *Source* Driven design applied to land administration system development)

The Netherlands implemented a new data model to match the LADM standard of capturing history and logging transactions in relation with the land. They ran their system in a semi-production environment for a year while the old system kept running in production. The new system mirrored all transactions in the old system, and after a year, they went through a "big bang" switchover. In a single weekend, they shut down the old system and replaced it with the new one. Issues appeared but they were handled efficiently by the IT team. This case shows how the development of a land registry can be approached in different ways (Oukes et al. 2021).

3D Cadastre implementation in Shenzhen in China using LADM
Shenzhen, China, faces an issue with its land registry as it resides on the border between Mainland China and Hong Kong. At the crossing point, an issue occurred where both Hong Kong and China's Mainland border control existed. This created an issue on how to manage RRR's on the land (Guo et al. n.d.). The Chinese looked for the international standard "LADM" for the solution. LADM offers 3D modeling as an option to be used when necessary (Oosterom et al. 2013; Guo et al. n.d.). gave both jurisdictions the ability to impose their rights, restrictions, and responsibilities according to their own respective laws as shown in Fig. 13.6 (Guo et al. n.d.).

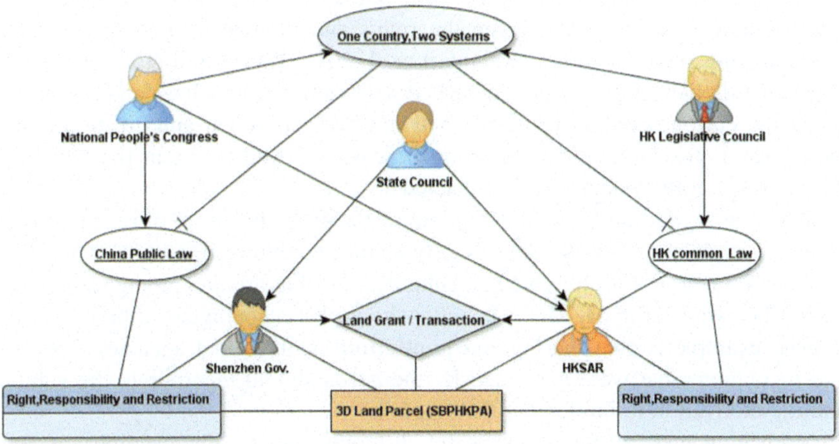

Fig. 13.6 Structure of RRRs on a land split between two jurisdictions using a 3D parcel model. *Source* Case study of 3D Cadastre in Shenzhen, China as experiment using the LADM

13.3 The Proposed Framework

The proposed framework is developed and implemented for a digital land registry in the Kingdom of Saudi Arabia. The framework consists of five stages that happen in sequence. These five stages were suggested based on the previous studies of land registry development and LADM adaptations. The stages are as follows:

Recognizing Needs and Creating Motivation
Institutional Agreements
Analysis of Process, Data, and Stakeholders to Define a "To – Be" Status
System Development of the Real Estate Registry Platform
Going Live and Capacity Building

Each of the above stages will be detailed in the following section.

13.3.1 Recognizing Needs and Creating Motivation

The organization motivation should be generated by the desire to fulfill a need that was created by either an existing problem or a new emerging opportunity. Needs of individuals, land, society, commercial entities, and government agencies should be taken into consideration. When the need is identified, multiple solutions should be considered out of which one of them will be recognized as the most suitable. Laws and regulations should be drafted and put into effect to carry out this solution, yet a solution on paper will remain on paper unless an enforcer of that solution is assigned and given the rights and privileges to carry out a plan implementing such a solution (Kalantari et al. 2015; Kalogianni et al. 2021).

Required changes need to be identified. How will the new laws and regulations affect the current existing process? Will it solve a problem and create another? Will related stakeholders benefit from the change? And how can the benefit be maximized for all if such a law is enacted?

When laws get drafted in the Kingdom of Saudi Arabia, a set of regulations are attached, explaining how the law exactly should be enforced, so one of the main motivation creators for all related stakeholders is what are they going to benefit if such a law was enforced. Those benefits should be encouraged in the regulation. A clear example is data sharing and joint efforts with related agencies. The laws and regulations that were generated by the needs should incorporate the solution and the motivation to implement the solution.

Motivation takes different forms for different entities. For example, an individual can be motived to participate in the process of developing a land registry by promising them that their rights will indisputably be guaranteed by the land registry. Another example is a government entity that can be motived to help by a clear direction from the head of the government and by the benefits that they will get such as fees and fines collected and data shared (Kalantari et al. 2015; Kalogianni et al. 2021).

13.3.2 Institutional Agreements

Implementing a new land registry in any jurisdiction will be considered a complex and daring undertaking. Implementing a new land registry in a country the size of the Kingdom of Saudi Arabia will be a mega project that will span years and will involve change management and the collaboration of different agencies and administrations (Kalantari et al. 2015; Kalogianni et al. 2021).

The second stage of this framework should focus on the agreement required to carry out such an implementation. The institutional agreements refer to the (formal and informal, state and non-state) institutions, their mandates, procedures, and capacities for implementing a country's policies, laws, and regulations. First, is the agreement required to share knowledge and data about the existing land registry, which is managed by the Ministry of Justice. Understanding the full current status "As – Is" is crucial to move towards the desired status "To – Be." The desired status was specified in the new laws and regulations but how to move there remains the challenge (Kalantari et al. 2015; Oukes et al. 2021; Guo et al. n.d.).

Other agreements must come into effect that will affect related stakeholders. The real estate domain is complicated; different parties play different roles and have different interests. All need to be accommodated into a development and operating model that will offer a seamless transition to work under the new law and regulations.

The main stakeholders in the Kingdom of Saudi Arabia are the regulator for the real estate domain the Real Estate General Authority; the jurisdictional authority and current regulator of the real estate domain the Ministry of Justice; the responsible agency for planning and administrating urban and rural areas the Ministry of Municipalities and Rural Affairs and Housing under which the Municipalities and Amanahs fall; and other parties such as the national housing company, WAFI the responsible agency for off-plan property management, Real Estate Developers, the Ministry of Environment, Water and Agriculture.

All of the above-mentioned agencies are included in the real estate domain. This indicates the deep complexity and interconnected processes that need robust agreements and a clear on-boarding change management process.

13.3.3 Analysis Exiting Processes, Data and Stakeholders to Define a "To – Be" Status

13.3.3.1 Understand the "As – Is"

Almost every country has a standing land registry of one form or another. Land registries differ in the amount and format of details that they record about property so, a detailed analysis of the process data and the involved stakeholders—with their process and data—needs to be carried out. This analysis is required to bring forth such a detailed description and deep understanding of the current status (Lemmen et al. 2015).

Building the "To – Be" should not be all about adherence to the new law or accommodating LADM. Unless you fill the existing gaps and solve the current problems, the new land registry has failed to achieve its purpose. Subject Matters Experts (SMEs) from the existing regulating body should collaborate with the current parties running the land registry to achieve the full understanding required of the current status. Weaknesses in the process, inconsistency in the data and inefficient execution by the responsible parties should be identified and documented through a gap analysis report, which should be referred back to later when structuring the "To – Be" status (Beck and Moss n.d.; Lemmen et al. 2015).

13.3.3.2 Define the "To - Be"

Uses the law and regulation as a base and the LADM and the current "As – Is" situation as a reference for a gap to be filled. A new process, data model, and RACI matrix should be defined.

New Processes

It is expected that any land registry in the world will have different processes out of which many will include different parties and agencies involved. Such processes need to be defined and modeled at the appropriate levels. The researcher recommends levels 1 and 2 "Strategic Level" and Level 3 "Functional Level" (Association of Business Process Management Professionals 2013).

The strategic level makes sure that the business process helps the overall strategy goals, while the functional level helps understand what will be done and by whom. This makes sure all that needs to be done is being done, but there is no need to dive into the operational level, as this will come in the next stage of the framework. Figure 13.7 helps illustrate the details contained on each level (Association of Business Process Management Professionals 2013).

13.3.3.3 Data Model

It is essential for any land registry to capture the correct dataset about land. LADM gives essential data that needs to be gathered in a land registry, such as:

Party—Person—Organization
Legal—Administrative
Spatial Data—Immovable Objects
Data on Surveying and Geometry Data "Meta Data"

These are the essential data required by LADM to be included in any land registry. Christiaan Lemmen and others suggest in their "Land Use Policy" the following model (Fig. 13.8).

The main observation is that the model has included all the essential information suggested by LADM and added some more. As it should be the case in the Kingdom of Saudi Arabia. LADM should be adopted not copied; what needs to exist should, but what does not should be excluded. Also, all land registries are unique thus giving the need to modify/add/remove any required data that exists in the LADM or in the current land registry.

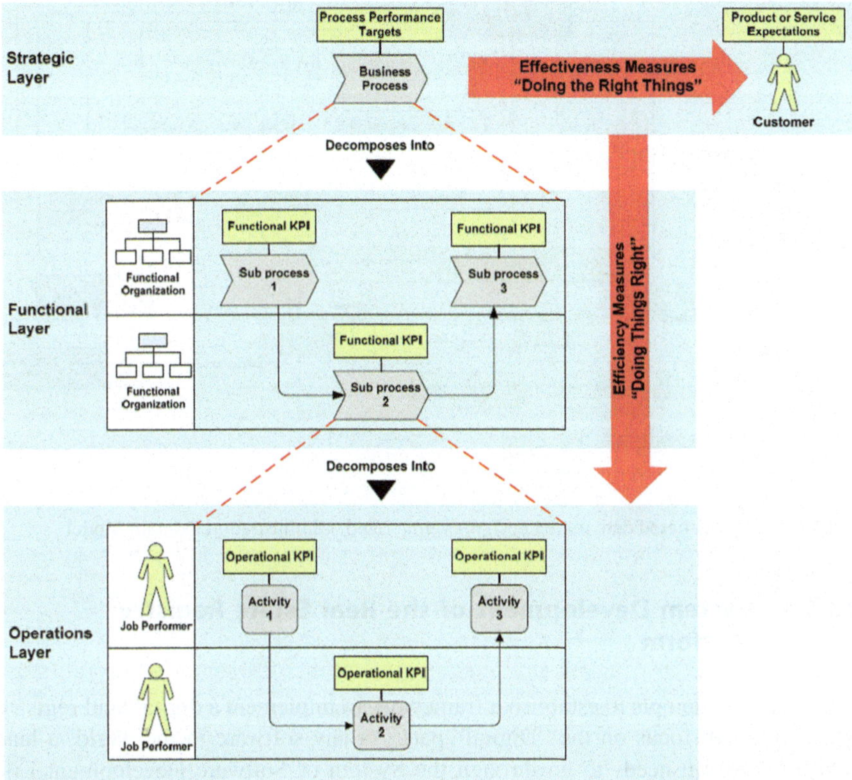

Fig. 13.7 Process levels as defined by the assassination of business process management professionals "ABPMP." *Source* Association of Business Process Management Professionals (2013). BPM CBOK: version 3.0ABPMP

13.3.3.4 Governance

A RACI matrix which is defined as "Responsible, Accountable, Consulted, informed (RACI), or often called the Responsibility Assignment Matrix (RAM), is a helpful tool for describing the work responsibility assignment matrix" should be designed on an organization level and within each organization should be defined as part of drafting the "To – Be" process. Different processes will be carried out and different datasets will be collected, to have a clear matrix describing who is doing what, who is overseeing, and who is being informed is important to have a governed seamless process flow (Dwi et al. n.d.).

Each process should have an owner and any change should go through a clear change management process. The governance framework should be flexible enough to incorporate any changes as a new land registry will take some time to reach high maturity and change is inevitable.

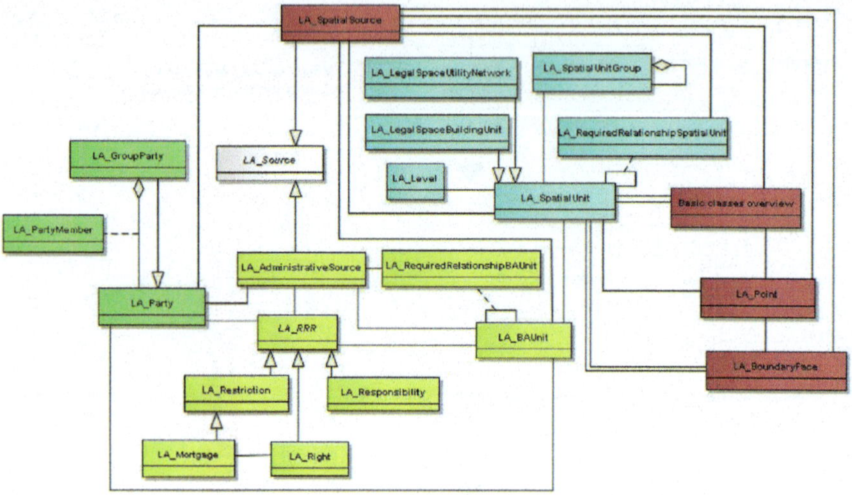

Fig. 13.8 LADM-based data model design. *Source* Land Administration Domain Model

13.3.4 System Development of the Real Estate Registry Platform

The research attempts to establish a framework to implement a digital land registry. This stage will focus on the "Digital" part. As any software in the world, a land registry platform needs to go through the System or Software Development Life Cycle (SDLC). SDLC was first described as a structured approach in 1970 by Winston W. Royce in what is known today as the waterfall model that contains the stages (Choudhary 2018) represented in Fig. 13.9. Stages may differ from one organization to another, and stages may be taken multiple times iteratively or one time based on the methodology used. Stages should happen in sequence for a functionality to be built or enhanced. This does not mean that analysis of a functionality cannot happen while another functionality is being developed. SDLC stages briefly are (Tilley 2020):

Planning: The scope of the system is determined and a baseline is established to build the functionality.

Requirements Analysis: The functionalities are broken down into smaller components, analyzed and assigned proper specifications.

Design: The creation of the blueprint of the software and the design of its architecture that is appropriate to the functionalities that need to be developed.

Implement: The development of functionalities starts in adherence to requirement specification and system design.

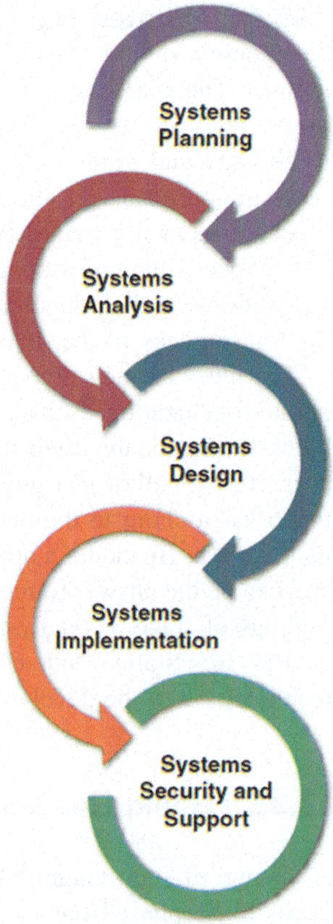

Fig. 13.9 The stages of a system development life cycle (SDLC). *Source* Scott Tilley, System Analysis and Design, 12th edition

Support: Any obstacle to the daily operation of the system is removed to support intended users.

13.3.4.1 Suggested Approach to Developing a System for the Real Estate Register in the Kingdom of Saudi Arabia

The suggested approach to develop a digital land registry needs both components from the waterfall and agile methodologies. With a hybrid approach, you can benefit from the flexibility of multiple releases as well as the thorough analysis and design that occurs in each release. As the land registry is a critical function that touches the public directly, an implementation needs to have both. Usually, hybrid approaches are customized to the organization's needs and do not follow a predefined framework such as the waterfall and agile, yet researchers have tried to put it into some sort of a shape called the waterfall-agile approach or waterfall-agile methodology (Prenner et al. 2020).

13.3.4.2 Adopted Technology

Technology should always be chosen to serve the purpose, not the other way around. There are multiple current technologies that can be utilized to build a digital land registry. A simple two- or three-tier architecture using any available technology and programming languages for frontend, backend, and database will serve the purpose. Yet, there is a case to be made for newer technologies such as blockchains as it provides the increased security and reliability that comes from a peer to peer authentications and smart contracts. A private blockchain between certain entities can help in streamlining the land registry processes as well as increase the transparency of the registry (Wu et al. 2023).

The introduction of blockchains is a must in case there is ever the need to incorporate tokenization. Converting the value of land assets into digital tokens on the blockchain to enable their digital transfer and ownership. The token represents the property with all of its rights and obligations. This means anyone who owns a land token has a claim to the underlying land property as well as any associated profits or losses. Blockchain applications in the land registry are novel and interesting and can be the answer to many problems. The framework suggests to capitalize the findings of stage 3 "Analysis Exiting Processes, Data and Stakeholders to Define a "To – Be" Status" and see if there is an actual need for the introduction of such technology into the development of a new digital land registry (Wu et al. 2023).

13.3.5 Going Live and Capacity Building

A change of such magnitude that will re-structure the land registry of a country needs to be phased out in stages. Different approaches may be taken, a parallel approach in which the new system is launched while the current one is still up and running. This approach will cost more and may create data inconsistency, but it will minimize the risks taken and will ease any rollbacks. A big bang approach can be taken where the old system is switched off and the new system is deployed at a single point. This approach increases the risks significantly but cuts down the cost of running two systems in parallel. Other approaches that may be taken include the migration of service from the old system to the new or a zone-by-zone migration, which may result in a less drastic result from an operational and cost point of view.

Before, during and following the go-live, capacity building to all aspects of the new land registry's operation needs to be scaled up. Business operations, training and support, IT operation, and infrastructure all need to be scaled up according to the approach taken to go live with the new land registry.

Suggested Approach to going live with a digital land registry in the Kingdom of Saudi Arabia

Going live in such a project is a delicate matter. A phased approach seems the most reasonable, as big bangs usually tend to cause a lot of problems and the real estate domain has a low tolerance for issues (Oukes et al. 2021). Areas should be defined

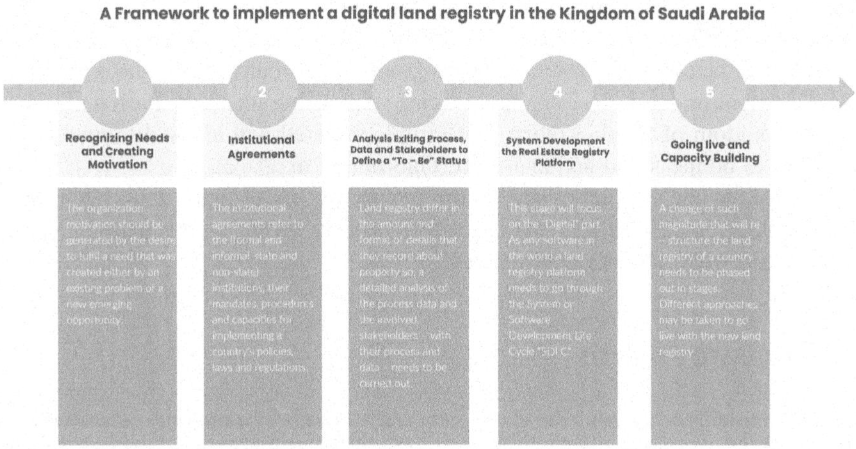

Fig. 13.10 The proposed framework

to start the implementation of the new system while maintaining the old system running outside these areas. This approach enables the piloting of the software as well as an easy rollback if anything goes wrong and it requires reverting back to the old system.

13.3.6 A Summary of the Framework Suggested

Figure 13.10 represents a summary of the framework discussed in this chapter.

13.4 Research Findings and Analysis

For this research, qualitative data will be collected through interviews, as this is the best technique for research using an exploratory approach. Such interviews are carried out with subject matter experts working in entities in the real estate domain in the Kingdom of Saudi Arabia. Interviews were chosen as well to give a wider perspective from a different a point of view of the current problem and how the suggested framework will contribute to enhancing the land registry domain in the country (Bhattacherjee n.d.).

13.4.1 Dataset Details

As mentioned in the previous section, the technique that will be used is unstructured interviews as data collection. Interviews provide in-depth knowledge of the subject, which is required for this research. In addition, the study will allow a selection of a few individuals who are considered SMEs of the topic to discuss the research questions (Dunwoodie et al. 2022).

Data will be collected from a selected group of 6–8 individuals who work in 3–4 government or semi-government agencies responsible for the real estate domain in the Kingdom of Saudi Arabia. The sample selected will allow different points of views to be measured and analyzed (Gubrium et al. 2012).

The role of the researcher was as the interviewer in all interviews. Each interview took about 25–30 min. All participants were aware of the data collection process and what the data will be used for.

13.4.2 Data Analysis

In the following chapters, the data gathered through several interviews will be analyzed. All questions asked in the interview were designed to test the suggested framework discussed in chapter three. The interviews were carried out in Arabic, as not all participants were proficient English speakers. In Table 13.1, all participants who have been interviewed will be listed according to their level and organization.

The transcribed interviews were analyzed with content analysis using a deductive approach as the research is testing a pre-defined framework using data collected to prove its viability. Codes were assigned to identify key data repeated in the transcripts. The following is the analysis of the data collected (Woiceshyn and Daellenbach 2018) (Table 13.2).

Analysis will be conducted in each aspect of the framework from which the questions for the interviews derived.

13.4.2.1 First Stage: Recognizing Needs and Creating Motivation

Percipients have agreed that that there are multiple needs that are not satisfied due to certain factors such as lack of alignment, outdated laws and regulation, no governance for investment in the real estate sector and problems that were inherited from legacy systems. Participants recognized the need for a broad change across strategic, tactical and operational levels to enable a solution to take effect to fill the existing gaps.

Table 13.1 Participants who contributed to the data collection

#	Management level	Agency
	Middle management	Ministry of Justice
	First—line management	Ministry of Justice
	Senior management	Economics Cities and special Zones Authority
	Middle management	Economics Cities and special Zones Authority
	Middle management	Real Estate Registry
	Analyst	Al Madina's Amanah
	Analyst	Ministry of Justice

Table 13.2 Questions and responses log

#	Question	Participant						
		1	2	3	4	5	6	7
	Do you think there are gaps in the land registry system in Saudi Arabia? If yes, how it can be best filled from regulation and operations point of view?	✓	✓	✓	✓	✓		✓
	Do you think all agencies (public or private) are fully trained and aware of the process? Do you think more and better alignment can happen between them and how?	✓	✓	✓	✓	✓	✓	✓
	Is all data necessary in a real estate registry is captured and in the correct form? What do you think is missing?	✓	✓	✓	✓	✓	✓	✓
	What in your opinion is the degree of digitization possible that could happen in a land registry in Saudi Arabia?	✓	✓	✓	✓	✓	✓	✓
	Do you think the better approach to develop a digital land registry should focus on precise planning, or higher flexibility, or the need for both is equal?	✓	✓	✓	✓	✓	✓	✓
	What is the best way to go-live in a brand-new digital land registry in Saudi Arabia?	✓	✓	✓	✓	✓	✓	✓

As one participant has stated: "*Distributing responsibility randomly across different agencies has created ambiguity, I think there is a lot of miss alignment. The process is for sure is capable of enhancing.*"

Another participant also said: "*There are two main gaps in our current system. Firstly, is the absence of judicial body that is specialized in real estate cases and the lack of regulation that allow the quick processing of such cases, Secondly, there is no clear mechanism for foreign investment in the country, this deters foreign investors of entering the Saudi market.*"

13.4.2.2 Second Stage: Institutional Agreements

Participants have mentioned that there is a lack of alignment among the different entities related to the real estate system. As one participant stated: "*There is simply not enough awareness shared among agencies, and with the new services coming live quickly the lack of shared understanding has become a nightmare.*"

Another participant said: "*Agencies are working as silos. Municipalities, the Ministry of Justice, and the courts are not aligned and there is big room for enhancements.*" Nonetheless, some have mentioned that this lack of alignment is not just related to lack of communication or understanding, but with the current state of the Kingdom of Saudi Arabia fast changes and quick development to achieve their 2030 goals have made some agencies move faster than others to meet their demands.

One participant said: "*Agencies have good understanding of processes but with the changing demands they need to adapt which is their right but it causes gaps in the knowledge of others involved.*"

13.4.2.3 Third Stage: Analysis Exiting Processes, Data, and Stakeholders to Define a "To - Be" Status

Participants have agreed that the current processes have gaps, which resulted in missing data captured by the real estate system. One participant has stated: "*Current data miss accuracy and completeness. Things such as the real estate type, real estate historical record, maintenance and utilities are all missing.*"

Another participant has said: "*Although a lot of information is being captured, I think the most important piece that is missing is accurate location information and its relationship with the ownership record.*" Almost all participants mentioned the lack of quality of geospatial information, either by being missing completely or outdated and out of sync with other agencies. Other percipients mentioned a lack of other data such as utilities in the real estate as they are maintained by other agencies, which are completely disintegrated from the real estate system.

One participant has mentioned that the current real estate system lacks the history of the property, he stated the following: "*Right now the deeds does not show who owned the property and when, some state the previous deed number but even if that is mentioned it can be not digitized and kept in the archives of the notary that issued it, and when you find it, you will not find the prior ones to that.*"

13.4.2.4 Fourth Stage: System Development of the Real Estate Registry Platform

Participants were divided with regard to the approach needed to develop a new real estate registration system. There was a segment asking for adapting an agile approach to the development of the new system, yet others preferred precise planning and to take a waterfall approach. Others stated the need for both is required, as the processes of real estate is too complex yet quickly changing; or as he stated: "*Mixing both adaptability and planning is required while always keeping in mind the big picture of a full comprehensive system that serves the current needs and has the scalability to enhance in the future to incorporate any new processes or changes.*"

But the general direction of the data leaned towards agile as another percipient stated: "*You will not believe the odd situation we face and need to handle, if an inheritor of an inheritor of the owner who passed away an 80 years ago comes to you to wanting to do any transaction on his old paper deed how can you process his request, is he even allowed to do anything as the property is most likely owned by a 100 individual at this point.*"

13.4.2.5 Fifth Stage: Going Live and Capacity Building

Most participants have agreed that a phased launch of a new real estate system is the suitable way to go, as a full cut-off from the old system to the new will create disruption that the real estate domain cannot be tolerated. One participant has stated: "*Based on best practices for such a project, it is absolutely need to be taken on in phases. A big bang approach will just disturb the market creating problems that could go on for years.*"

One participant argued in favor of a nation-wide release and said: "*A single release to unify all process across the country will make more sense in my point of view. It will help minimize confusion as the market will be affected by inconsistencies in data and process if two systems were running in parallel.*"

13.5 Results and Discussion

This research aims are:

To evaluate the current status of the real estate current status in the Kingdom of Saudi Arabia
To define best practices to establish a new land registry or enhance an existing one
To design the framework required to implement a digital land register in the Kingdom of Saudi Arabia
These objectives were driven from the main research question, which is: **How to implement a digital land registry in the Kingdom of Saudi Arabia?**
Extensive literature review was carried out and based on that a framework was designed to describe the implementation of a digital land registry in the Kingdom of Saudi Arabia. Based on this, a set of questions were put to a selected sample to collect the required data to test the framework, which were analyzed in the previous chapter. This chapter will discuss and elaborate the results based on the literature review and the collected data.

13.5.1 The Viability of the Framework

According to the data analyzed, the framework seems to capture the required stages for its purpose. Some interesting points were discussed that might be taken into consideration to better refine the framework such as:

- In the first stage of the framework, the data collected strongly suggested that there is a significant need to implement a new land registry in the Kingdom of Saudi Arabia. Most participants expressed their dissatisfaction with the current system, emphasizing the need for a change to solve the different existing problems.
- In the second stage, the framework suggests an institutional agreement across all responsible stakeholders in the land registry. This stage was intended to describe the agreements required by agencies to draw and design processes to

carry out real estate operations. Yet the data suggested that the challenge also lies in the fast-moving changes in the country and how organizations need to find a way to manage these quickly moving markets and not just focus on agreements and contracts. Change management and clear governance need to be introduced to the framework as the data collected emphasizes its importance.
- In the fourth stage of the framework, the data collected indirectly agreed that a hybrid model needs to be adopted, as the data suggest the need for flexibility and precise planning.
- In the fifth stage of the framework, the data collected shed light on an extremely important factor that might create problems if the suggested approach in the framework was adapted, which is data inconsistency and process ambiguity. As two real estate systems will be running in parallel but in different areas, the migration of data will happen over a long period of time. The data model for both systems will be evolving, thus the need for constant updating and incorporating of new processes in the two systems will remain.

The literature provides great insight into the findings from the data collected, as described in the study carried out by Kalantari and others in 2015, where they discuss a road map to adapt the Land Administration Domain Model in cadastral information systems. They have assigned one of their 6 stages to "Governance and Engagement" where they manage changes between processes and stakeholders as no two implementations of the LADM are the same.

On other aspects, the literature agrees with the data collected that a hybrid approach for software development is possible to be used from the framework suggested. Such an approach sounds reasonable for an LADM-based land registry in the Kingdom of Saudi Arabia. As the Kingdom is going through mega real estate development projects as well as their existing land registry system, this demands the integration of both flexibility due to these mega projects and planning and analysis of their existing system. A hybrid approach can provide both to a good extent, allowing the Kingdom to adapt LADM without major disruption to its current development or legacy system.

Another intersection between the literature and the data collected is the launch of a new land registry. The literature mentions the case of the Netherlands discussed by Oukes and others (2021) in their paper Domain-Driven Design applied to land administration system development: Lessons from the Netherlands Land Use Policy, where they discuss the deployment of a new land registry system, how it was carried out and what was learned. In their study, it was described that the Netherlands opted for a parallel operation of two systems but in a different way than was suggested in the framework. The land registry department ran the new system for an entire year as a copy of the old one. When a transaction would occur in the old system, it was reflected in the new without the new being exposed to end users. This can eliminate the fear that the requirements were not captured correctly, and it makes sure that the data model used can serve the intended purpose to its fullest eliminating the risk of significant data inconsistencies when moving from the old system to the new.

The data collected suggested that the implementation of a new land system should depend on a hybrid approach, but leaning toward flexibility. This can be analyzed against the case discussed in the literature where Guo described in his paper 3D Cadastre Implementation in Shenzhen China using LADM (2011), a scenario where agility served to provide a quick solution between China's and Hong Kong's governments and implement LADM in a disputed border in the province of Shenzhen.

The data collected suggested that the current situation of the real estate market does not support foreign investment in the country. This affects the entire country's vision and direction to diversify the economy. The framework should incorporate guidelines to enable the real estate market to be a global player.

It seems that the generality of the framework gives it an advantage from the point of view of letting the circumstances of the situation form stages into detailed execution plans. Yet, it also creates a disadvantage by allowing certain aspects broadly discussed such as a full stage of change management and governance, as it might allow people to fill these gaps with poor practices not specified in the LADM or any tested standard (Table 13.3).

13.5.2 Conclusion

Land management has existed since the dawn of humanity, and with the surge of industrial and technological advances, the administration of land has become more complex and essential. The Land Administration Domain Model (LADM) or ISO19152 was born in 2012 to bring more discipline and order to the chaos. This research examined closely LADM and the case studies of its implementation across the world. It also studied different concepts of real estate and the current situation of the Kingdom of Saudi Arabia. The research then went on to examine the current approaches to develop a software for such a land registry. This enabled the design of a specific framework related to the Kingdom. The data collected and the literature was analyzed to test the framework designed. The research confirmed that the framework was viable to be used for its purpose to a certain extent. Nonetheless, the research also suggests that there are certain gaps that can be covered to better allow the framework to function properly which were discussed previously. It is expected that this research will offer insight into the land registry system in Saudi Arabia and how to enhance it. The findings of this research can be utilized in different fields such as real estate transaction management, software development, and Institutional governance.

This research was faced with different limitations and challenges. First, the time of the study was limited to a single academic semester, which put a constraint on the analysis of the literature, design of the framework, and data collection and analysis. Another limitation was the sample selected. It limited the research from two different perspectives. First, the size of the sample was not as large as desired, which may have limited the point of view and the quality of the data collected. Also, the number of agencies that the sample was drawn from was too few. The

Table 13.3 Summary of research findings

Framework stage	Literature findings	Interview findings	Similarities and differences
	"Land administration is widely recognized as a critical component of sustainable development, as it affects a wide range of economic, social and environmental outcomes. The Land Administration Domain Model (LADM) provides a common conceptual framework for land administration, which can assist in improving the efficiency, transparency and accountability of land administration processes and procedures" (Oosterom et al. 2013)	The data collected has strongly suggested that there is a significant need to implement a new land registry in the Kingdom of Saudi Arabia based on the new best practices and standards	The data analysis found that stage one is similar between the interview and literature findings as there is a need to be fulfilled
	"Integration and collaboration between organizations, leading the land administration in a country is critical to improve efficiency, accuracy and consistency in land administration" (Beck n.d.)	The data collected suggests institutional agreements must come in place with the room for fast base changes	The data analysis found that stage two is similar in suggesting the agreements between different stakeholders. Yet different in allowing a fast base change to occur
	"Analysis and interpretation of information in addition to data organization are critical components in understanding the existing process. A well-structured and organized process, combined with a thorough analysis of relevant information, provides valuable insights into the current state of the process, identifies areas for improvement, and helps in determining the right course of action" (Association of Business Process Management Professionals 2013)	The data collected suggests the importance of defining the "As – Is" and building the "To – Be" form process and data point of view	The data analysis found that stage three is similar in suggesting the deep analysis and data organization. Yet different in emphasizing the design of the "To – Be"

(continued)

Table 13.3 (continued)

Framework stage	Literature findings	Interview findings	Similarities and differences
	"Hybrid software development models are possible and could be useful in providing insights into the organization and implementation of these models. Such a review would help identify best practices, common challenges, and potential benefits of hybrid development approaches" (Prenner et al. 2020)	The data collected suggests that a hybrid model of software development is most suitable to develop a new real estate system	The data analysis found that stage four is similar in suggesting a hybrid model to software development
	"Phased deployment is an important aspect of successful land administration system development. By gradually implementing new functionality, organizations can test and refine the system before full-scale deployment, reducing the risk of failure and ensuring a more seamless transition to the new system" (Oukes et al. 2021)	The data collected suggests an area-by-area approach is most suitable to go live in a new land registry	The data analysis found that stage five is similar in suggesting a phased (area by area) approach to go live in a new land registry

real estate domain has many stakeholders. This research collected data from people in the main agencies working on the sector but did not include some of the other agencies that can give viable input to the framework. A big limitation that faced by the author was gaining access to the sample required, as they include people who are in key positions in agencies related to the real estate domain and are not easily reachable. This research was designed for the implementation of a land registry system in the Kingdom of Saudi Arabia, thus creating a lack of generalizability as the result may not be applicable for other regions. Yet, a future study can try to generalize the framework enough to be applicable at least to regions with similar demographic, geographic, and social attributes.

Future studies could focus on refining the framework as well. As the analysis of the collected data and literature found the lack of dedicated governance and change management stage could result in some problems, as the new land registry will be working in close contact to the legacy system. Another study can dive deeper into data management, migration, and governance in real estate digital system implementations. In addition, it would allow research on how to enable the land registry to activate global investments in the Saudi market.

References

Aborujilah A, Yatim MNM, Al-Othmani A (2021) Blockchain-based adoption framework for authentic land registry system in Malaysia. Telkomnika (Telecommun Comput Electron Control) 19(6):2038–2049. https://doi.org/10.12928/TELKOMNIKA.v19i6.19276

Ali Mohammed H, Zeebaree S, Mujdat Tiryaki V, Sadeeq M (2021) Web-based land registration management system: Iraq/Duhok case study. J Appl Sci Technol Trends 2(04):113–119

Association of Business Process Management Professionals (2013) BPM CBOK: version 3.0ABPMP

Barbieri M (2017) Blockchain-can this new technology really revolutionize the land registry system? http://www.forbes.com/sites/laurashin/#1143578f655d

Beck A (n.d.) Developing a Hohfeldian theory of real rights to support LADM modelling

Beck A, Moss D (n.d.) LADM patterns to support the efficient modelling of cooperative (community and strata) titles for land registers

Bhattacherjee A (n.d.) Scholar commons social science research: principles, methods, and practices

Choudhary S (2018) Evolution of system development life cycle (SDLC). https://www.linkedin.com/pulse/evolution-system-development-life-cycle-sdlc-shantanu-choudhary/

Deininger K, Goyal A (2012) Going digital: credit effects of land registry computerization in India. J Dev Econ 99(2):236–243. https://doi.org/10.1016/j.jdeveco.2012.02.007

Dunwoodie K, Macaulay L, Newman A (2022) Qualitative interviewing in the field of work and organisational psychology: benefits, challenges and guidelines for researchers and reviewers. Appl Psychol. https://doi.org/10.1111/apps.12414

Dwi R, Suhanda P, Pratami D (n.d.) RACI matrix design for managing stakeholders in project case study of PT. Int J Innov Enterp Syst

Gubrium JF, Holstein JA, Marvasti AB, McKinney KD (2012) The SAGE handbook of interview research: the complexity of the craft

Guo R, Ying S (n.d.) A multi-jurisdiction case study of 3D Cadastre in Shenzhen, China as experiment using the LADM

Hayes A (2023) Blockchain facts: what is it, how it works, and how it can be used

Kalantari M, Lester C, Boyle D, Coupar N (2009) Towards eLand administration–electronic plans of subdivision in Victoria. In: Ostendorf B, Baldock P, Bruce D, Burdett M, Corcoran P (eds) Surveying & spatial sciences institute biennial international conference. Surveying & Spatial Sciences Institute, Adelaide, pp 155–162

Kalantari M, Dinsmore K, Urban-Karr J, Rajabifard A (2015) A roadmap to adopt the land administration domain model in cadastral information systems. Land Use Policy 49:552–564. https://doi.org/10.1016/j.landusepol.2014.12.019

Kalogianni E, Janečka K, Kalantari M, Dimopoulou E, Bydłosz J, Radulović A, Vučić N, Sladić D, Govedarica M, Lemmen C, van Oosterom P (2021) Methodology for the development of LADM country profiles. Land Use Policy 105(2021):105380

Kommana K (n.d.) Implementation of a geoserver application for GIS data distribution and manipulation

Lazăr I (n.d.) The administration right and the entry in the land register of the ownership over the real estate subject to public propriety

Lemmen C, van Oosterom P, Bennett R (2015) The land administration domain model. Land Use Policy 49:535–545

McCormick M (n.d.) Waterfall vs. Agile methodology (2012)

Mithöfer K (2008) Development of a GIS-based land registry for Tanzania

Nasorudin NN, Hassan MI, Zulkifli NA, Rahman AA (2016) Geospatial database for strata objects based on land administration domain model (LADM). Int Arch Photogramm Remote Sens Spat Inf Sci-ISPRS Arch 42(4W1):329–334. https://doi.org/10.5194/isprs-archives-XLII-4-W1-329-2016

Oosterom P, Lemmen C, Uitermark H (2013) ISO 19152:2012, Land administration domain model published by ISO. ISO, The Netherlands, pp 1–8

Oukes P, Van Andel M, Folmer E, Bennett R, Lemmen C (2021) Domain-driven design applied to land administration system development: lessons from the Netherlands. Land Use Policy 104. https://doi.org/10.1016/j.landusepol.2021.105379

Prenner N, Unger-Windeler C, Schneider K (2020) How are hybrid development approaches organized? A systematic literature reviews. In: Proceedings—2020 IEEE/ACM international conference on software and system processes, ICSSP 2020145–154. https://doi.org/10.1145/3379177.3388907

Simsion GC, Witt GC (2005) Data modelling essentials, 3rd edn. Elsevier Inc., San Francisco

Steudler D, Williamson I, Rajabifard A (2004) A world-wide comparison of cadastral systems, cadastral template. GIM International

The World Bank (n.d.) Innovations in land rights recognition, administration, and governance

Tilley S (2020) System analysis and design, 12th edn

Woiceshyn J, Daellenbach U (2018) Evaluating inductive vs deductive research in management studies: implications for authors, editors, and reviewers. Qual Res Organ Manag: Int J 13(2):183–195. https://doi.org/10.1108/QROM-06-2017-1538

Wu R, Abdullah N, Ismail AR, Ali M (2023) Blockchain-based framework for secure and reliable land registry system. Int J Inf Manag 55:102412. https://doi.org/10.1016/j.ijinfomgt.2022.102412

Blockchain for Secure Payments: A Bibliometric Review

14

Vibhuti Jain, Ruchi Mehrotra, Jose Arturo Garza-Reyes, Rajesh Tiwari, and Khem Chand

Abstract

Blockchain has gained the attention of stakeholders as a potential technology for peer-to-peer transactions. Traditional online financial transactions are susceptible to cyberattacks. The poor adoption of online payments has been a concern due to security issues. Blockchain offers a secure alternative for banks and non-bank entities engaged in the payment sector. The chapter explores the literature on blockchain for secure transactions. A bibliometric analysis was done on the Scopus database. 311 documents were used for the analysis. The VOS viewer software version 1.6.19 was used for the analysis. 187 documents were published in the Computer Science domain, 94 documents were published in the business and management domain, 86 documents in engineering, 58 in economics, econometrics, and finance, and 51 documents were published in the social science domain. The findings have implications for researchers as they explore major themes for research in secure payments using blockchain.

Keywords

Blockchain • Security • Payments • Scopus • Banks • Fintech • Bibliometric analysis

V. Jain · R. Mehrotra (✉) · R. Tiwari
Department of Management Studies, Graphic Era (Deemed to Be University), Dehradun, India
e-mail: uniqueruchi@rediffmail.com

J. A. Garza-Reyes
Head of Centre of Supply Chain Department, University of Derby, Derby, UK
e-mail: J.Reyes@derby.ac.uk

K. Chand
Mittal School of Business, Lovely Professional University, Jalandhar-Delhi, G.T. Road, Phagwara, India

A. Kumar et al. (eds.), *Sustainable Security Practices Using Blockchain, Quantum and Post-quantum Technologies for Real Time Applications*, Contributions to Environmental Sciences & Innovative Business Technology,
https://doi.org/10.1007/978-981-97-0088-2_14

14.1 Introduction

Technological advancements and innovations have revolutionized the financial services system in the whole world which today, it is referred to as FinTech. The innovative Fintech applications developed by Startups have also contributed majorly toward the Fintech Revolution (Gomber et al. 2018). The rising use of Mobile payments through UPI, the growth of microfinance institutions which are adopting innovative digital ways for providing services, P2P lending, Crowdfunding, and Blockchain, all are new trends in the Finance market and are the future of Fintech (Liu et al. 2020).

Whenever there is a centralized system of control, there is a risk of information hacking and here a decentralized system like Blockchain comes to the rescue (Li et al. 2019a, b). The decentralization of the information is what makes Blockchain a secure system (Thakor 2019). It is highly helpful in increasing the efficacy of all Financial Institutions, especially in the payment and credit information system in banks. It simplifies the payment procedures by using a decentralized ledger as a regulator and by eliminating unwanted intermediaries (Wu and Liang 2017).

Digitalization and Automation are the current trends for which Blockchain technology is a path to be followed. It is a unique technology as it is immutable and backed by a cryptographic protection system. It is being used for banking services, payment settlement, IOT devices, car sharing, and so on (Vasilievich Babkin et al. 2018). Peer-to-Peer lending, UPI, Cryptocurrencies, Robo-advisory, InsurTech (Insurance & Technology), and many more are all part of Fintech, which is backed by Blockchain (Thakor 2019). Thus, the introduction of Blockchain technology is said to bring Disruption in the market (Guo and Liang 2016).

The worrisome conditions during global financial crises have made the banks more vigilant and hence more regulative but these regulations cannot fully save the financial services and thus, there arises the need for a more transparent system rather than a rigid one for which Blockchain now is considered to be a better and sustainable choice (Nguyen 2016).

Blockchain technology automates the assessment of customers' credit information, which improves the credit system as it reduces credit risk by having all information beforehand safe and secure. It even shortens the timing for cross-border payments making them quick and efficient (Zhang et al. 2020).

With this changing era, financial institution intends to replace RTGS system with a Blockchain-enabled system, which will be trustworthy, decentralized, privacy-preserving, liquidity-saving saving, and efficient. For this purpose, a prototype based on Hyperledger Fabric backed by a blockchain platform for an end-to-end IBPS system is introduced which provides a high level of payment settlement (Wang et al. 2018). The emergence of P2P is a good competition for banks and thus, to compete with it, banks can either acquire or may come in partnership with P2P lending firms (Thakor 2019). Constructive engagement of stakeholders is crucial for the success of reforms (Tiwari et al. 2020).

In this era of smart devices used at workstations, for traveling and even at homes the blockchain-enabled IOT (Internet of Things) devices embedded with edge computing are the need of the hour for eradicating security issues (Li et al. 2019a, b). The impact or the value of a network in a Blockchain system increases or grows with the increase in the number of active users which is modeled by Metcalfe's law (Alabi 2017). If data analytics methods such as machine learning and deep learning are combined with blockchain technology it will make data more trustworthy, valid, transparent, suitable, and standardized (Swan 2018).

To improve and upgrade the IOT devices, machine learning training has to be done but if combined with blockchain, decentralized edge notes are useful in securing the privacy of the data sets and these devices can enjoy the benefit of automation and also make the payment system quick and cost-efficient (Fan et al. 2020).

The use of AI (Artificial Intelligence) and Blockchain has fully evolved the financial market. It has fully automated trading and various other finance activities and led to the progress of various organizations, entities, and so on. It has brought so many opportunities that various Educational Institutions have started introducing Fintech in their curriculum (Hendershott et al. 2021).

Blockchain is going to be a trend in the future but its acceptance by the users of technology is necessary to leverage its benefits. The user's perception of the usefulness of a technology and the ease of its use decides whether the user will accept the technology or not. Penetrating the idea of the technology gradually in a social system over time helps the individuals to understand the need for the idea to be able to accept and adopt it (Lou and Li 2017). Security is a mediator between perception and satisfaction from technology-enabled financial services (Chand et al. 2023a, b). Satisfaction from technology-enabled payment systems influences satisfaction, which is a determinant of continuous usage in the long term (Chand et al. 2022). Blockchain technology doesn't include any intermediaries, which makes it a secure and reliable system. The public blockchains are not trustworthy as compared to the private Blockchains (Swan 2018). The major problem in its implementation is the legal and institutional norms of a country (Vasilievich Babkin et al. 2018). The involvement of employees plays an important role in outcomes for an organization (Sapna et al. 2023). The adverse impact of technology initiatives on carbon dioxide emission and the sustainability of the environment needs to be examined (Kohli et al. 2023).

Easy communication facilities, the omnipresence of the Internet, and the urge to bring automation in all the processes paved the way for accepting and opting the Fintech services (Milian et al. 2019). The fusion of technology resources and human resources brings optimal results (Tiwari et al. 2023). Blockchain can be of great importance to almost all sectors if and only if they are ready to accept and keep up with this change and able to exploit and embrace the opportunities it brings (Li et al. 2019a, b).

14.2 Significance for Payment Settlement in Various Sectors

(a) Construction Sector

The opportunities for utilization of Blockchain technology are not only limited to the banking sector but are also useful in various other industries like the construction industry in which it helps by reducing cost and settling payments (Li et al. 2019a, b). Delayed payments and at times insolvency are the problems associated with the construction sector but by going digital and making use of Blockchain technology this problem is being resolved. Through a smart contract-based secured payment system a decentralized, transparent, and efficient system for secure payment is possible in this sector (Ahmadisheykhsarmast and Sonmez 2020).

There are three models for Supply Chain Management (SCM) application of Blockchain Technology in the Construction sector which are; Project Bank Accounts (PBAs) Model, Reverse Auction, and Crowdfunding. The prototypes for testing these models were developed by using Ethereum. The result of the test was that the use of Blockchain in payment settlement systems was easier than in supply chain management systems but it surely has a future in this sector (Tezel et al. 2021).

(b) Insurance Sector

Blockchain network shows its usefulness in the insurance process as well as it implements the whole process of insurance as a smart contract. Right resources reduce stress (Karthick et al. 2023). Blockchain maintains the information of the insurance policy transactions and also the result of these transactions, which will help the insurance company in case there will be any false claim or manipulation by customers and also can be used as proof for accountability of the company (Raikwar et al. 2018).

(c) Trading

Data trading is done in two modes; the hosting mode in which the data trading center retains the data and the aggregation mode in which the data trading center can retain data but do not retain it. These two modes are not very reliable, thus there is a need for changing the data trading modes to the ones, which are based on smart contracts backed by Blockchain Technology and Machine Learning. This mechanism significantly provides authenticity to both the data owner and the data purchaser for secure data trading (Xiong and Xiong 2019).

(d) Smart Vehicles

With the increase in technological advancements, there has been the development of smart vehicles but the major concern of manufacturers is privacy issues. To resolve this issue, a Blockchain-based strategy is suggested for fuelling smart vehicles which is known as Petro-Block. In this strategy, there is the use of Hyperledger Fabric and Smart Contracts for resolving privacy and security issues (Jamil et al. 2021). The fear of Sybil attacks attached to charging stations of EVs can be prevented by embedding it with Blockchain technology to secure the privacy of the real EV user (Baza et al. 2021).

(e) Electricity Sector

Blockchain can have a future in the electricity sector as well in the coming years. It will help in decentralizing the whole system, encouraging P2P trading in microgrids, automating payment and billing systems, and so on (Hohne and Tiberius 2020). Blockchain enables P2P electricity trading and settlement. The settlement system which is used for the centralized market cannot be applied to Blockchain thus, electricity exchange settlement with the help of Blockchain can be possible through two settlements which are; Global Balancing Settlement (GBS) and Splitting Settlement (SS) (Oprea et al. 2020).

(f) Other Sectors

E-tourism which includes online booking and payments through a centralized system becomes an easy mode of online fraud therefore a decentralised system like Blockchain becomes a saviour (Bodkhe et al. 2019). Digitalization of transactions has made them easier to audit through computer systems and reduces all the unnecessary hustle in tax collection as well. BPCSS (Bitcoin Collection Supervision System) is also introduced for cost-effective payment transactions between a buyer and a seller in a merchandise store (Chen et al. 2017).

14.3 Digital Currency

Bitcoin is a digital currency or cryptocurrency based on Blockchain technology and cryptography. Mining of Bitcoin makes less use of resources and more use of energy but still less than what is needed for gold mining. Though it needs less energy as compared to gold mining, coins and currency notes to transform the monetary system into a virtual monetary system by using cryptocurrencies, will still be a challenging task (Giungato et al. 2017). This gaining popularity of Bitcoin has led to the introduction of CBDC (Central Bank Digital Currency) by Central Banks (Sun et al. 2017). Digitalization of the monetary system prevents it from the issue of duplicate currency notes and provides us with a transparent system in which fraud is easily identifiable. This digitalization is not only cost-saving but it also saves customer's interest (Chen et al. 2017). Optimal utilization of resources enhances efficiency (Padmalosan et al. 2023).

To digitalize its functioning Central Bank needs a Distributive Ledger Technology like Blockchain to grab all the opportunities and deal with the challenges coming ahead. It needs Blockchain for Central Bank Digital Currency, Payment Clearing and Settlement Systems, Assets Transfer, Regulatory Compliance, and Audit Trail System (Dashkevich et al. 2020). Using Blockchain Technology enables the system to be decentralized, tamper-proof, and detectable, which is what makes it useful for CBDC. But in the coming time, it will be more effective to use multi-blockchain technology, which will further improve the system's adaptability and make payment procedure even faster. Political support is needed for major technological reforms and adaptation (Tiwari et al. 2021). This system will enable Central banks to keep the user's identity and their transaction information separate, hence securing privacy (Sun et al. 2017). Though the awareness and use of cryptocurrency are increasing day by day its acceptance as Fiat Money is ambiguous. Rather, it is considered as Foreign Money because it is decentralized and unregulated (Prochazka 2018).

14.4 Method

The bibliometric analysis was done using the Scopus database. Keywords used were "blockchain" AND "payments" AND "fintech" OR "bank." 320 documents were obtained from the Scopus database. Only documents in the English language were included. Out of 320, 9 documents were excluded (Russian: 2, Spanish:3, German: 2, Portugese:1, Chinese: 1). 311 documents were included after exclusion based on language. The bibliometric analysis was performed on 311 documents using VOS viewer software.

14.5 Results

14.5.1 Citation Analysis

A paper having high citations means that the published paper has research significance. By keeping a minimum no. of documents 5, out of 239 only 6 met the limit. IEEE ACCESS is the most cited source with 144 citations from 7 documents. The next analysis was focused on leading cited documents. The threshold was 50 citations and 15 met the threshold. "On the Fintech Revolution: Interpreting the Forces of Innovation, Disruption, and Transformation in Financial Services," is the highest-cited paper with no. of citations at 536. The top cited countries were examined by keeping the minimum no. of documents to 5, out of 73 only 25 met the limit. The top three countries are the United States with 42 documents and 1208 citations China with 43 documents and 1158 citations and the third is Singapore with 8 documents and 650 citations.

14.5.2 Co-Citation Analysis

Figure 14.1 represents co-citation through cited authors in which by keeping 20 minimum no. of documents, out of 12,190, only 48 met the limit. Figure 14.2 represents co-citation through cited references in which by keeping 5 minimum

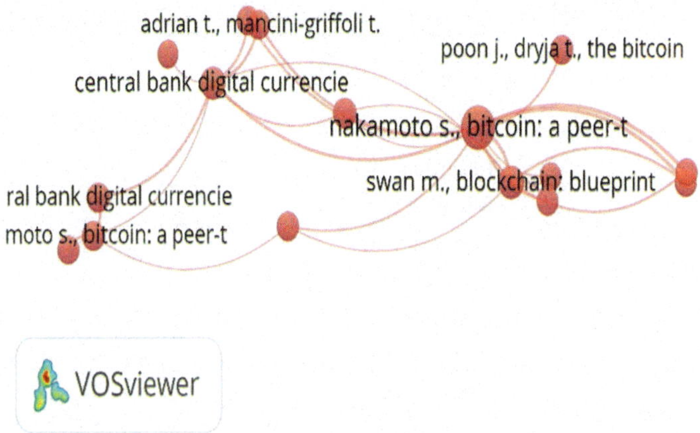

Fig. 14.1 Co-citation through cited authors

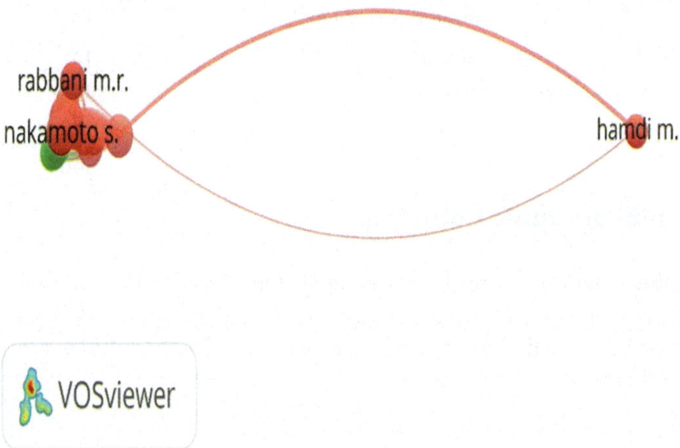

Fig. 14.2 Co-citation through cited references

no. of documents, out of 10,071, only 17 met the limit and only 16 out of which are connected.

14.5.3 Co-Occurrence Analysis

Table 14.1 represents the most occurred keyword in which Blockchain was the most occurred word at 221 times followed by Fintech at 69 times by keeping a minimum no. of occurrences at 15.

Table 14.1 Most frequent keywords

Keyword	Occurrences
Blockchain	221
Fintech	69
Block-Chain	65
Electronic Money	48
Bitcoin	44
Cryptocurrency	40
Smart contract	34
Finance	33
Payment systems	25
Central bank	21
Distributed ledger	19
Financial service	19
Central bank digital currency	17
Digital currency	16
Peer to peer networks	16
Banking	15
Decentralized	15
Security	15

14.5.4 Bibliographic Coupling

Bibliographic coupling is useful to identify the clusters. Figure 14.3 shows the major themes of research through clusters using a VOS viewer. A total 14 clusters were formed out of which 6 clusters were selected which includes a minimum of 10 no. of articles.

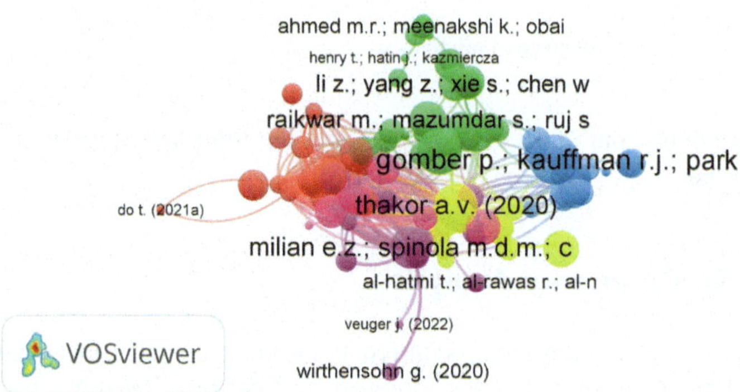

Fig. 14.3 Major themes of research through clusters

Cluster 1: Cluster 1 is shown in red color. It contains 48 articles with 314 total citations. The cluster was focused on security and payment settlement issues. The highly cited articles are Ahmadisheykhsarmast and Sommez (2020) with 81 citations and 61 total link strength, Xiong and Xiong (2019) with 56 citations and 67 total link strength and Sun et al. (2018) with total link strength of 1 and 30 citations.

Cluster 2: It is shown as green color. The cluster had 38 articles. A total of 814 citations were found in the cluster. The theme of the second cluster focuses on the characteristics of Blockchain and suggests applications of Blockchain. The leading cited documents in this cluster are Li et al. (2019a, b) with 261 citations and 11 total link strength, Giungato et al. (2017) with 116 citations and 2 total link strength and Raikwar et al. (2018) with 96 citations and 7 total link strength.

Cluster 3: It is shown in blue color. It contains 35 articles with 698 total citations. The theme of this cluster focuses on how Blockchain is transforming and innovating financial services. The highly cited articles in this cluster are Gomber et al. (2018) with 516 citations and 9 total link strengths, Wang et al. (2018) with 25 citations and 5 total link strengths, and Soloviev (2018) with 22 citations and 4 total link strengths.

Cluster 4: It is shown in purple. It contains 30 articles with 280 total citations. The theme of this cluster focuses on filling the gaps in research on Blockchain and proposing new models. The highly cited articles in this cluster were Milian et al. (2019) with 169 citations and 3 total link strength, Fan et al. (2021) with 34 citations and 68 total link strength and Lee et al. (2021) with 16 citations and 9 total link strength.

Cluster 5: It is shown in yellow. Cluster had 25 articles with 452 total citations. The cluster focused on fintech. The leading cited articles in this cluster were Thakor (2020) with 246 citations and 14 total link strengths, Liu et al. (2020) with 74 citations and 2 total link strengths and Lou et al. (2017) with 32 citations and 1 total link strength.

Cluster 6: It is shown in pink. Cluster had 12 articles with 230 total citations. The cluster-focused challenges. The leading cited articles in this cluster were Zhang et al. (2020) with 68 citations and 14 total link strength, Swan (2018) with 47 citations and 71 total link strength and Tezel et al. (2021) with 42 citations and 72 total link strength.

14.6 Conclusion

The research initiative for using blockchain for secure payments has been taken by the engineering domain researchers. IEEE ACCESS has emerged as the most cited source. On the Fintech Revolution: Interpreting the Forces of Innovation, Disruption, and Transformation in Financial Services" was the most cited paper with

536 citations in Scopus and 1188 citations in Google Scholar. The dominance of established large players in financial services is challenged by new fintech startups. Singapore has emerged as the third top-cited country. The US and China are the top two cited countries with more than a thousand publications. Asian economies and emerging economies are showing keen interest in research on blockchain technology for secure payments. Blockchain is the most frequently used keyword appearing 221 times. Fintech is the second most frequently occurring keyword with 69 occurrences. Six major research themes were identified after bibliographic coupling. The themes are security, services, innovation, new business models, Fintech and Challenges of the financial sector. Blockchain offers the opportunity to leverage technology to eliminate blockages in credit flow, enhance operational efficiency, enhance financial inclusion and support economic growth and development. However, the stakeholders need to focus on the energy efficiency of using blockchain. High energy usage and capital expenditure for blockchain systems and transactions is a deterrent for organizations. Government support to provide a national blockchain support system will go a long way to make blockchain a viable alternative to traditional payment systems and financial services. Regulatory agencies need to take precautionary measures to avoid misuse of blockchain by terrorists and anti-social elements. Blockchain needs a robust regulatory framework to make it a tool for sustainable and inclusive growth.

References

Ahmadisheykhsarmast S, Sonmez R (2020) Smart contract system for security of payment of construction contracts. Autom Constr 120:103401. https://doi.org/10.1016/j.autcon.2020.103401

Alabi K (2017) Digital blockchain networks appear to be following Metcalfe's law. Electron Commer Res Appl 24:23–29

Baza M, Sherif A, Mahmoud MMEA, Bakiras S, Alasmary W, Abdallah M, Lin X (2021) Privacy-preserving blockchain-based energy trading schemes for electric vehicles. IEEE Trans Veh Technol 70(9):9369–9384. https://doi.org/10.1109/tvt.2021.3098188

Bodkhe U, Bhattacharya P, Tanwar S, Tyagi S, Kumar N, Obaidat MS (2019) BloHosT: blockchain enabled smart tourism and hospitality management. IEEE Xplore. https://doi.org/10.1109/CITS.2019.8862001

Chand K, Tiwari R, Sapna (2022) Effect of perception and satisfaction on preference for mobile wallet. FIIB Bus Rev 231971452210773. https://doi.org/10.1177/23197145221077365

Chand K, Jangra S, Singh PC, Tiwari RS, Tiwari R (2023) Sustainability of mobile wallets in the era of low-cost data usage in India. In: Data-driven approaches for effective managerial decision making. IGI Global, pp 102–126

Chand K, Tiwari R, Bhardwaj K (2023) Impact of welfare measures on job satisfaction of employees in the industrial sector of Northern India. Financ India 37(2)

Chen PW, Jiang BS, Wang CH (2017) Blockchain-based payment collection supervision system using pervasive Bitcoin digital wallet. In: 2017 IEEE 13th international conference on wireless and mobile computing, networking and communications (WiMob). https://doi.org/10.1109/wimob.2017.8115844

Dashkevich N, Counsell S, Destefanis G (2020) Blockchain application for central banks: a systematic mapping study. IEEE Access 8:139918–139952. https://doi.org/10.1109/access.2020.3012295

Fan S, Zhang H, Zeng Y, Cai W (2020) Hybrid blockchain-based resource trading system for federated learning in edge computing. IEEE Internet Things J 1. https://doi.org/10.1109/jiot.2020.3028101

Giungato P, Rana R, Tarabella A, Tricase C (2017) Current trends in sustainability of bitcoins and related blockchain technology. Sustainability 9(12):2214. https://doi.org/10.3390/su9122214

Gomber P, Kauffman RJ, Parker C, Weber BW (2018) On the fintech revolution: interpreting the forces of innovation, disruption, and transformation in financial services. J Manag Inf Syst 35(1):220–265

Guo Y, Liang C (2016) Blockchain application and outlook in the banking industry. Financ Innov 2:1–12

Hendershott T, Zhang X (Michael), Zhao JL, Zheng Z (Eric) (2021) FinTech as a game changer: overview of research frontiers. Inf Syst Res 32(1):1–17. https://doi.org/10.1287/isre.2021.0997

Hohne S, Tiberius V (2020) Powered by blockchain: forecasting blockchain use in the electricity market. Int J Energy Sector Manag (ahead-of-print). https://doi.org/10.1108/ijesm-10-2019-0002

Jamil F, Cheikhrouhou O, Jamil H, Koubaa A, Derhab A, Ferrag MA (2021) PetroBlock: a blockchain-based payment mechanism for fueling smart vehicles. Appl Sci 11(7):3055. https://doi.org/10.3390/app11073055

Karthick M, Somesh V, Gudadhe N, Boopathi B, Padmanabhan S, Tiwari R, Sharma A (2023) Structural analysis of motorcycle spokes design using finite element analysis with alloy materials. Mater Today: Proc

Kohli S, Srikavya BP, Dhapekar NK, Tiwari R, Kumar R, Yadav AS, Sharma N, Sharma A (2023) Impact of nano materials on engine performance run on biofuels. Mater Today: Proc

Li J, Greenwood D, Kassem M (2019a) Blockchain in the built environment and construction industry: a systematic review, conceptual models and practical use cases. Autom Constr 102:288–307

Li Z, Yang Z, Xie S, Chen W, Liu K (2019b) Credit-based payments for fast computing resource trading in edge-assisted internet of things. IEEE Internet Things J 6(4):6606–6617. https://doi.org/10.1109/jiot.2019.2908861

Liu J, Li X, Wang S (2020) What have we learnt from 10 years of fintech research? a scientometric analysis. Technol Forecast Soc Chang 155:120022. https://doi.org/10.1016/j.techfore.2020.120022

Lou AT, Li EY (2017) Integrating innovation diffusion theory and the technology acceptance model: the adoption of blockchain technology from business managers' perspective

Milian EZ, Spinola MDM, Carvalho MMD (2019) Fintechs: a literature review and research agenda. Electron Commer Res Appl 34:100833. https://doi.org/10.1016/j.elerap.2019.100833

Nguyen QK (2016) Blockchain a financial technology for future sustainable development. In: 3rd international conference on green technology and sustainable development (GTSD). https://doi.org/10.1109/gtsd.2016.22

Oprea SV, Bara A, Andreescu AI (2020) Two novel blockchain-based market settlement mechanisms embedded into smart contracts for securely trading renewable energy. IEEE Access 8:212548–212556. https://doi.org/10.1109/access.2020.3040764

Padmalosan P, Vanitha S, Kumar VS, Anish M, Tiwari R, Dhapekar NK, Yadav AS (2023) An investigation on the use of waste materials from industrial processes in clay brick production. Mater Today: Proc

Prochazka D (2018) Accounting for bitcoin and other cryptocurrencies under IFRS: a comparison and assessment of competing models. Int J Digit Account Res 18:161–188. https://doi.org/10.4192/1577-8517-v18_7

Raikwar M, Mazumdar S, Ruj S, Sen Gupta S, Chattopadhyay A, Lam KY (2018) A blockchain framework for insurance processes. In: 2018 9th IFIP international conference on new technologies, mobility and security (NTMS). https://doi.org/10.1109/ntms.2018.8328731

Sun H, Mao H, Bai X, Chen Z, Hu K, Yu W (2017) Multi-blockchain model for central bank digital currency. IEEE Xplore. https://doi.org/10.1109/PDCAT.2017.00066

Swan M (2018) 'Blockchain for business: next-generation enterprise artificial intelligence systems. Adv Comput 121–162. https://doi.org/10.1016/bs.adcom.2018.03.013

Tezel A, Febrero P, Papadonikolaki E, Yitmen I (2021) Insights into blockchain implementation in construction: models for supply chain management. J Manag Eng 37(4). https://doi.org/10.1061/(asce)me.1943-5479.0000939

Thakor AV (2019) Fintech and banking: what do we know? J Financ Intermediation 41:100833. https://doi.org/10.1016/j.jfi.2019.100833

Tiwari R, Chand K, Anjum B (2020) Crop insurance in India: a review of Pradhan Mantri Fasal Bima Yojana (PMFBY). FIIB Bus Rev 9(4):249–255. https://doi.org/10.1177/2319714520966084

Tiwari R, Chand K, Bhatt A, Anjum B, Thirunavukkarasu K (2021) Agriculture 5.0 in India: opportunities and challenges of technology adoption. A Step Towards Soc 5:179–198

Tiwari R, Agrawal P, Singh P, Bajaj S, Verma V, Chauhan AS (2023) Technology enabled integrated fusion teaching for enhancing learning outcomes in higher education. 18(7):243–249. https://doi.org/10.3991/ijet.v18i07.36799

Vasilievich Babkin A, Dmitrievna Burkaltseva D, Viktorovich Betskov A, Shapievich Kilyaskhanov H, Sergeevich Tyulin A, Vladimirovna Kurianova I (2018) Automation digitalization blockchain: trends and implementation problems. Int J Eng Technol 7(3.14):254. https://doi.org/10.14419/ijet.v7i3.14.16903

Wang X, Xu X, Feagan L, Huang S, Jiao L, Zhao W (2018) Inter-bank payment system on enterprise blockchain platform. IEEE Xplore. https://doi.org/10.1109/CLOUD.2018.0008

Wu T, Liang X (2017) Exploration and practice of inter-bank application based on blockchain. In: 2017 12th international conference on computer science and education (ICCSE). https://doi.org/10.1109/iccse.2017.8085492

Xiong W, Xiong L (2019) Smart contract based data trading mode using blockchain and machine learning. IEEE Access 7:102331–102344. https://doi.org/10.1109/access.2019.2928325

Zhang L, Xie Y, Zheng Y, Xue W, Zheng X, Xu X (2020) The challenges and countermeasures of blockchain in finance and economics. Syst Res Behav Sci 37(4):691–698. https://doi.org/10.1002/sres.2710

Formalized Overview of ZX-Calculus, the Notion of Completeness Clifford Computation-Based and One Representative Application Case

15

Luis Gerardo Ayala Bertel

Abstract

This paper aims to consider the importance and use of ZX-calculus taking into account the solid structure of this branch to give a representation of quantum algorithms by finding the function's properties pictorially using logic and category theory. It's desired to show this topic formally, establishing axioms and definitions that support which foundations will be operating, explaining the processes and steps followed. Will start with the notion of what a ZX-diagram is, what spiders are, symmetries, and transposes. After that, get into ZX-calculus regarding spider fusion, π-commutation, color changing, and algebraic structures; it will show reasoning about the influence of connectivity. After this, their advantages in simplification, and, finally, how a ZX-diagram is translated into a quantum circuit.

Keywords

Complex space · ZX-calculus · Graphical representation · Linear transformation · Quantum computation

15.1 Introduction

The ZX-calculus is a graphical language that contributes to giving an axiomatic path of how circuits behave when moving within the Z, X basis using logic and mathematical properties such as commutation, association, distribution to show all the connections between the input values and the quantum gates diagrammatically until reaching their output value. One of the essential characteristics of ZX language

L. G. Ayala Bertel (✉)
Faculty of Exact and Natural Sciences, Department of Mathematics, Cartagena, Bolivar, Colombia
e-mail: ayalabluis08@gmail.com

A. Kumar et al. (eds.), *Sustainable Security Practices Using Blockchain, Quantum and Post-quantum Technologies for Real Time Applications*, Contributions to Environmental Sciences & Innovative Business Technology,
https://doi.org/10.1007/978-981-97-0088-2_15

is that all the graphs are undirected, isomorphic, and any deformation can be returned to its original state, yielding the same output.

It's formalized that quantum theory is used precisely in Hilbert spaces; this means, natural behavior of these phenomena can be modeled mathematically enclosed by linear algebra and n-dimensional calculus within a complete metric space under an inner product. Alluding to the fact that despite the advances or concepts in this field today, the theoretical foundations remain as those postulated several decades ago. Exist a universality criterion in this language where any $2^n \times 2^m$ matrix can have its diagrammatic representation as a ZX-diagram, thus having real applications in the field of quantum computing (Coecke and Kissinger 2017). There is a particular algebraic structure called monoidal categories that can interact with computational processes in algebraic systems as linear logic, i.e., it describes the parallel combination of objects and morphism when it is possible (Heunen and Vicary 2019). Since the linear map algebras in finite-dimensional Hilbert spaces are significantly connected to this pictorial viewpoint used in quantum theory, the ZX-calculus helps to describe composite quantum systems; therefore, the quantum mechanics representation diagrammatically is suitable (Coecke 2010).

Definition 15.1 The monoidal structure of the Hilb category has the following properties:

- The tensor product $\otimes$: $Hilb \times Hilb \rightarrow$ Hilb is the Hilbert space tensor product.
- The unit object I is the one-dimensional Hilbert space C.
- Associator $(H \otimes J) \otimes K \rightarrow H \otimes (J \otimes K)$ is the single linear map structure which satisfies $(\mathrm{u} \otimes \mathrm{v}) \otimes \mathrm{w} \rightarrow \mathrm{u} \otimes (\mathrm{v} \otimes \mathrm{w})$ for all $\mathrm{u} \in H$, $\mathrm{v} \in J$ and $\mathrm{w} \in K$.
- Left unitor $C \otimes H \rightarrow H$ is the single linear map structure which satisfies $1 \otimes \mathrm{u} \rightarrow \mathrm{u}$ for all $\mathrm{u} \in H$.
- Right unitor $H \otimes C \rightarrow H$ is the single linear map structure which satisfies $\mathrm{u} \otimes 1 \rightarrow \mathrm{u}$ for all $\mathrm{u} \in H$.

There exists another operator satisfying these properties called direct sum ($\oplus$), which means, all elements on the Hilbert space under this operator are also monoidal structure. The reason for defining only the tensor above is because it has a selective consideration in view of the fact that correctly describes in quantum theory the space over the states of a system when it is composed (Heunen and Vicary 2019).

Definition 15.2 ZX-diagram is a compact dagger category, meaning, it has a symmetrical monoid structure, it is closed compact, and it comes provided with a dagger, so all these structures interact adequately.

One of the foundations in ZX-diagram is its notation, which makes its representation logical and provides rigor through the diagrammatic structure which can be combined in order to simplifying the operations, thing that is not possible in quantum circuits; its diagrams are intuitive to use and can generate significant improvement by providing a syntax to manipulate the linear operations that are the essential elements of quantum mechanics (Bergholm and Biamonte 2011). It should be noted that the ZX-diagrams given below involve angles multiples of $\pi/2$ only, this group in the ZX-calculus is called *stabilizer*, and there is where the completeness criterion is met (Gottesman 1997).

15.2 ZX-Diagram Terminology

ZX-diagrams provide a more visually pragmatic representation of a quantum calculation than a circuit; this satisfies the necessary condition for quantum circuits but not the sufficient condition, meaning that every circuit is represented as a ZX-diagram, but not all ZX-diagrams have model representation as a quantum circuit, thus in the ZX-diagrams, it is possible to be able to develop simplifications and operations that haven't been an analogous quantum circuit (Duncan et al. 2020).

Definition 15.3 A diagram of the ZX-calculus is a finite undirected framed open graph whose interior vertices are of the following types:

- $Z(\alpha)$ vertices, designated by α angle where $0 \leq \alpha < 2\pi$. These are framed color as green circles, internal label is erased when the angle is zero.
- $X(\beta)$ vertices, designated by β angle, where $0 \leq \beta < 2\pi$. These are framed color as red circles, internal label is erased when the angle is zero.
- H vertices, appointed to be represented $\frac{1}{\sqrt{2}}\begin{pmatrix} 1 & 1 \\ 1 & -1 \end{pmatrix}$, which rotate the angles along Z- and X-axis. Are depicted as yellow color in a square;

Proposition 15.1 *Given a diagram $D : n \to m$, the adjoint $D\dagger : m \to n$ is interpreted to be the diagram acquired by negating all the angles and reflecting the diagram around the vertical axis.* When two diagrams are called isomorphic meaning that every operation identifies the two objects as if they were really the same object with the equal properties and vector structure (Fagan and Duncan 2019).

15.3 Clifford Group

A Clifford algebra is operated in vector space with a quadratic form called bilinear, the focus that will be given is related to objects called spinors. These objects arise within physics in quantum mechanics and Dirac's theory as entities that represent the quantum state of a system. Is characterized to be a unitary group acting on C^{2^n} which is generated finitely using these two matrices:

$$S = \begin{pmatrix} 1 & 0 \\ 0 & i \end{pmatrix}; V = \frac{1}{\sqrt{2}} \begin{pmatrix} v & \hat{v} \\ \hat{v} & v \end{pmatrix}$$

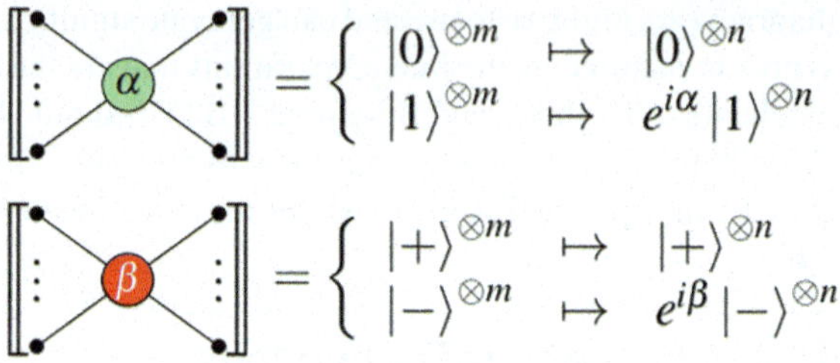

Fig. 15.1 m inputs, n outputs diagram, and its zx-calculus condition representation

where $v = e^{i\pi/4}$; it's possible to construct the Pauli matrices with $Z = S^2$ and $X = V^2$ defined as Pauli group normalizer, and this group generates ZX-calculus completeness structure.

Definition 15.4 In a zx-diagram D : m → n, its linear function map $[[D]] : (C^2)^{\otimes m} \to (C^2)^{\otimes n}$ on single vertices are represented for green or red dot attached with wires to the left and wires to the right.

Figure 15.1 shows in a generalized form in the inputs and outputs representation holding the fact ZX-diagrams are universal, that's it; $\forall A \in (C^2)^{\otimes m} \times (C^2)^{\otimes n}$, $\exists$ D : m → n, $[[D]] = A$.

Note that above condition is only for a finite set of angles, as same the completeness is valid inside the group of angles that are multiples of $\pi/2$. Already observed what ZX are, to start reasoning a set of axioms is need, being those the base for further arguments.

15.4 ZX-Calculus Axioms

An essential requirement to be able to understand is internalize the basic rules, these are fundamental principles that allow mathematics to be developed within a coherent ontological framework, those basic rules are called axioms or postulates, here will be the relevant ones to cover this topic.

Axiom 15.1 A single wire-diagram is the 2×2 identity matrix, two cross wires is the 4×4 swap gate matrix.

Axiom 15.2 A zx-diagram with one green color vertex and one input, one output is the $R_Z(\alpha)$ phase gate. A zx-diagram with one red color vertex and one input, one output is $R_X(\beta)$ phase gate (Coecke et al. 2009) (Figs. 15.2 and 15.3).

Definition 15.5 Let ZX_s be a product and permutation sub-category of ZX obtained by restricting Z- and X- vertices only by the angles $\alpha, \beta \in \{0, \pi/2, \pi, 3\pi/2\}$, then the functor on the generators $f : Cliff \to ZX_s$ are represented by

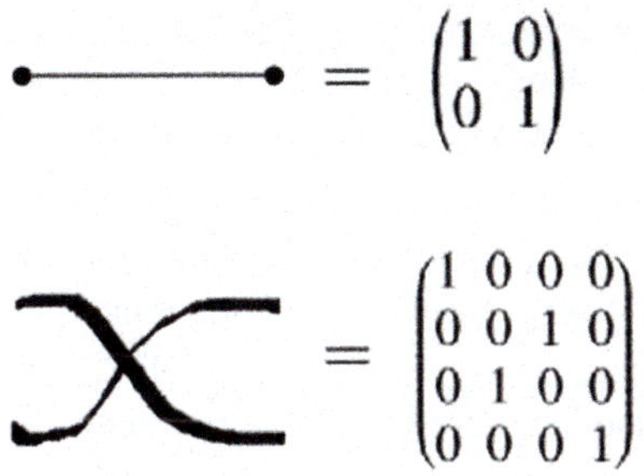

Fig. 15.2 Any single wire-diagram added doesn't affect the quantum state. A swap decomposition in the summation form is $\frac{I \otimes I + X \otimes X + Z \otimes Z}{2}$

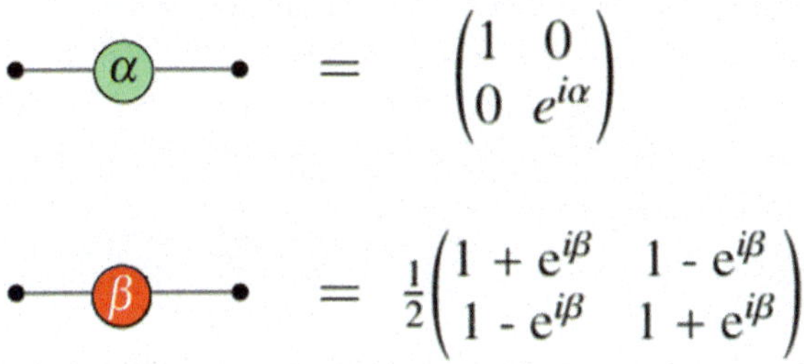

Fig. 15.3 For one input, one output green color vertex the matrix is the $R_Z(\alpha)$ phase gate. If $\alpha = \pi$ it will be the Pauli Z-gate. For one input, one output red color vertex the matrix is the $R_X(\beta)$ phase gate. If $\beta = \pi$ it will be the Pauli X-gate. When the angle $\alpha = \beta = 0$ by default it removes the Greek symbol as it becomes phaseless, i.e., $\frac{1}{2}\begin{pmatrix} 1+e^{i(0)} & 1-e^{i(0)} \\ 1-e^{i(0)} & 1+e^{i(0)} \end{pmatrix} = \frac{1}{2}\begin{pmatrix} 1+1 & 1-1 \\ 1-1 & 1+1 \end{pmatrix} = \begin{pmatrix} 1 & 0 \\ 0 & 1 \end{pmatrix} = \begin{pmatrix} 1 & 0 \\ 0 & e^{i(0)} \end{pmatrix}$ which is the identity matrix.

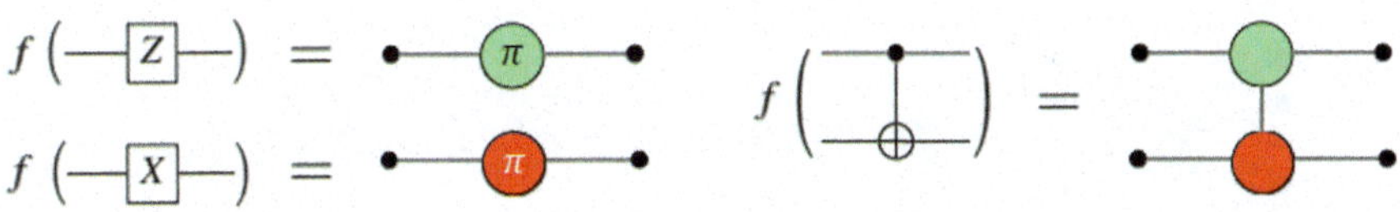

Axiom 15.3 A zx-diagram with one phaseless green color vertex as input and one single wire output is the eigenstate $|0\rangle$; with one phase π green color vertex as input and one single wire output is the eigenstate $|1\rangle$; with one phaseless red color vertex as input and one single wire output is the eigenstate $|+\rangle$; with one phase π red color vertex as input and one single wire output is the eigenstate $|-\rangle$.

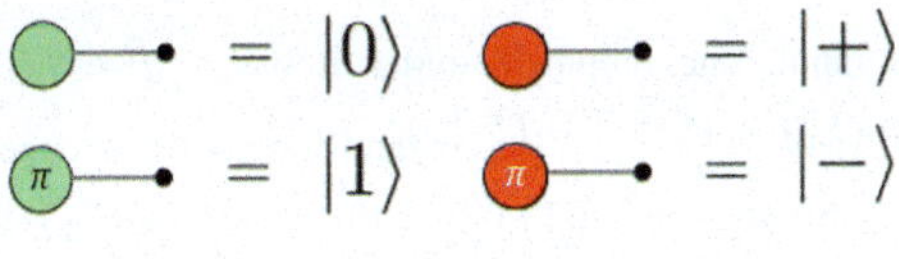

Definition 15.6 Any unitary single-qubit gate can be represented as Z-rotation followed by an X-rotation, followed again by a Z-rotation using the Euler angles. Which is equal to the Bloch sphere rotation over some axis (van de Wetering, 2020).

Axiom 15.4 Exist a derived generator from the unitary single-qubit that maps the basis eigenstates into a superposition, called Hadamard gate.

Since any single-qubit can be represented diagrammatically as

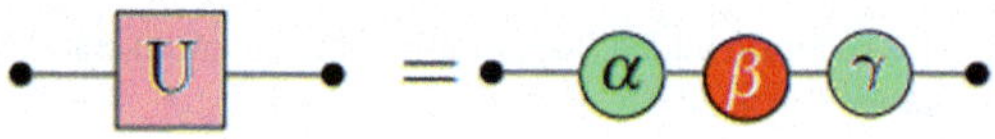

The particular instance U = H is given when the angles $\alpha = \beta = \gamma = \frac{\pi}{2}$ (Ambos-Spies et al., 2009), i.e.,

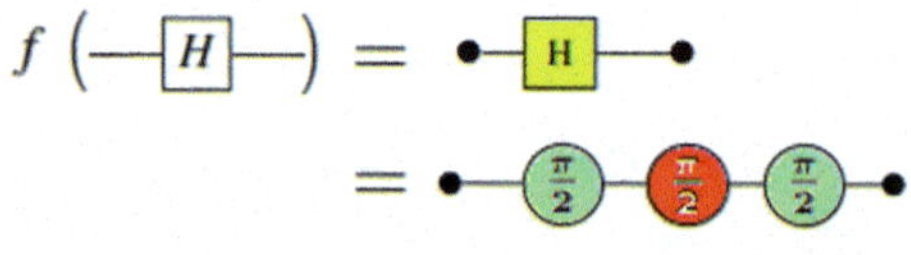

meaning H = $Z(\gamma) \circ X(\beta) \circ Z(\alpha)$. Other types of generators that are used are the Z-spider and X-spider; these can have whichever number of wires as inputs from left and whichever number of wires as outputs from right; they have this linear map representation (Fig. 15.4).

In ZX-spiders, each wire corresponds to a two-dimensional index; if there are more or equal to two wires, both input and output, the ZX-diagram has the same behavior as tensor networks having this similarity with the standard graphical tensor network postulated by Penrose (van de Wetering 2020).

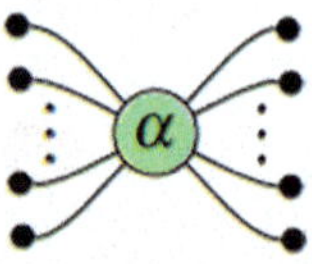

$$= |0\cdots 0\rangle\langle 0\cdots 0| + e^{i\alpha}\,|1\cdots 1\rangle\langle 1\cdots 1|$$

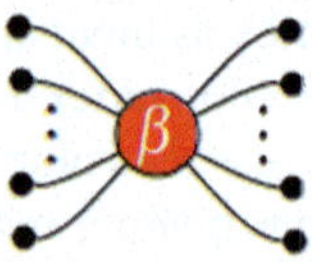

$$= |+\cdots +\rangle\langle +\cdots +| + e^{i\alpha}\,|-\cdots -\rangle\langle -\cdots -|$$

Fig. 15.4 Let's announce that $|0\rangle = \binom{1}{0}$, $|1\rangle = \binom{0}{1}$in Z basis, and $|+\rangle = \frac{1}{\sqrt{2}}\binom{1}{1}$, $|-\rangle = \frac{1}{\sqrt{2}}\binom{1}{-1}$in X basis then for one input, one output Z-spider it has; $R_Z(\alpha) = |0\rangle\langle 0| + e^{i\alpha}|1\rangle\langle 1| = \binom{1}{0}(1\ 0) + \binom{0}{e^{i\alpha}}(0\ e^{i\alpha}) = \begin{pmatrix}1 & 0\\ 0 & 0\end{pmatrix} + \begin{pmatrix}0 & 0\\ 0 & e^{i\alpha}\end{pmatrix} = \begin{pmatrix}1 & 0\\ 0 & e^{i\alpha}\end{pmatrix}$and for one input, one output X-spider it has; $R_X(\alpha) = |+\rangle\langle +| + e^{i\alpha}|-\rangle\langle -| = \frac{1}{\sqrt{2}}\binom{1}{1}\frac{1}{\sqrt{2}}(1\ 1) + e^{i\alpha}\left(\frac{1}{\sqrt{2}}\binom{1}{-1}\frac{1}{\sqrt{2}}(1\ {-1})\right) = \frac{1}{2}\begin{pmatrix}1 & 1\\ 1 & 1\end{pmatrix} + \frac{1}{2}e^{i\alpha}\begin{pmatrix}1 & -1\\ -1 & 1\end{pmatrix} = \frac{1}{2}\begin{pmatrix}1+e^{i\alpha} & 1-e^{i\alpha}\\ 1-e^{i\alpha} & 1+e^{i\alpha}\end{pmatrix}$.

As remark the notation $|00\rangle = |0\rangle \otimes |0\rangle$ meaning the tensor product and $X|0\rangle = \left(\begin{smallmatrix} 0 & 1 \\ 1 & 0 \end{smallmatrix}\right)\left(\begin{smallmatrix} 1 \\ 0 \end{smallmatrix}\right)$ the matrix multiplication, thus for multi-qubits is possible to represent synchronic operations Kronecker product based, i.e., $X|q_1\rangle \otimes H|q_0\rangle = (X \otimes H)|q_1 q_0\rangle$, gathering all this is possible to describe the state of the certain system as state vectors. Let's consider two more generators called *Controlled-gates (CX, CZ)*, both act in two-qubits dimension, the *CX* apply *X-gate* the second qubit iff the first qubit be on state $|1\rangle$. Its matrix representation is given by $CX_{q_0,q_1} = I \otimes |0\rangle\langle 0| + X \otimes |1\rangle\langle 1|$, if the control is the qubit two and the target qubit one then $CX_{q_1,q_0} = |0\rangle\langle 0| \otimes I + |1\rangle\langle 1| \otimes X$, the *CZ* is analogous instead apply *Z-gate* to the target if control is $|1\rangle$. Those generators play a fundamental role in the construction of gate-based quantum operations; there is a fact that by making simple multiplications is possible to find that $HXH = Z$ and $HZH = X$.

Definition 15.7 The spiders are symmetric tensors. Under permutations are invariant, which means that it's possible to interchange their wires with each other (Dixon and Kissinger, 2014). If we made diagrammatically the representation of this definition then

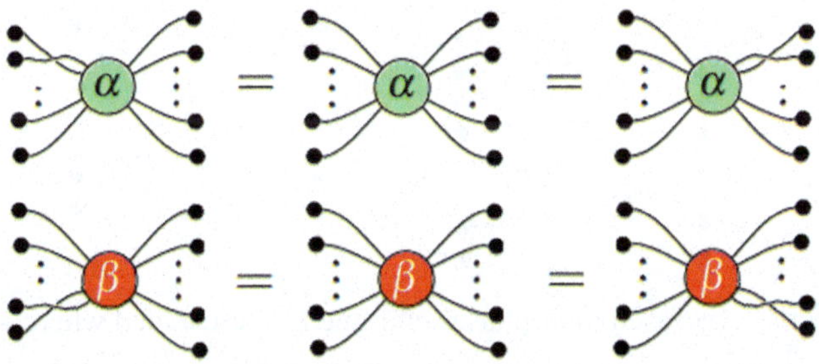

this permutation constructed by three CX-gate interleaving the controller and target from first qubit to second qubit so-called *swap*, which have this matrix representation; $swap = |00\rangle\langle 00| + |01\rangle\langle 10| + |10\rangle\langle 01| + |11\rangle\langle 11|$.

Now combining the CX-gate with swap, we diagrammatically have this symmetry;

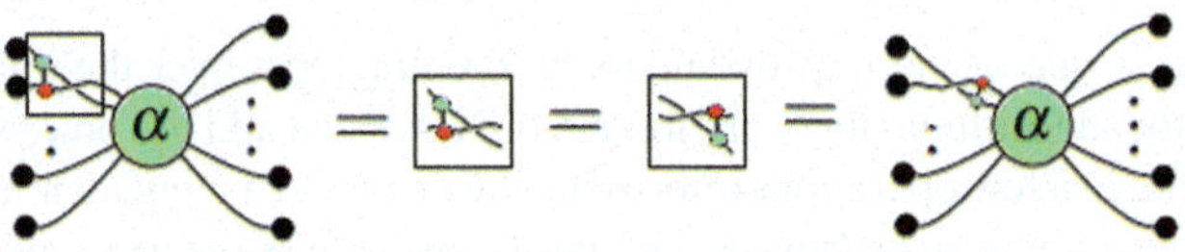

other generators in ZX-diagrams that have this symmetry, which are represented as bent parts of wires called *cap* and *cup*, they have this matrix representation;

$$\underset{|00\rangle + |11\rangle}{\frown = (} := \begin{pmatrix} 1 \\ 0 \\ 0 \\ 1 \end{pmatrix} \qquad \underset{\langle 00| + \langle 11|}{\smile =)} := (1\,0\,0\,1)$$

this piece of wires satisfies the yanking equations (van de Wetering, 2020). Also, it is possible to represent the transpose of the matrix using this generator as follows:

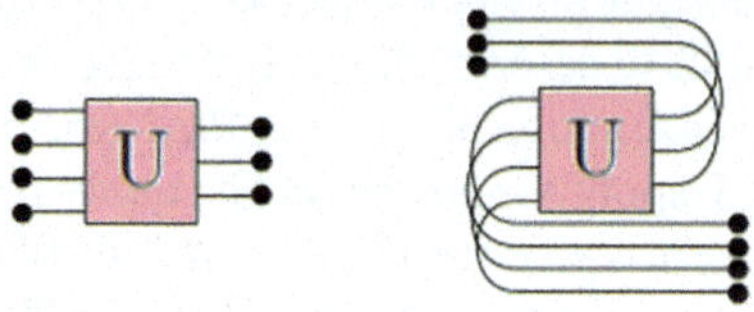

where the *cup* changes every output to input and the *cap* changes every input to output.

15.5 Spider fusion and π-commutation

Definition 15.8 Two same color spiders connected by one or more wires are possible to fuse altogether.

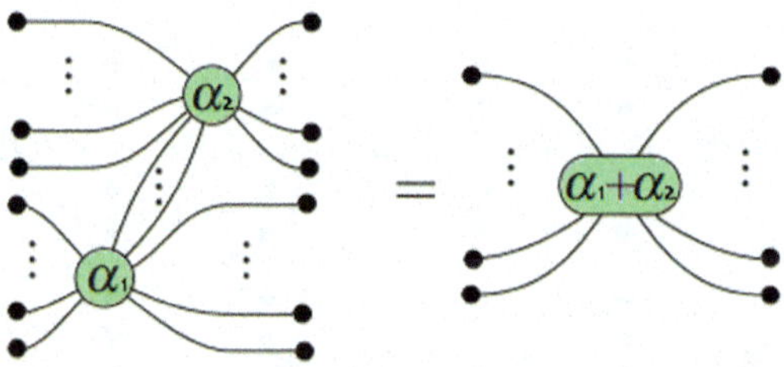

As the spider α_1, α_2 fused their phases $e^{i\alpha_1}$ and $e^{i\alpha_2}$ are added where $\alpha = \frac{\pi n}{2}$, with n $\in Z$.

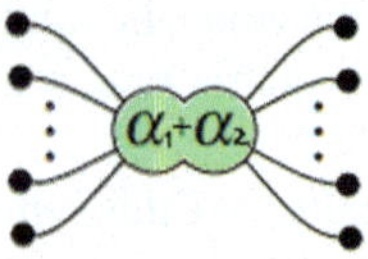

$$= |0\cdots 0\rangle\langle 0\cdots 0| + e^{i(\alpha_1+\alpha_2)}|1\cdots 1\rangle\langle 1\cdots 1|$$

This feature is one of the improvements of ZX-diagrams over the quantum circuit since if all the algorithms run over either Z or X, we can add the phases on the same axis, resulting in a few operations improving the error and having identical outcomes. It satisfies the symmetry properties, i.e., let $\rho_1, \rho_2, \rho_3 \in Z$, $\rho_1 \sim \rho_1$, if $\rho_1 \sim \rho_2$, then $\rho_2 \sim \rho_1$; if $\rho_1 \sim \rho_2$ and $\rho_2 \sim \rho_3$, then $\rho_1 \sim \rho_3$, meaning that we can fuse many spiders as we can (Fagan and Duncan, 2019). Analogous with X.

Now it is possible to perform linear operations between the two bases if we have this circuit,

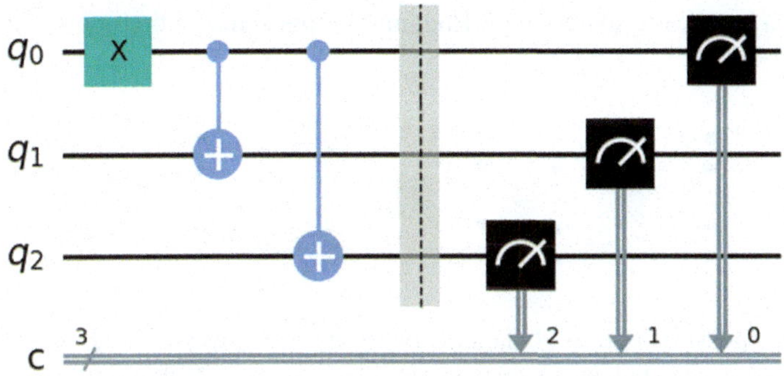

which consists of one input and three outputs, then in ZX-diagram, we have

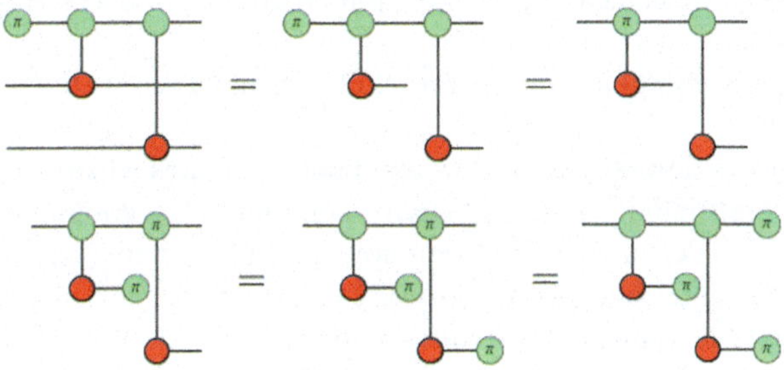

In the above diagrammatic representation, there are two essential features, elements of the same color basis commute and elements of different color basis distribute (π-commutation), which means a linear combination that each output is equivalent to $(X \otimes \cdots \otimes X)|1 \cdots 1\rangle\langle 1| + (X \otimes \cdots \otimes X)|0 \cdots 0\rangle\langle 0| = |0 \cdots 0\rangle\langle 1| + |1 \cdots 1\rangle\langle 0|$; this is for phaseless states if we have a phase α it's necessary to solve the equation $X(|0\rangle + e^{i\beta}|1\rangle)$, i.e.,

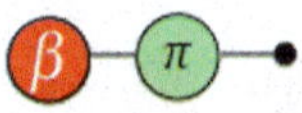

$$= X|0\rangle + e^{i\beta} X|1\rangle = |1\rangle + e^{i\beta}|0\rangle$$
$$= e^{i\beta}(|0\rangle + e^{-i\beta}|1\rangle) = (|0\rangle + e^{-i\beta}|1\rangle)$$

$=$ 

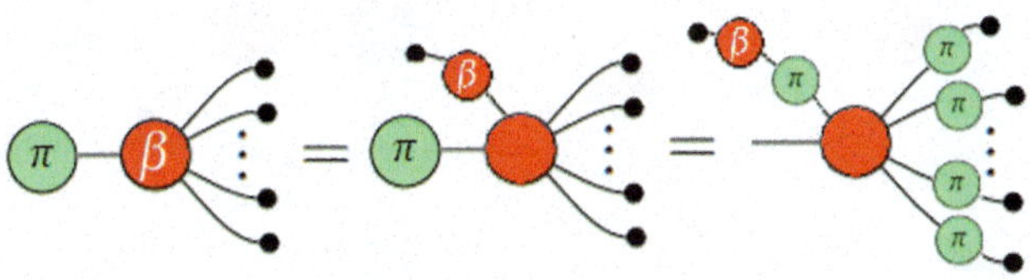

also it is possible to switch from one basis to another one by the Hadamard property;

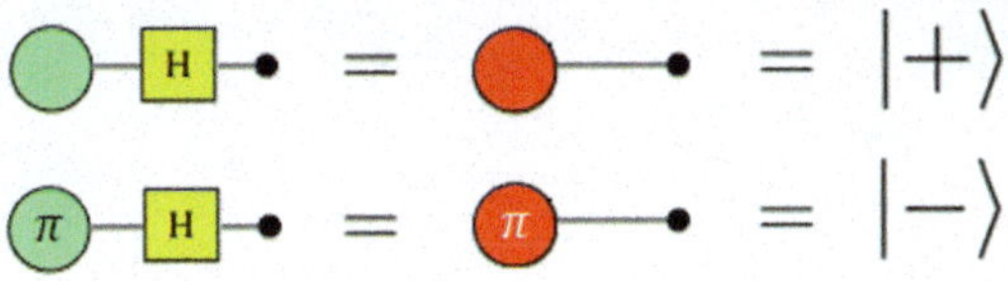

since $HZH = X$, the spiders with Hadamard inputs change their color through this process;

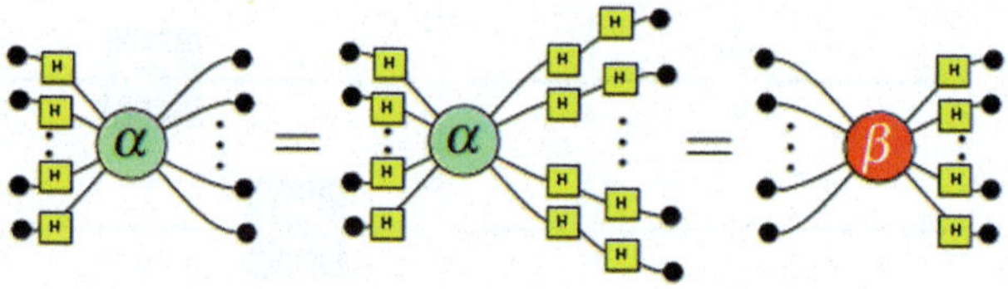

being this possible due to Hadamard being its own inverse, $HH = I$. Therefore, we can place two Hadamard gates in any place of any ZX-diagram yielding by the morphism. Morphisms are linear map functions between sets in vector spaces, then for three objects, we can have a composition function i ∘ (j ∘ k) = (i ∘ j) ∘ k, and for tensor products

I ⊗ J is always isomorphic to $J \otimes I$ for all objects in those sets (Fagan and Duncan, 2019).

The spiders, by definition, also follow algebraic structures where they represent the properties of associativity, co-associativity, commutativity, co-commutativity, if the Z-spiders or X-spiders are under a comonoid, i.e., a linear map $\delta : S \rightarrow S \otimes S$ and $\epsilon : S \rightarrow I$, satisfying $(\delta \otimes id_S) \circ \delta = (id_S \otimes \delta)$ and $(\epsilon \otimes id_S) \circ \delta = id_S = (id_S \otimes \epsilon) \circ \delta$. Here δ map is denominated co-multiplication and ϵ is the co-unit. In other words, Z- and X-spiders operate on algebra, coalgebra, and bialgebra over a certain field in vector spaces (Coecke and Pavlovic, 2008).

Grammatically this properties are represented as follows:

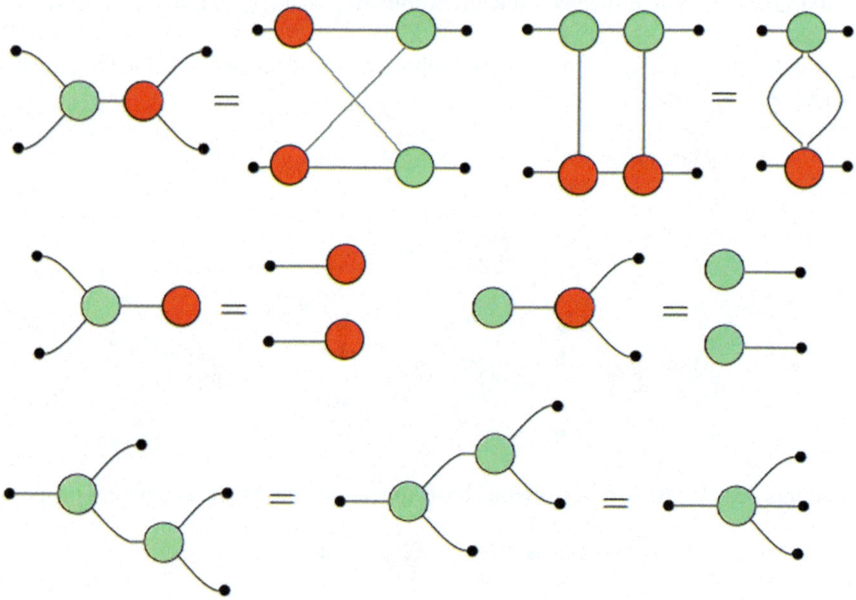

An important behavior of Z-spider and X-spider in a single interaction is when there is one phaseless dot of a different color; this cancels each other, allowing the connection between opposite color spiders to detach.

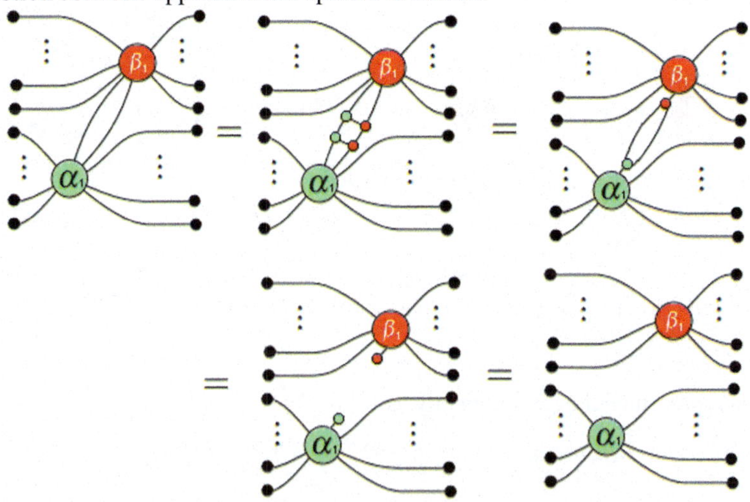

Is essential to say that any diagram can be topologically deformed whenever the input and output orders in the graph are maintained, which means that the bent and unfolded wires will represent the same matrix.

15.6 ZX-Quantum Algorithm

The algorithms are the way of performing operations acting as a finite organized series of steps which describes the process that must be followed to solve a specific task; in quantum, this process can be defined by using a black box denominated oracle, viewed as the encoding of functions. For ZX-calculus, these oracles can be decomposed into diagram components that have algebraic properties, which can provide more understanding.

Definition 15.9 Given a function by $f : 2^n \to 2^m$ exist a linear map $f : C^{2^n} \to C^{2^m}$ where the function *arrow* is defined as

$$\left[\!\left[\xrightarrow{f} \right]\!\right] = |x\rangle \mapsto 2^{(m-n)/4}\, |f(x)\rangle$$

and any function arrow satisfies the following attributes:

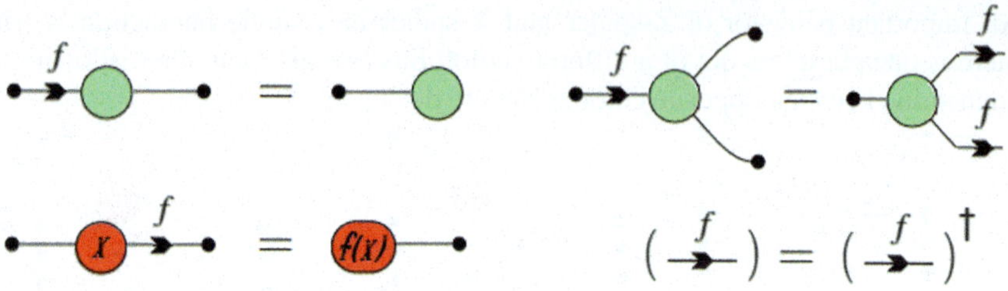

the function arrow f has these two properties:
- f is *balanced* iff $\forall x, y \in 2^m$ $|f^{-1}(\{x\})| = |f^{-1}(\{y\})|$ iff $\sum_{x \in 2^n} |f(x)\rangle = \sum_{y \in 2^m} 2^{(n-m)/2} |y\rangle$.

- f is *injective* iff $\forall x, y \in 2^n$ it has $f(x) = f(y) \rightarrow x = y$ iff $\delta_{x=y} |f(x)\rangle = \delta_{f(x)=f(y)} |f(x)\rangle$.

This is represented diagrammatically as

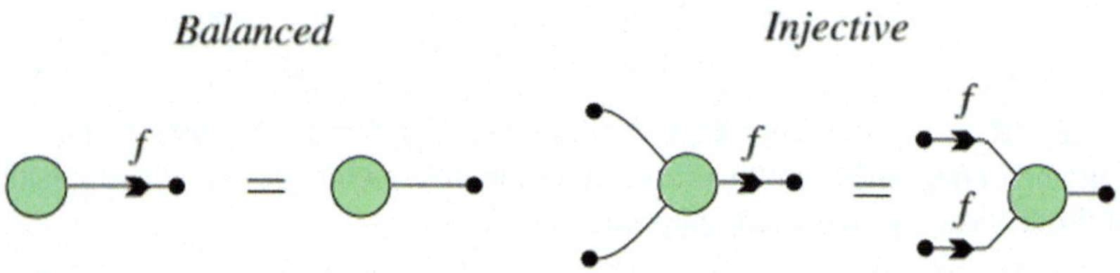

The arrow function is unitary if and only if f is bijective. Given any function f: $2^n \rightarrow 2^m$ the unitary transformation is denoted $U_f = |x\rangle|y\rangle \rightarrow |x\rangle|f(x) \oplus y\rangle$, called quantum oracle (Carette, 2021), constructed diagrammatically like this:

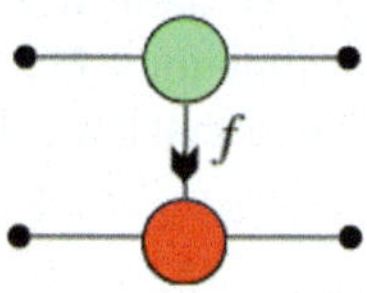

$[n] \otimes [m]$, then

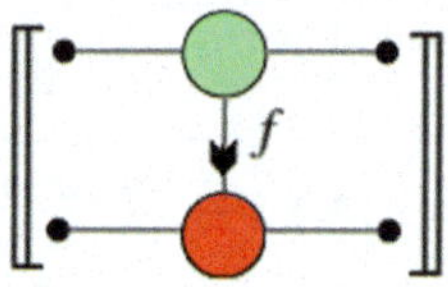

$[n] \otimes [m] = |x\rangle\ |y\rangle \rightarrow |x\rangle\ |f(x) \oplus y\rangle$.

For Boolean function $f : 2^n \rightarrow 2$, then the oracle of f is represented as

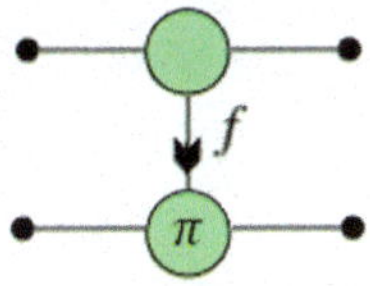

$[n] \to [n]$, then,

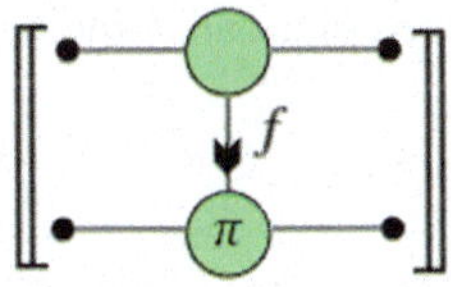

$[n] \to [n] = |x\rangle \to (-1)^{f(x)}|x\rangle$.

When the oracle acts on the wires with inputs, it has this property:

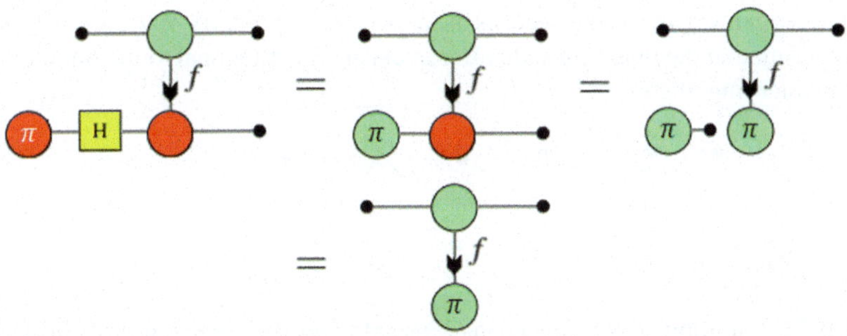

Now representing the first quantum algorithm released called Deutsch-Jozsa, let us consider the function $f(x_0, x1) = x_0 \otimes x_1$ the set of classical values that applying the function will be $f(0{,}0) = 0$; $f(0{,}1) = 1$; $f(1{,}0) = 1$; $f(1{,}1) = 0$, then its quantum state is given by $|\phi_0\rangle = |00\rangle_{01} \otimes |1\rangle_2$, where the subscript meaning the index of the qubits, and the oracle for this two inputs will be $U_f|x_1, x_0\rangle = (-1)^{f(x_1, x_0}|x\rangle$, which is the same as the *function arrow* representation. Since the goal is to evaluate the function f to determine if constant or balanced without the needing for an exponential number of queries, diagrammatically it be

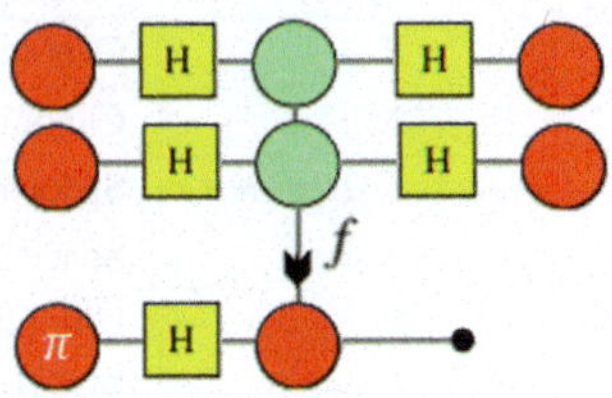

check that mathematically there are many things here; the qubits with Hadamard is $|\phi\rangle = \frac{1}{2}(|00\rangle + |01\rangle + |10\rangle + |11\rangle) \otimes \frac{1}{\sqrt{2}}(|0\rangle - |1\rangle)$, now that we know the rules of ZX we can graphically simplify, by the change color rule;

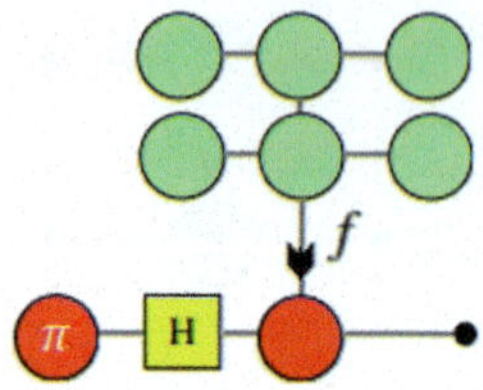

since the spider fuse the same color can combine, then

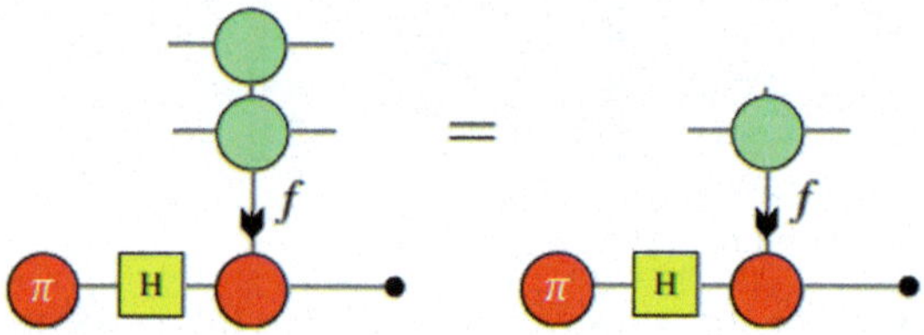

then as we can see, this is the same as the above oracle property of the arrow function, considering that the quantum state after querying is orthogonal to the quantum state before querying, therefore

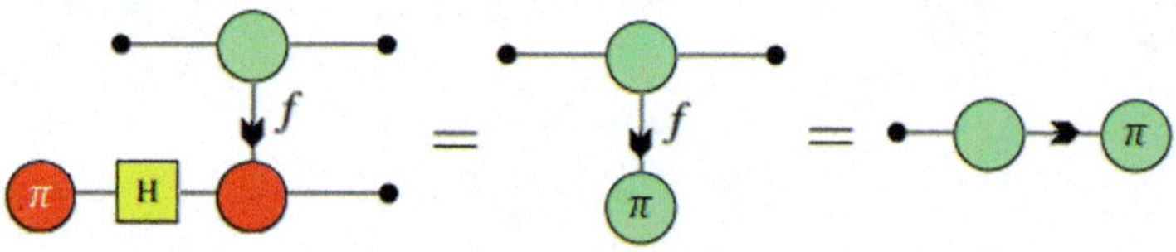

which, by definition, f is balanced, as we want to prove. This example is provided to show a basic understanding of the simplification theory by using the ZX-diagrams; for further or profound explanation can check the open-access archive *Wielding the ZX-calculus, Flexsymmetry, Mixed States, and Scalable Notations* (Carette 2021).

15.7 Symbol Table

See Table 15.1.

Table 15.1 Information about meaning of letter and symbol representation

Symbol letter	Description	Space
Z	Canonical axis	Complex
X	Canonical axis	Real
C–H–J–K	Space representation	Complex
u–v–w	Space elements	Complex
I	Unit object	Complex
n–m	Matrix dimensionality	Real
α–β	Phase angles	Real
f	Linear map	Real-Complex
U–H–S–V	Matrix representation	Real-Complex

15.8 Conclusion

In this work, the goal is to show in pictorial fashion the basic structure of ZX-diagrams and the compound field explaining theoretically, and like this, the reader once sees another paper to know about what the representation means, from where all this comes by explaining in a surface level that ZX-calculus will have too much to say in the following years, highlighting categorical axiomatization of quantum theory can provide a step toward diagrammatic formalism; the paper *categorical semantics of quantum protocols* (Abramsky and Coecke 2007) was pioneered through the theorem on schematic reasoning for protocols such as teleportation (Coecke 2004), one of the current developments is the automated reasoning, meaning, construct a software framework for graphical reasoning without doing manually, there are many aspects that in few pages are not possible to include, then is wanting in a humble way communicate the interest of this area that has the potential to unify the quantum theory with other theories, there exists extension of the zx-calculus which can represent mixed states that are very wide and have open degrees of freedom to develop high-dimensional study cases, which can have multiple applications in many areas as optimization, error correction, and natural language processing (Ganguly et al. 2022), where the idea is to give another perspective of the quantum computation circuit based with formalism, diagrammatic reasoning, and techniques from monoidal categories, among others, having as well solid structure for making or discover new axiomatic foundation in quantum to solve problems that classical computers are very hard or not possible to do.

References

Abramsky S, Coecke B (2007) A categorical semantics of quantum protocols

Ambos-Spies K, Löwe B, Merkle W (eds) Mathematical theory and computational practice. Springer, Berlin. https://doi.org/10.1007/978-3-642-03073-4

Bergholm V, Biamonte JD (2011) Categorical quantum circuits. J Phys A: Math Theor 44(24):245304. https://doi.org/10.1088/1751-8113/44/24/245304

Carette T (2021) Wielding the ZX-calculus, flexsymmetry, mixed states, and scalable notations. PhD thesis, HAL CCSD, arXiv, U. Lorraine

Coecke B (2004) The logic of entanglement

Coecke B (2010) Quantum picturalism. Contemp Phys 51(1):59–83. https://doi.org/10.1080/00107510903257624

Coecke B, Kissinger A (2017) Picturing quantum processes: a first course in quantum theory and diagrammatic reasoning. Cambridge University Press, Cambridge. https://doi.org/10.1017/9781316219317

Coecke B, Paquette EO, Pavlovic D (2009) Classical and quantum structuralism

Coecke B, Pavlovic D (2008) Quantum measurements without sums

Dixon L, Kissinger A (2014) Monoidal categories, graphical reasoning, and quantum computation

Duncan R, Kissinger A, Perdrix S, Wetering J (2020) Graph-theoretic simplification of quantum circuits with the ZX-calculus. Quantum 4:279. https://doi.org/10.22331/q-2020-06-04-279

Fagan A, Duncan R (2019) Optimising clifford circuits with quantomatic. Electron Proc Theor Comput Sci 287:85–105. https://doi.org/10.4204/eptcs.287.5

Ganguly S, Morapakula SN, Bertel LGA (2022) An introduction to quantum natural language processing (qnlp). CRC Press eBooks. https://doi.org/10.1201/9781003244660-1

Gottesman D (1997) Stabilizer codes and quantum error correction
Heunen C, Vicary J (2019) Categories for quantum theory: an introduction. Oxford University Press, United Kingdom
van de Wetering J (2020) ZX-calculus for the working quantum computer scientist

The manufacturer's authorised representative in the EU is Springer Nature Customer Service Centre GmbH, Europaplatz 3, 69115 Heidelberg, Germany. If you have any concerns regarding our products, please contact ProductSafety@springernature.com

Printed and bound by CPI Group (UK) Ltd, Croydon, CR0 4YY
01/10/2026
02225870-0006